Interventions in Smallholder Agriculture

Implications for Extension in Zimbabwe

Editors: **A. Bolding, J. Mutimba, P. van der Zaag**

UNIVERSITY OF
ZIMBABWE
Publications

First published in 2003 by
University of Zimbabwe Publications
P.O. Box MP 203
Mount Pleasant
Harare
Zimbabwe

ISBN 0–908307–52–7

Front cover by Inkspots based on a photograph by E. D. Alford in 1928, National Archives of Zimbabwe

Typeset by the University of Zimbabwe Publications.

Contents

Notes on Contributors v

Introduction 1
    Alex Bolding, Jeff Mutimba, Pieter van der Zaag and Jens Andersson

PART I: PAST AND PRESENT OF AGRICULTURAL EXTENSION

Chapter 1
Alvord and the demonstration concept: Origins and consequences of the agricultural
demonstrator scheme, 1920-44 35
    Alex Bolding

Chapter 2
Towards an understanding of technology needs of smallholder farmers 84
    Jeff Mutimba

Chapter 3
Of science and livelihood strategies: Two sides of the commercialisation
debate in smallholder irrigation schemes 110
    Emmanuel Manzungu

Chapter 4
Contextualising communal agriculture: Observations on labour migration
and farming in Save Communal Land 131
    Jens Andersson

PART II: TECHNOLOGY TRANSFER

Chapter 5
Of trees and grassroots interactions: Forestry extension in
Chinyika Resettlement Scheme, Zimbabwe 157
    Benjamin Hanyani-Mlambo

Chapter 6
The bench terrace between invention and intervention: Physical and
political aspects of a conservation technology 184
    Pieter van der Zaag

PART III: LOCAL KNOWLEDGE AND NETWORKS

Chapter 7
Women farmers make the markets: Gender, value and performance
of a smallholder irrigation scheme 209
Carin Vijfhuizen

Chapter 8
Knowledge is like live coals, it can be borrowed: How indigenous
knowledge and Shona proverbs can be used in agricultural extension 228
Olivia N. Muchena

PART IV: ENCOUNTERS

Chapter 9
Farmer-extensionist encounters in Nyamaropa: From clashes to a
negotiated order? 247
Dumisani Magadlela

Chapter 10
The uncooperative statistic: Farmers' response to experts' fact-finding
mission 271
Emmanuel Manzungu

Chapter 11
Getting back to the source: farmer participatory research at the
University of Zimbabwe 282
Jeff Mutimba

Chapter 12
Conclusion 308
Alex Bolding, Jeff Mutimba and Pieter van der Zaag

Bibliography 318

Index 339

Notes on Contributors

Jens Andersson is a research associate with the Department of Development Sociology at Wageningen University, The Netherlands. His PhD research on the development and social organisation of rural-urban linkages between Buhera district and Harare in Zimbabwe was hosted by the Department of Sociology, University of Zimbabwe, and funded by the Netherlands Foundation for the Advancement of Tropical Research (WOTRO).

Alex Bolding is a PhD researcher with the Irrigation and Water Engineering Group at Wageningen University, The Netherlands. His research on the models and practices around socio-technical interventions in the Nyanyadzi catchment, Chimanimani district was hosted by the Department of Soil Science and Agricultural Engineering of the University of Zimbabwe. His research interests focus on agricultural extension and soil and water management, encompassing both dryland and irrigated agriculture.

Dumisani Magadlela obtained his PhD thesis on irrigation interventions in Nyamaropa communal area in 2000 from Wageningen University, The Netherlands. He lectured social theory in the Department of Sociology at the University of Zimbabwe. Presently he is with the Department of Water Affairs and Forestry in South Africa.

Emmanuel Manzungu completed his PhD research with the Irrigation and Water Engineering Group at the Wageningen University in 1999. His research on the interplay between technical infrastructure and management in smallholder irrigation schemes was hosted by the Department of Soil Science and Agricultural Engineering, University of Zimbabwe. A crop scientist by training, his interests have broadened towards water management in irrigated agriculture and to related policy issues. Together with Pieter van der Zaag, he edited a book on *The Practice of Smallholder Irrigation: Case Studies from Zimbabwe*, published in 1996 by University of Zimbabwe Publications (UZP). He also edited a book together with Aidan Senzanje and Pieter van der Zaag on *Water for Agriculture in Zimbabwe: Policy and Management Options for the Smallholder Sector*, which was published by UZP in 1999. Presently he is engaged as a post-doctoral researcher on catchment management agencies in Zimbabwe and South Africa, hosted by the Irrigation and Water Engineering Group at Wageningen University.

Benjamin Hanyani-Mlambo is a lecturer and PhD researcher with the Department of Agricultural Economics and Extension, University of Zimbabwe. He is an agricultural economist, who obtained an MSc degree in agricultural extension. Current areas of interest include farming systems, farmer participation and other approaches for improved interventions within rural societies.

Olivia N. Muchena received her PhD degree from Iowa State University on the implications of indigenous knowledge systems for agricultural extension education. She was a senior lecturer in agricultural extension with the Department of Agricultural Economics and Extension, University of Zimbabwe, and Deputy Minister of Lands and Agriculture. Presently she is Deputy Minister responsible for the accelerated land resettlement programme in Zimbabwe.

Jeff Mutimba is an agricultural extension specialist with the Alemaya University of Agriculture in Ethiopia. In 1997 he obtained a DPhil degree at the University of Zimbabwe on the subject of 'Farmer participation in research: an analysis of resource-poor farmer involvement in, and contribution to, the agricultural research process in Zimbabwe.' He also lectured in the Department of Agricultural Economics and Extension at the University of Zimbabwe.

Carin Vijfhuizen concluded her PhD research with the Department of Rural Development Sociology, Wageningen University, in 1998. Her research on gender, networks and ideology in a communal area was hosted by the Department of Sociology at the University of Zimbabwe. Her main research interests focus on the power dynamics between and among women and men in everyday life. Presently she is a lecturer of sociology at the Eduardo Mondlane University in Maputo, Mozambique.

Pieter van der Zaag is a senior lecturer of IHE Delft, the Netherlands, and currently seconded to the Department of Civil Engineering, University of Zimbabwe. An irrigation engineer from Wageningen University, he wrote his PhD dissertation on formal and informal organising practices in irrigation management in Mexico. From 1993 to 1996, he coordinated the linkage programme between the University of Zimbabwe and Wageningen University.

Introduction

ALEX BOLDING, JEFF MUTIMBA, PIETER VAN DER ZAAG AND
JENS A. ANDERSSON[1]

This book is the result of research conducted under the auspices of the
Zimbabwe Programme on Women's Studies, Extension, Sociology and
Irrigation (ZIMWESI). The first phase of this inter-university exchange
programme on teaching and research between the University of Zimbabwe
(UZ), Wageningen Agricultural University (WAU) and the Department of
Agricultural, Technical and Extension Services (Agritex), got under way in
1993. One of the objectives of the project was a critical understanding of
intervention issues and practices in rural areas. To capture the complexity
and dynamics of smallholder agriculture in Zimbabwe, a team of researchers
of technical and social backgrounds was assembled. Academically the aim
was to provide interdisciplinary insights by means of doing in-depth case
studies that are concerned with day-to-day practices of the actors involved.
This resulted in a trilogy of books that reflect the multidisciplinary ambitions
of the team members. These books share a sense of collective authorship:
the authors extensively reviewed each other's work, and we have all benefited
from each other's insights and disciplinary specialisations. Whereas the first
two books cover issues related to smallholder irrigation (Manzungu and Van
der Zaag, 1996) and future options for water management in smallholder
agriculture (Manzungu, Senzanje and Van der Zaag, 1999), this book concerns
itself with the theory and practice of agricultural extension in the smallholder
sector. It aims to contribute insights and options for future extension
initiatives in Zimbabwe. To do so it draws on the work of nine authors from
various backgrounds, who present their case studies in 12 chapters. Before
going into the subject matter of each chapter, an overview is given of
international developments in agricultural extension. Thereafter a brief history
of agricultural extension in Zimbabwe is presented by sketching salient
developments in policy, research, extension and their effects on farming
opportunities.

---

1. We gratefully acknowledge the valuable comments made by Professor Linden Vincent on
   drafts of this chapter.

FOUR PARADIGMS OF AGRICULTURAL EXTENSION

The term agricultural extension normally invokes a number of images related to agricultural prosperity, progress, improvement and modernity in general. Another common association is that of state or public intervention: agricultural extension being one of the prime movers of agricultural development that is part and parcel of the fabric of a modern nation. In this book a number of case studies are presented that look into the workings of agricultural extension, and its impact on the smallholder sector of Zimbabwe.

Extension refers to an extrapolation, a movement or flow of knowledge, information, technology and services from a particular source to a receiver or client. The first use of extension in the sense we know it today was made in England in the 1840s. It was meant to be an extension of the university beyond its walls to the doorstep of the common people. In 1873 Cambridge University became the first to formally adopt a system of organising centres for extension lectures (Van den Ban and Hawkins, 1988, 7-8). Since then a variety of agricultural extension systems has been established across the globe. These systems generally aim at the promotion of agricultural development, commonly conceived as an increase in agricultural production and/or agricultural productivity. In 1914 the United States of America were the first to formalise the co-operative extension system organised by land-grant colleges as a public responsibility of the state. After World War II the promotion of public extension systems assumed a special significance. These were seen as an effective way to promote agricultural growth and trigger a process of modernisation in support of import substitution and industrialisation policies. It was believed that there was a considerable amount of readily available technologies on the shelf that, once disseminated in a systematic way, would trigger a process of modernisation of the national economy and the society as a whole (Rivera, 1991, 3). Recently, the conception of public extension's role has expanded to cover sustainable rural development of which agricultural production is only one facet. Considerable cross-fertilisation has taken place between contemporary ideas on development policies and how public extension systems can be used as an instrument in supporting these policies. Over time a number of different 'schools of thought' or paradigms of public extension have emerged responding to new development theories. These paradigms differ in their conception of what is to be extended (message), from where (source), to whom (receiver) and by which means (method). Before dealing with the salient features of Zimbabwe's public extension system, first four paradigms of public extension are described.

The dominant and most influential paradigm on public extension is based on the **Transfer of Technology (TOT)** model. Many national extension services have been designed and drafted on the basis of this model. The basic premises of the model is that tremendous increases in agricultural productivity can be achieved by means of the transfer and dissemination of (new) technologies. The successful introduction of new high yielding varieties of many food and export crops in the 1960s (Green Revolution) has reinforced the belief in a state-led process of agricultural development based on scientific innovations. The TOT model consists of three components that play distinctive roles in a linear process of technology generation, transfer and utilisation. Research produces technology. Extension transfers this technology to farmers. And finally farmers adopt the technology. The sequential, one-directional flow of innovations was to go from a research institute to the experiment station to a subject matter specialist to an extension worker to a contact farmer. Farmers were not considered experimenters and technology developers, but rather passive recipients and users. Innovations are spread over the farming community through a process of diffusion. Research in the 1940s and 1950s showed that three adopter categories could be distinguished: *innovators* who were fast and active in taking up innovations, *followers*, and finally *laggards* who were slow in taking up innovations (Rogers 1962; 1971; 1983). The overall success of the TOT system could be measured in terms of adoption rates. Non-adoption of innovations was explained in terms of 'resistance to change' or inherent conservatism of the farming community. The scientific basis of innovations was supposed to guarantee the appropriateness of extension messages.

However, after the initial successes of Green Revolution technology in intensive farming contexts (mostly in irrigated agriculture) the adoption rates started to drop. The initial response to this slow-down in technology transfer was to strengthen the intensity of interaction between researchers and extensionists, and extension agents and innovative farmers. Benor and Harrison (1977) devised the Training and Visit (T&V) system as a model for better extension services through better management (regular training of extension agents and farmers). The T&V approach proved to work in green revolution settings in India and was subsequently adopted by the World Bank as the ideal way of reviving ailing TOT systems (Hulme 1991; Bagchee, 1994).

Even with the high intensity of interaction between the various actors, weaknesses of the TOT model continued to emerge. Questions were raised on the appropriateness or profitability of some innovations for different categories of farmers. It was discovered that the farming community was not homogenous. Women, resource rich, resource poor, large and small scale farmers were found to have different needs and means at their disposal to incorporate innovations in their farming operations. A move was made from distinguishing adopter categories to distinguishing target categories (Röling, 1988, 29). Extension agencies took up the task of translating research findings into tailor-made packages of innovations that would suit different target categories. A two-way model of the research-extension-farmer continuum was developed by Havelock (1969) stressing the need for a feed-back mechanism from farmers via extensionists to researchers to articulate the needs of different target categories.

When it was discovered that the needs of some target groups (notably women and resource poor dryland farmers) were systematically not addressed by the TOT system, it was argued that these groups had to be mobilised in farmer organisations that could exert 'countervailing power' on research and extension agencies (Röling, 1988). An international research institute (ISNAR) was established in the early 1980s to develop effective linkage mechanisms between the main components of the TOT system (Kaimowitz et al, 1989a; 1989b). The 1980s also witnessed a new approach in research being promoted: Farming System Research (FSR). This approach acknowledged that optimal technologies developed in ideal circumstances (research stations) requiring high levels of inputs might not work in the context of the farming system. On-farm research and analysis of different farming systems were used to produce more appropriate innovations that would fit in the existing system of farming and be compatible with the level of inputs available. Thus, instead of simulating farming circumstances at the research station the reverse had to take place: bringing the research station to the farm(er)s (Chambers, 1993, 67).

From the 1980s onwards criticism on the merits of the TOT system increased: research priorities and locations were considered to be wrong, extension messages were inappropriate, and packages were rejected by farmers. In response a new paradigm emerged: **Farmer First**. This paradigm comprises a host of different approaches all stressing the need for, and benefits of, farmer participation and empowerment. Farmer Participatory

Research (FPR); Participatory Technology Development (PTD); Participatory Rural Appraisal (PRA); Farmer First and Last (Chambers, 1983; 1993; Chambers *et al,* 1989) are all concepts that have emerged in a concerted effort to ensure that the farmer takes centre stage. The paradigm stresses that the reasons why farmers do not adopt innovations lie not in their supposed ignorance or farm-level constraints, but in deficiencies in the technology and the process that generates it. To overcome these deficiencies roles have to be reversed: researchers and extensionists have to learn from farmers. Contrary to the universalist claims of 'scientific' knowledge and technology from research stations in the TOT system, farmer-led innovations rely on local 'indigenous' knowledge and technology development that suits the specific circumstances and conditions of the locality (Brokensha *et al.*, 1980; Richards, 1985). The aim of the approach is to empower farmers to capitalise on their inherent ability to experiment with a 'basket of technology choices' and enhance farmers' capacity to adapt existing technologies to suit their circumstances. The main mode of extension is from farmer to farmer. The extension agent acts as a catalyst of this process. Outsiders like researchers, extensionists and non-governmental organisations (NGOs) have to develop and transfer principles and methods of farmer-led experimentation and innovation. All actors in the TOT system have to be mobilised to contribute to the basket of choices (Chambers, 1993).

From the start this approach has been promoted vigorously by aid agencies and NGOs. The main emphasis so far has been on developing participatory methods and establishing principles of farmer experimentation. An IFAD study observed that despite the enormous amount of literature on Farmer First approaches there has been limited impact on the ground. Farmer-led extension programmes still are 'rare islands in a sea of conventional programmes' (IFAD, 1996). Critics of the approach have pointed at the exorbitant expense in human and financial resources that is required to make it work, the piecemeal effects produced by isolated pilot projects and the difficulties in scaling-up successful examples by transforming institutions and policies (Scoones and Thompson, 1994). In most cases Farmer First approaches end up prescribing policy strategies which advocate what ought to happen, without specifying necessary and sufficient conditions for a true Farmer First approach. The initiative for technology development in Farmer First approaches often continues to originate from outsiders (NGOs, donor agencies) rather than from the farmers themselves. The fact that most Farmer

First projects treat farmers, households and communities as undifferentiated and harmonious entities has not helped in producing the tailor-made farming options for different segments of the farming population (Gubbels, 1994, 241-242).

During the 1990s the contours of two more paradigms involving agricultural extension emerged. One paradigm more or less evolved from the Farmer First movement and focuses on communicative support for collective action aiming at developing ecologically sound and sustainable farming systems. As such it rallies against the environmental and social effects produced by modern agro-technologies and the invisible hand of the market. The other paradigm embraces neo-liberal approaches in favour of market mechanisms and is generally known as new institutionalism.

The **collective action for sustainable agriculture** approach has emerged in response to increasing concerns over persistent degradation of the resource base, negative environmental impact of high input agricultural modernisation and the resulting precariousness of livelihoods in (semi-)arid environments. On the one hand rising population pressures combined with unscrupulous exploitation of natural resources by transnational corporations have led to increased competition over scarce resources between different stakeholders. This in turn has led to further degradation, poverty and conflicts over the distribution of scarce resources (Röling, 1994; Röling and Pretty, 1997). The realisation that many smallholders in Africa live a precarious life due to limited access to natural resources, vague property rights and a degrading resource base, has led to a re-discovery of common property management regimes (Bromley *et al.*, 1992). Secondly, the paradigm postulates that sustainable agriculture is about the effective use of internal resources. Low External Input Sustainable Agriculture (LEISA) initiatives seek to minimise the use of environmentally destructive external inputs by fostering regeneration of internal resources (Reijntjes *et al.*, 1992). Unlike Farmer First approaches, LEISA proponents not only seek sustainable options at farm level, but also at higher levels in the agri-business complex by means of collective management arrangements (Röling and Pretty, 1997, 184). Finally the approach draws its aims from the realisation that smallholder farmers' strategies are mainly informed by risk considerations. Risk and uncertainty dominate people's lives in dryland areas. People's livelihood strategies have to cope with micro-level variabilities like rainfall patterns, crop pests, heterogeneity of soil types as well as with macro-level variabilities in market conditions, wage levels and environmental hazards (Scoones *et al.*, 1996).

Social and environmental sustainability are the main concern of this approach. The emphasis lies on management of natural resources rather than on technology transfer or adaptation. The answer for a more sustainable agricultural livelihood is sought in collective action by groups or communities at the local level. However, the problem is that such platforms for resource use negotiation do not exist and need to be created and facilitated by outsiders. Besides using participatory techniques as developed by Farmer First approaches, the paradigm relies heavily on soft systems methodology (Checkland, 1981; Checkland and Scholes, 1990). Platform building starts with an exercise in shared problem appreciation by all stakeholders in the natural resource. An agricultural knowledge and information system (AKIS) has to be developed to support decision making processes. On the basis of this information a consensus can be established through negotiation and accommodation between the various stakeholders. The end result is a process of shared learning which can lead to effective collective action (Röling and Pretty, 1997). Despite the young age of the paradigm, some successes have been reported. The land care programme in Australia led to increased care for natural resources by both farmers and policy makers (Pretty, 1995). Equally, the Integrated Pest Management approach in Indonesia has led to a decrease of pesticide use, whilst crop yields increased (Röling *et al.*, 1995). Critics to the approach have pointed at the difficulties of consensus building (Leeuwis, 1995). Stakeholders (farmers, officials, academics) differ in learning approach and capacity to articulate their views and problem analyses (Bawden, 1994). In the absence of a common understanding and perspective for action, local platforms for resource use negotiation often fail to produce sustainable action. Furthermore, doubts have been raised whether low external input agriculture can be sustained in semi-arid areas with nutrient-poor soils, since it tends to produce high cost unsustainable livelihoods (Low, 1994).

The fourth approach, **new institutionalism**, is a product of the neo-liberal economic adjustment policies that have been promoted by the International Monetary Fund and World Bank since the mid-1980s. These structural adjustment policies resulted in a reduction in public spending, leading to reduced budgets for public research and extension services and a general retreat of the state (Farrington, 1994; Beynon *et al.*, 1998). Furthermore, attention has been drawn to the inefficient nature of public institutions, which has been attributed to a lack of incentives for public researchers and extension agents to perform better. In response to this criticism, a group of new institutionalists has emerged. This group seeks to establish new institutions according to market principles, reform existing institutions and promote a more active role of private (agri-business) agencies (Rivera, 1991). The central concepts in the new institutionalist approach are cost recovery, financial autonomy, accountability and partnership (Beynon *et al.*, 1998). *Cost recovery* depicts a process where public research and extension agencies sell their services to their clientele. *Financial autonomy* implies that these agencies rely more and more on client fees, whilst retaining control over the funds thus generated. Cost recovery and financial autonomy together lead to increased *accountability*, by which is meant that the performance of public staff can be monitored and controlled by clients (e.g. farmers). Accountability can be further fostered by means of strong agency-client linkages and decentralisation of the policy making process (e.g. farmers setting research priorities). Those services that can be provided by private agencies have to be relinquished by public agencies. However, since not all tasks of public agencies (e.g. regulatory tasks) can be relinquished *partnerships* are to be established with the private sector (Carney, 1998). Christopolos and Nitsch (1996) argue that it is no longer a question of how the state can provide for a well-functioning national research and extension system, but rather how the state can support the agricultural development of specific target groups and meet specific objectives. Within the new institutionalist model a multi-plural complex of public agencies, privately funded agencies, agri-business firms, NGOs, farmer associations and private consulting firms delivers services to a variety of clients with different needs (Haug, 1999). During the 1990s most, if not all, national extension agencies have been forced to go through some degree of reform inspired by the above principles. Market-oriented and large-scale farmers have found their way to privately funded research and extension services. The big question that remains is how one can ensure that smallholder farmers with little political leverage or economic 'buying power' are not further marginalised by these reforms (Rukuni *et al.*, 1998).

After a relatively short life-span of trying to improve the fate of rural producers in developed and developing countries alike, extension finds itself the subject of 'improvement', as reflected in the title of the most recent FAO reference manual on agricultural extension (Swanson *et al.*, 1997). Others prefer to speak of 'critical turning points' (Rivera, 1991), public extension agencies facing the 'abyss' (Christopolos, 1999), and the need for a 'New Professionalism' (Pretty and Chambers, 1993). The final analysis as to why public extension is facing a crisis depends on the paradigm one advocates. The crisis in agricultural extension has been attributed to a lack of suitable technologies for smallholder farmers (Bagchee, 1994), a lack of effective linkages between the three sub-systems (Swanson *et al.*, 1997), a lack of incentives for researchers to deliver appropriate technologies (Beynon *et al.*, 1998), or a failure to transform public agencies into facilitators of farmer-led approaches (Scoones and Thompson, 1994). Whatever the reasons may be on the international front, Zimbabwe has not been able to evade this world-wide crisis in agricultural extension, as is succinctly expressed in the main report of the land tenure commission:

> Today, however, farmers are extremely unhappy about the performance of all major institutions, particularly research, extension and credit. The Commission was indeed shocked to find that the whole country and all ... farming sectors believe these lay institutions to be almost irrelevant to their situation (Zimbabwe, 1994, 137).

AGRICULTURAL EXTENSION IN ZIMBABWE

How did this situation come into being? As a first attempt to answer this question, this section provides an overview of major developments in agricultural extension in Zimbabwe.

Roughly speaking four different periods of agricultural interventions in Zimbabwe can be distinguished

— the period up to 1945, characterised by settlement of white farmers and segregated agricultural development;
— the period between 1945 and 1965, characterised by the first agricultural revolution in the commercial farming sector, and state control over African agricultural development;
— the period between 1965 and 1980, characterised by the need for crop diversification and the consolidation of segregated development;
— the period since 1980, characterised by an attempt to commercialise the smallholder farmer.

Below, each period is characterised in terms of the dominant intervention rationale, research efforts, agricultural extension approaches and their effects on farming opportunities.

Period to 1945: Settlement and segregated agricultural development

Intervention rationale

After the pioneer column of 1890 and their followers failed to find the mineral wealth that the British South Africa company had hoped to find in what was then known as Rhodesia, agriculture became the mainstay of the white settlers. The initial emphasis was on opening up the territory and building of communications (railways, roads) to facilitate commercial agricultural production and export, whilst simultaneously securing settler control over land and African labour. By carving up the available land resources along racial lines, the indigenous inhabitants of Zimbabwe were allocated land of inferior quality to be occupied under a communal tenure regime. The boundaries of these so called 'Reserves' were redrawn by the Reserves Commission of 1914-15 in reaction to increases in the number of white settlers. The Commission's proposals cut down earlier established reserves, especially where they were within easy access of main markets (Palmer, 1977b, 104-130). European settlers occupied most of the best land on freehold terms, covering almost one third of the total available land by the early 1920s. This racial division of land was accompanied by other measures that aimed to undercut African farmers' productive capacity. For instance, the settlers' company government tried to force the inhabitants of the Reserves to sell their labour in the rapidly expanding white capitalist sector. Forced labour and the imposition of taxes constituted the two prongs of this policy of labour mobilisation, and although the impact of these measures on African agriculture varied, circulatory labour migration became firmly rooted in social life in many reserves, turning them simultaneously into labour reserves (van Onselen, 1976; Ranger, 1985, 41-43; Johnson, 1992; Andersson, 2001a, b). Besides the extraction of labour and the racial land policy aimed at undermining African peasant agriculture, the sector was increasingly faced with unfair competition in markets for agricultural produce. From 1908 onwards white agriculture was actively fostered, at the neglect of African agriculture.

In 1923 Rhodesia became a self-governing colony, which strengthened the settler governments' ability to pursue its ideal of segregated development of the races, each at their own pace, using their own resource bases. The settler government felt a moral obligation to guide the African masses into the modern world of Christian ethics, improved agricultural techniques and general education, but without allowing them to compete with settler interests. The Morris Carter Land Commission of 1925 endorsed the principle of segregation in spite of the so-called Cape clause in the Rhodesian constitution allowing the British colonial office to veto discriminatory legislation against Africans. In exchange for the right to 7.5 million acres of – generally ill-situated — Native Purchase Areas (NPAs), which would be made available to a class of loyal, progressive yeoman farmers, the area available for white settler occupation was increased from some 31 million acres in 1925 to 49 million acres in 1930 (Palmer, 1977b, 185; Ranger, 1970, 115-117; Phimister, 1984).

The Land Commissions' findings formed the basis of Southern Rhodesia's most contentious legislation, the Land Apportionment Act of 1930. Brought into effect in 1931, and amended in 1940 and 1950, the Act triggered a massive translocation of Africans who, until then, had often been allowed to remain on their ancestral lands in areas assigned to the white settlers. Thus, the specific agricultural economy, so characteristic for colonial and post-colonial Zimbabwe, was born. In 1930 the 48,000 white settlers were given on average some 1,000 acres per person, whereas the one million Africans living in Reserves and Native Purchase Areas had, on average, access to a mere 29 acres per person (Palmer, 1977b, 185-6). The Land Apportionment Act became the cornerstone of subsequent settler policies and determined a course of African smallholder development that was more constrained than in other British colonies (Ranger, 1985, 68).

The economic depression of the early 1930s saw the nascent commercial farming sector on the brink of collapse, despite its privileged access to the natural resources of the country. Starting in 1931 the government actively intervened in the agricultural industry by controlling prices, extending subsidies to commercial farmers and controlling international trade through agricultural parastatals like the Maize Control Board (1931), Dairy Marketing Board (1931), Tobacco Marketing Board (1936), Cotton Research and Industry Board (1936), Cold Storage Commission (1937), Pig Industry Board (1937), Natural Resources Board (1941), and Sugar Industry Board (1944) (Weinrich, 1975). All these boards, in one form or another, have remained a characteristic feature of the agricultural industry up to very recent times (Rukuni, 1994).

State interventions for the development of African Reserves were part and parcel of the larger policy framework of segregated development. They aimed at improving agricultural practices to enable the Reserves to carry a larger human and livestock population and thus avoid the necessity to acquire more land for African occupation (Phimister, 1988). Missionaries and agricultural demonstrators instructed Africans on modern methods of agriculture that allowed for intensive rather than extensive use of the land. Indigenous agricultural practices, like shifting cultivation and intensive cultivation of streambeds, were considered primitive and environmentally destructive and had to be replaced by a system of permanent farming, based on crop rotations on fixed plots, integrating both crop and livestock production. During the 1930s and 1940s intervention policies became increasingly informed by conservationist concerns (Anderson, 1984; Beinart, 1984; Beinart and Coates, 1995). Policies of land rationalisation, through the separation of arable and grazing land (centralisation), were implemented to stem the dangers of erosion, to increase state control over African producers and to refine carrying capacity notions (Drinkwater, 1989; 1991).

Research

The Department of Agriculture started in 1897 with the broad mandate to promote the interests of commercial farmers and the progress of the agricultural industry of the territory. After the appointment of a specialist from the Cape colony (Dr Nobbs) the department got its commodity based organisation structure to realise its mission of identifying new commercial crops. By 1922 the Department consisted of a staff of 36 technical officers and 24 administrative and clerical officers. Salisbury experiment station and Gwebi experiment farm were opened in 1909 to test crops such as tobacco, cotton, rubber and coffee. The crop that proved most viable for commercial farmers was maize and in 1909 the first significant amount of maize was exported through the Salisbury farmers' co-operative society. Different varieties of white and yellow maize were tested and in 1919 the maize breeders' association was formed to cooperate closely with the department on such issues as methods of seed selection and production of pure-bred seed (Weinmann, 1972). However, the chief agriculturist (Mundy) soon raised alarm on the dangers of maize mono-cropping, threatening the viability of the commercial farming sector. Henceforth the department paid considerable attention to soil fertility management by experimenting with suitable systems of crop rotations and integrated 'mixed' farming (Wolmer and Scoones, 2000). Mundy postulated his evolutionary view of agricultural development, from pioneer farming characterised by destructive mono-cropping and land mining to a system of mixed farming, integrating separate cropping and livestock components in a sustainable farming system (Mundy, 1921). Trials were done on crop rotations, green manuring, application of kraal manure and application of chemical fertilisers (see Mundy and Walters, 1919; Weinmann, 1972).

In 1930 the Department of Agriculture was considerably expanded with separate divisions on animal industry, plant industry, agricultural engineering, forestry, entomology, chemistry and the lands department, office of the surveyor general, and veterinary services. During the ensuing decades a range of research stations were opened, mostly in favourable natural regions close to commercial farming centres, to host research that reflected settler agriculture interests. Some of these stations have remained important research centres to the present day, like the Cotton Breeding Station in Kadoma (1925), the Grasslands Experiment Station in Marondera (1930), the Tobacco Research Station in Trelawny (1935), and the Henderson Research Station north of Harare, devoted to crop, pasture and animal husbandry research (1947) (Weinmann, 1975). Research efforts were thus fully geared towards the development of the commercial agriculture. A demand-led approach was taken, characterised by close interactions between individual farmers and scarce researchers, later brought together in commodity boards (Scoones et al., 1996, 129-130). In sharp contrast, the only source of innovations for the smallholder sector were mission stations, neighbouring commercial farms and the odd agricultural demonstrator, as well as smallholders' own experiments. For instance the spread of the plough among African cultivators was largely precipitated by returning labour migrants, mission converts and white traders (Rennie, 1973; Ranger, 1985).

Extension

For the commercial farming sector no official extension service was established until 1948. The close interaction between farmers and researchers, as well as the publication of experiment results by both farmers themselves and officials of the Department of Agriculture in the *Rhodesia Agricultural Journal* (started in 1903) made a public extension service unnecessary (Scoones et al., 1996, 129). Mechanical protection of soils and dam development was done on a voluntary basis with help of government subsidies and advice from the irrigation branch in the Department of Agriculture (Weinmann, 1972; 1975; Phimister, 1986).

Extension in African areas started in 1927 when the first agricultural demonstrators were deployed at the request of Native Commissioners, under the guidance of E.D. Alvord (see Bolding, this volume). Demonstrators assisted willing farmers to implement a package of improved farming methods like crop rotations, heavy manuring, use of improved seed, weeding, ploughing and mono-cropping. This package of mixed, permanent farming was to sell itself by demonstrating its resulting high yields at pre-harvest meetings and agricultural shows. Thus an evolutionary pattern of development and modernisation of the African farmer was established, from follower (adopting some improved methods of farming), to plotholder (working closely with the demonstrator), to Master Farmer, who had a proven record of good farming practices. Reserve modernisation and development could be achieved with the deployment of different types of demonstrators. First an agricultural demonstrator; next a community demonstrator to help building improved, square brick houses and assist in centralisation of the area; a home demonstrator to teach proper health, nursing and household chores; and finally soil erosion, forestry, livestock or irrigation demonstrators to work in their respective areas of training. By 1948, 163 agricultural, 86 community, 25 erosion control, 34 forestry and 20 livestock demonstrators were operating in the Reserves. In spite of initiatives to increase smallholder production by means of irrigation and introduction of new cash crops like cotton, the programme shifted its emphasis to conservation during the mid-1930s. By 1944, the voluntary demonstration approach had reached its limits. Coercion was deemed to be required to save the Reserves from environmental destruction due to over-population, over-stocking and continued use of 'wasteful' agricultural practices (Alvord, 1948).

Effects on farming opportunities

During the first two decades of settler rule European farming remained near subsistence level, amounting to little more than market gardening for urban and mining centres (Palmer, 1977a, 228). However on the back of favourable policies, commercial agriculture expanded with three main export concerns: maize, tobacco and beef. By 1924, 78% of the land cultivated by Europeans was under maize, and its production expanded steadily until 1935, when the collapse of global maize markets forced many farmers to switch to tobacco (Palmer, 1977a, 232). Tobacco production went through several cycles of boom and bust during the periods of 1909-1913 and 1925-1928, before settling on a steady course of growth after the establishment of the Tobacco Marketing Board in 1936. From 1939 to 1948 the number of tobacco growers grew from 638 to 1,600, whilst export earnings increased from just over 1 million pounds to more than 11 million pounds sterling (Phimister, 1988, 225). The shift to tobacco combined with a rapid growth of the domestic economy resulted in a loss of food self-sufficiency (Phimister, 1988, 227). By 1942 maize production reached its lowest output in 16 years and a government bonus scheme was introduced for maize grown under sound farming practices (using green manure and compost on land protected from erosion by contour ridges) . Through this scheme the growing problem of under-capitalised farms mining the land under steady maize mono-cropping in an 'after me, the desert' fashion, was addressed in earnest for the first time (Phimister, 1986).

When the settlers arrived in Rhodesia they found a variety of indigenous farming systems like cattle ranching, intensive streambed (dambo) cultivation, and extensive shifting cultivation of the upper lands, facilitated by the increased availability of ploughs (Mtetwa, 1976; Palmer, 1977a). With increased security and the availability of new marketing opportunities provided by the nascent settler state, a vibrant African peasantry emerged, reaching its peak in the 1920s. A category of plough entrepreneurs, cultivating large tracts of land under maize mono-cropping, emerged and was further stimulated by missionaries like Alvord whose demonstrator programme provided further incentives to improve production. However, the drought of 1920-21, the implementation of the Land Apportionment Act of 1931 combined with the effects of a world-wide economic crisis and discriminatory price policies in maize and cattle marketing during the early 1930s put a severe constraint on the viability of the 'peasant option' in the Reserves (Palmer, 1977a). Centralisation, by restricting plot sizes, severely hampered the plough entrepreneurs of the 1920s. Furthermore, marketing opportunities for African farmers were drastically reduced, forcing many more Africans to work as wage labourers for longer periods of time. But as the economic crisis eased, the late 1930s and early 1940s saw the emergence of an African elite consisting of Master Farmers, Native Purchase Area yeomen and regimented settlers on irrigation schemes. Much more aware of the role of the state in restricting marketing, the non-provision of facilities in the Native Purchase Areas and the curbing of possibilities for Master Farmers and irrigation plot holders to expand, prevented them from becoming a loyal class of African farmers. In 1938, the Southern Rhodesia Bantu Farmers' Congress was established by disgruntled farmers in the Marirangwe Native Purchase Area (Ranger, 1985, 59-77).

1945-1965: First agricultural revolution and state control over African development

Intervention rationale

The post-war influx of settlers (up from c. 80,000 in 1945 to c. 125,000 in 1950), combined with the tobacco boom and growing manufacturing industry (the number of factories went up from 294 in 1939, to 473 in 1948), allowed a liberal minded government to pursue ideals of racial partnership and laissez-faire policies towards the commercial agricultural sector (Phimister, 1986; 1988). At the same time conservationist concerns of over-population and over-stocking in reserves culminated in increased technical control over African rural and agricultural development. The Natural Resources Board (1941), whilst advocating a voluntary change in farming practices and mechanical conservation of commercial farming areas through enhanced state subsidies and the formation of Intensive Conservation Area committees, prescribed forced de-stocking and compulsory contour ridging in African Reserves. The Native Production and Trade Commission (1944) confirmed that agricultural modernisation would be achieved through compulsion, which led to the passing of the Good Husbandry Act (1948) culminating in the Native Land Husbandry Act (1951). Forced de-stocking, contour ridging, crop rotation and individual tenure of land and grazing rights on the basis of rational planning of production and carrying capacity were the result. The interventionists were confident that those Africans who could not be accommodated in the Reserves would be absorbed as permanent labourers in the expanding industrial sector in the urban areas.

Research

The establishment in 1948 of the Department of Research and Specialist Services (DR&SS) introduced a shift from demand-led to supply-led research, where scientists identify research needs, whilst designing complex research trials at favourably located research stations, with emphasis on high input technology solutions (Scoones *et al.*, 1996, 128). Thus the vestiges of the Transfer of Technology model in agricultural development were established in Zimbabwe. The result became known as the first agricultural revolution in the commercial farming sector. A breakthrough in the production of superior hybrid maize varieties available for commercial production in 1948/ 49, combined with the availability of cheap fertiliser, resulted in tremendous increases in yields (Weinmann, 1975). Over the period 1948-1965, maize yields increased by 155%, cotton by 100%, tobacco by 300% and wheat by 185%, compared to a mere 25% yield increase in sorghum. In the same period livestock research focused on improvement of dry season feeding, resulting in a 150% increase in beef offtake per herd (Tawonezvi, 1994, 93). Increased efforts in land surveying, mapping, and climatologic research led to the development of a system of agro-ecological classification in relation to farming potential. This identification of so-called Natural Regions has remained influential in agricultural extension advice ever since (Box 1).

Box 1: The system of agro-ecological classification of Zimbabwe in 5 Natural Regions (Vincent and Thomas, 1957; Ivy, 1978)

Natural Region 1: Specialised and Diversified farming region: High rainfall (more than 1000 mm per annum) normally during all months of the year. Afforestation, fruit and intensive livestock production, as well as tea and coffee production in frost-free areas, can be practised.

Natural Region 2: Intensive farming region: Rainfall is confined to summer and is moderately high (750-1000 mm). The region is suitable for intensive systems of farming based on crops/and or livestock.

Natural Region 3: Semi-intensive farming region: Moderate rainfall (650-800 mm), but with fairly severe mid-season dry spells. Suitable for mixed farming systems based on both livestock and crop production.

Natural Region 4: Semi-extensive farming region: Fairly low rainfall (450-650 mm), which is subject to periodic seasonal droughts and severe dry spells during the rainy season. The region is suitable for livestock production combined with drought-resistant crop production.

Natural Region 5: Extensive farming region: The rainfall is low and erratic (below 600 mm). This region is considered only suitable for extensive cattle ranching or game ranching.

In the context of the British colonial policy of divestment and preparation for Independence, the predecessor of the University of Zimbabwe was opened in 1957 as a multi-racial institute of higher learning. Critical studies on the sociological and economic aspects of smallholder farming (Aquina, 1963), and the relationship between population pressure and changing patterns of African land use (Allan, 1945; Floyd, 1959; Jordan, 1963; Hamilton, 1964), were undertaken by the Rhodes-Livingstone Institute.

Extension

In 1950 the Soil Conservation Branch of DR&SS was upgraded to become the Department of Conservation and Extension Services (CONEX) with the mandate to give technical and extension advice to white farmers, though CONEX's primary concern was mechanical conservation (Kennan, 1980, 184). The number of conservation officers increased from 16 in 1944, to 69 in 1949, most of them being ex-servicemen trained at the University of Witwatersrand, Johannesburg, in a special soil conservation course (Weinmann, 1975, 147). In 1945 the government instituted a 50% subsidy for commercial farmers on the costs of construction of soil and water conservation works in Intensive Conservation Areas. As a consequence by 1950 most (91.5%) of the total cultivated area on commercial farms was protected by means of contour ridges (Phimister, 1986, 267). Attention within CONEX soon shifted to commodity-based extension and technical advice. Its field branch was organised through provincial and area officers that worked in close interaction with local farmer associations and researchers of DR&SS. By the late 1950s the number of ordinary field advisers was 87, providing one adviser to every 65 farmers (Dunlop, 1971, 38). Officers were trained at the Gwebi College of Agriculture (opened in 1950) or alternatively attained BSc degrees in agriculture at the University of Rhodesia. An in-service training programme was established, though not much attention was paid to extension methods. In 1962 a regional workshop on extension methods introduced the extension philosophy of the diffusion of innovations in the TOT fashion (Vincent, 1962). However, only by 1978 an extension philosophy similar to that of the United States Co-operative Extension Service was formalised in CONEX (Kennan, 1980, 185).

Agricultural extension in the Reserves in this period was characterised by swings between persuasion and coercion, production and conservation, and placement of the responsible department under either the Ministry of Agriculture or the Ministry of Native Affairs (later Internal Affairs) (Box 2).

Box 2: Public extension services for the smallholder and large-scale commercial sector

Public extension services for the smallholder sector

Years	Department	Ministry
- 1925	—	—
1926	Agriculturist for Natives in Department of Agriculture	Secretary for Lands and Agriculture
1927-30	Agriculturist for Natives in Dept of Native Education	Ministry of Education
1930-33	Agriculturist for Natives in Dept of Native Development	?
1933-44	Agriculturist for Natives in Native Affairs Dept.	Ministry of Native Affairs
1944-61	Dept of Native Agriculture	Ministry of Native Affairs
1962-63	Dept of Native Agriculture	Ministry of Agriculture
1963-68	CONEX	Ministry of Agriculture
1969-78	Dept of Tribal Agriculture	Ministry of Internal Affairs
1978-80	Dept of Agricultural Development (DEVAG)	Ministry of Lands, Natural Resources and Rural Development
1980-81	DEVAG	Ministry of Land, Resettlement and Rural Development
1981-2000	Dept of Agricultural and Technical Extension Services (AGRITEX)	Ministry of Land, Resettlement Rural Development

Public extension services for the large-scale commercial sector

Years	Department	Ministry
1897-1948	Dept of Agriculture	Secretary for Lands and Agriculture
1948-1950	Sub-dept of Conservation and Extension (CONEX)	Ministry of Agriculture
1950-1981	CONEX	Ministry of Agriculture
1981-	AGRITEX	Ministry of Land, Resettlement and Rural Development

The Department of Native Agriculture, formed in 1944 under the directorship of Alvord, steadily shifted its approach from persuasive extension focusing on production intensification with help of demonstration methods, to coercive extension approaches emphasising forced de-stocking, compulsory contour ridging and rationalisation of land use during the implementation of the Native Land Husbandry Act. After 1951 the implementation of the Act, supported by massive World Bank loans, became the Department of Native Agriculture's major focus. Only after 1961 was there a policy swing back to persuasive, production oriented, extension with more emphasis on Master Farmer training and group approaches under the community development approach imported from the United States of America (Kennan, 1980; Mutizwa-Mangiza, 1985). From 1963 to 1969, CONEX became responsible for agricultural extension in all farming areas, leading to a variety of new approaches (see next period).

Effects on farming opportunities

Commercial farming output increased tremendously from 1945 to 1957, after which production in some crops tailed off. Due to the availability of various export markets and a growing demand for food stuffs on the domestic market a specialisation of farming occurred. Industrial development increased on the back of the vibrant commercial farming sector (Dunlop, 1971).

A birds-eye view on Southern Rhodesia's African reserves in the early 1960s would immediately reveal the impact of the unprecedented level of state intervention in this period. Farming activities were forced into a rigid frame of consolidated blocks of grazing and arable lands, the latter being subdivided into plots marked by contour ridges. But on the ground the effect of this intensive planned modernisation programme was even more pronounced, as it met with growing African resistance. Whereas African protest against the settler state had been largely confined to urban areas in the 1940s – as the 1948 strikes exemplify (Vickery, 1996; Raftopoulos, 1999; van Velsen, 1964) – the massive state intervention in the rural areas in the 1950s helped to spread the nationalist opposition to the countryside. Thus, urban and rural opposition to the colonial state became linked, just as rural and urban livelihoods had been intimately connected since the time Africans entered into wage labour relations with white employers (see Andersson, 2001b). The opposition to the Native Land Husbandry Act of 1951 was broad based. The Act's levelling effect – limiting the cattle numbers and restricting access to land to 8-15 acres per individual farmer – threatened the Reserve entrepreneurs, who vehemently opposed the Act. But also the labour migrants working in town, running the risk of losing their claim to land in what they considered their home area, resisted the Act. Hence, opposition was not class-based.

Meanwhile, the state's agricultural intervention programmes sometimes seemed contradictory. Whereas the conservationist concerns and rationalisation of land use restricted farming opportunities in the Reserves, the food production drive policy of the early 1950s and the availability of hybrid maize seeds both encouraged and enabled African smallholders to grow more food. The widely held belief that the interventionist era of the 1940s and 1950s, and the implementation of the Native Land Husbandry Act in particular, marked the end of independent farming outside the African Purchase Areas and effectively halted rural differentiation, has recently been challenged. The slow progress in the implementation of the Native Land Husbandry Act – the Act was only implemented in 42% of the total Reserve area – and its short-lived enforcement as a result of increased opposition, has raised doubts on its impact on African farming opportunities (Phimister, 1993; Andersson, 2001a; Weinrich, 1973). As Phimister has argued, a 'large number of the better-off peasants came through the Land Husbandry Act, if not unscathed, then more or less intact' (Phimister, 1993, 237).

1965-1980: UDI, crop diversification and consolidation of segregated development
Intervention rationale

The Unilateral Declaration of Independence by Ian Smith in 1965 led to international isolation and triggered a policy of diversification of the commercial farming sector, away from tobacco to maize, cotton and wheat production (Muir and Blackie, 1994). Massive subsidies and public investment in irrigation development, particularly for wheat production, resulted in a total irrigated area on commercial farms of 139,000 hectares by 1982 (Rukuni, 1984). Ian Smith's Rhodesian Front party revived a policy of segregated development of the races. Under the broad banner of community development, 'traditional leaders' were given the lead in the development of Tribal Trust Lands (formerly Reserves). Chiefdoms were mapped and African communities delineated. The Tribal Trust Land Authorities Act (1967) subsequently gave chiefs and their land authorities rights to allocate land and supervise cultivation. The policy of reviving 'traditional' authorities resulted in the Tribal Courts Act of 1969 (Ranger, 1985). The Tribal Trust Land Development Corporation (TILCOR) was established in 1968 to foster economic development in Tribal Trust Lands by establishing a network of growth points, serviced by agricultural estates remote from current centres (Rukuni, 1994). Smallholder irrigation development, which had been halted between 1958 and 1965 on account of its unprofitability (Hunt, 1958), was organised around these growth points. However, due to chronic under-capitalisation of TILCOR only a limited number of growth points were established (Gasper, 1988; Rukuni, 1994). The liberation war during the late 1970s ultimately crippled state interventions in the Tribal Trust Lands. In a last ditch attempt to gain support from 'progressive' Africans, the new regime of Zimbabwe-Rhodesia under the leadership of Bishop Muzorewa (1979), implemented a policy of protected villages to prevent guerrilla insurgency, whilst at the same time providing credit facilities and other services to ailing African Purchase Areas (Ranger, 1985). Ultimately the Lancaster House agreement of 1980 put an end to UDI and inaugurated Zimbabwe.

Research

In 1970 the Agricultural Research Council was established to direct research efforts in close coordination with farmer representatives of the Rhodesia National Farmers' Union (RNFU). By 1976 it assumed direct responsibility for research programmes, bringing together top representatives from the RNFU, DR&SS, CONEX, Tobacco Research Board, University of Rhodesia and the agro-industry. Breakthroughs came in maize production through the further development of high yielding hybrid varieties; in cotton by the development of effective pest and disease control technologies in the mid-1960s; and in wheat due to short straw, high yielding varieties with good resistance to diseases and capable of utilising high levels of fertiliser. Yield increases in commercial agriculture over the period 1965-1980 amounted to 67% for maize, 48% for cotton and 138% for wheat. Self-sufficiency in wheat was reached in 1976. In livestock production, the increased availability of cottonseed and soyabean meal as sources of supplementary feeding during the dry season resulted in the adoption of intensified cattle fattening strategies (Tawonezvi, 1994, 94-96).

Extension

After a brief interlude of production-oriented extension, introducing tea, tobacco, cotton and cattle fattening, African agriculture was once more placed under Internal Affairs in 1969, leading to a renewed emphasis on mechanical conservation (Bates, 1980, 187). In order to prevent the bias in favour of few relatively well endowed progressive farmers, as had become the practice under Alvord's demonstrator programme, the group approach was adopted (Plowes, 1980).The trickle down effect of agricultural innovations was actively fostered by means of organised group work, co-operative marketing arrangements and pooling of resources. Master Farmer associations, Young Farmers Clubs, co-operative societies, savings clubs, women's clubs and commodity marketing groups became the mainstay of the community development programme (Weinmann, 1975; Bates, 1980; Plowes, 1980).

Effects on farming opportunities

In spite of the international trade sanctions and resulting market loss for tobacco, the policy of diversification resulted in sustained success of the commercial farming sector, particularly in wheat, cotton and maize. However, gross margins on commercial farms fell by 27% in real terms from 1973 to 1979 (Muir and Blackie, 1994). In addition, the ongoing liberation war severely constrained commercial farming operations. In some districts, e.g. Chimanimani, a majority of white farmers vacated their farms for fear of being killed by freedom fighters (Alexander, 1995).

In the African smallholder sector in this period, state intervention was concentrated in particular areas and its impact highly differentiated. For instance, in rural centres and irrigation schemes, a regimented, commercial type of smallholder producer emerged, whose farming operations were firmly under the control of white managers (Hughes, 1974). In Gokwe resettlement scheme, the state initiated an innovative cotton programme, thus triggering viable smallholder production in the late 1960s. By 1980 the smallholder sector produced almost 20% of the total cotton production in the country (Tawonezvi, 1994). Such intervention-driven agricultural development contrasted with the situation in other areas, such as Gokwe's neighbouring Sanyati district, where the implementation of the Native Land Husbandry Act during the 1950s only resulted in a decline of peasant production (Worby, 2000). Furthermore, the general policy shift from state-led agricultural modernisation in the 1950s towards decentralised community-based development in the 1960s, can be seen as a prelude to the state's loss of control over developments in the African areas in the 1970s when guerrillas gradually took control over the rural areas. With the Tribal Trust Land Authorities Act (1967) state control over land allocation and the supervision of cultivation diminished as traditional leaders were given the responsibility over these issues (Holleman, 1969). Hence, whereas in some areas state intervention opened up new opportunities for African smallholder farmers, its impact on farming in other areas was limited. While African Purchase Area farmers and Master Farmers in favourable regions continued to do relatively well, the majority of African farmers in the Tribal Trust Lands were merely scraping through, especially those without livestock (Weinrich, 1975). With the intensification of the guerrilla war in the mid-1970s the work of officers and extension assistants in Tribal Trust Lands came to a virtual standstill (Kennan, 1980; Plowes, 1980). Many Master Farmers, purchase area farmers and extension assistants were accused of being sell-outs and were killed by the freedom fighters (Ranger, 1985, 269-272). The policy of protected villages further alienated rural progressives from the Smith regime, since no compensation was given for any lost property (Ranger, 1985, 266). The protected villages themselves are an example of the contrasting developments in rural intervention in this period. An extreme product of state control, these villages formed the counterpoint of the surrounding rural areas where, by the late 1970s, state control had been lost completely.

1980-1996:[2]: Commercialising the smallholder farmer
Intervention rationale

After Independence a re-orientation to the smallholder agricultural sector was effected through increased public investments in credit facilities, research and extension services, as well as the provision of government controlled marketing facilities. The civil service was rapidly Africanised and expenditure on education soared. The national reconciliation policy, announced by Mugabe at Independence, allowed the large-scale commercial farming sector to re-establish itself and enter a period of sustained growth, though the number of commercial farmers dropped from around 6,000 in 1980, to some

---

2. Since the bulk of the empirical material presented in this book was collected during the mid-1990s, the historical coverage of this book effectively stops in 1996.

4,000 in 1990 (Muir and Blackie, 1994, 198). The resettlement programme, though initially ambitious, came to a virtual standstill towards the end of the 1980s due to a lack of public funds to honour the 'willing seller, willing buyer principle'. By 1989, 48,000 households had been resettled on 2.8 million hectares, whilst the original intention was to resettle 165,000 households on five million hectares of land (Moyo, 1995, 118-125). By 1986 a shift occurred in government interest from resettlement to restructuring of communal areas, in order to achieve production increases. This shift was caused by the high cost of drought relief: from 1982-85 the expenditure on drought-relief was comparable to that for resettlement during 1980-85. Thus improvement of land husbandry practices in communal areas became a distinct issue, separate from resettlement. Whilst the Riddell Commission of 1981 had been the first to raise this issue, the Chavunduka Commission (1984) pointed out that land husbandry was the main problem facing communal area production. The internal resettlement of communal areas was vigorously pursued from 1985-1990, on lines that were very similar to those of the Native Land Husbandry Act three decades earlier (Drinkwater, 1989). Thus rural interventions very much sustained the colonial rationale of realising the potential of communal areas by means of calculations of carrying capacities, grazing schemes, improvement of farming practices and the establishment of rural growth points.

From 1991 onwards Zimbabwe embarked on a programme of Economic Structural Adjustment, liberalising the economy. This policy resulted in a further decline in real terms of government funding for resettlement, research and extension services. A number of marketing boards were privatised and price policies were no longer controlled by the government. This policy has amongst other things resulted in a growing role of NGOs and donors in rural development initiatives.

Research

At Independence the mandate of DR&SS was re-oriented towards the smallholder sector. Farming Systems Research was established with a focus on on-farm research. Also, increased research efforts were made towards the improvement of millets and sorghum (often referred to as 'small grains'), horticultural crops promotion and conservation tillage (Tawonezvi, 1994). However, the bias remained on promoting high input agriculture in favourable Natural Regions through conventional on-station trials. Scoones *et al.* (1996, 133) remarks in this regard that '[a]gricultural research in Zimbabwe appears to have been seduced by scientific credibility and lost its way.' Some notable exceptions were the work on conservation tillage at the Makoholi Experiment Station in Natural Region 4 (Hagmann and Murwira, 1995) and the work on agronomic practices in low rainfall regions at the Chiredzi Research Station, situated in Natural Region 5 (Nyamudeza, 1999). The liberalisation of the economy after 1991 further contributed to a decline of the role of DR&SS, due to a sharp decline in government funding. As a direct result most research stations only had operating funds for the first 6 months into the financial year. A staggering 75% of DR&SS expenditure was on salaries and administration, leaving limited operating funds (Rukuni *et al.*, 1998, 1078).

The commercial farming sector more or less withdrew from DR&SS and the Agricultural Research Council, relying on its own privately funded research institutions, such as the Tobacco Research Board, Pig Industry Board, Rattray Arnold Research Station of Seed-co, Zimbabwe Sugar Association experimental station, and the Agricultural Research Trust (ART farm, established in 1982 and entirely funded by commercial producer organisations). By 1992 these semi-public research institutions accounted for 32% of the total agricultural research expenditure in Zimbabwe (Beynon *et al.*, 1998). Ever since the policy of liberalisation was adopted, international agro-chemical, fertiliser and seed producing companies, like Ciba-Geigy, Bayer, CARGILL, and Pioneer play an increasing role in producing and releasing new short season, drought tolerant varieties in maize, cotton and other commercial crops, though surprisingly little in millets and sorghum. This trend has fuelled fears about the lack of voice of smallholders in directing the research agenda (Rukuni *et al.*, 1998). An early attempt in 1986 to revive the Transfer of Technology model by linking smallholders' needs, extension advice and research efforts through the establishment of a Committee for On-farm Research and Extension (COFRE) failed to produce tangible results (Madondo, 1992; Tawonezvi, 1994).

Extension

A merger of CONEX and DEVAG resulted in the establishment of the Department of Agricultural, Technical and Extension Services (Agritex) in 1981. Many features of CONEX remained intact in Agritex, such as the in-service training for new officers and the commodity based structure of the technical division. Agritex's mandate steadily increased over the years to cover also regulatory tasks, the provision of drought relief, and the co-ordination of other development agencies operating in communal and resettlement areas (Madondo, 1992). Agritex's extension mandate committed itself towards commercialising the smallholder farmer by stimulating the adoption of 'proven' technologies, thus confirming a strong belief in the Transfer of Technology paradigm. This has remained so, in spite of the exposure of Agritex to a variety of donor funded experiments with other extension approaches. Thus Agritex has largely continued to build on previously established models of mixed smallholder farms, integrating livestock and high input crop production, increasingly seeking to simulate the large-scale farming model at a smallholder scale (Madondo, 1992; 1995; Wolmer and Scoones, 2000). The dominant extension methods in Agritex reflect a high resonance with the approaches developed under Alvord and his successors. Field days, agricultural shows, Master Farmer training, and group approaches for commodity groups comprise the main menu of the average front line extension worker. After a brief experiment in 1984 in Midlands province, the World Bank promoted Training and Visit system was discarded due to a lack of suitable technologies to be promoted, and to the drain on scarce human and financial resources (Pazvakavambwa, 1994, 106). The on-going professionalisation of officer staff, through the attainment of university degrees in specialised agricultural disciplines, has led to increased specialisation of services, possibly at the expense of a full appreciation of the day-to-day realities of communal and resettlement farmers (Madondo, 1992; Pazvakavambwa, 1994). Alternative extension approaches that based themselves on the realities at the grassroots have been tried by donor funded NGOs, such as ITDG (Scoones *et al.*, 1996), the GTZ funded Conservation Tillage programme, the programmes of ENDA Zimbabwe and others. However, the sometimes successful farmer innovations thus generated have proven to be resource intensive and *ad hoc*, prohibiting effective scaling up to the entire nation. After adopting the Economic Structural Adjustment Programme, government funding of Agritex declined affecting the operational activities of its staff (Madondo, 1992). In response, the Zimbabwe Farmers Union made an attempt to set up its own extension services. In addition, the importance of specialised extension advice provided by various agro-industry companies increased during the 1990s.

Effects on farming opportunities

In the early 1980s Zimbabwe's smallholder sector witnessed the so-called second agricultural revolution. Cotton and maize production rose dramatically as a consequence of access to government credit facilities and opening up of new marketing depots in rural areas. By 1985 communal farmers produced and sold more cotton to the Cotton Marketing Board than their large-scale counterparts (Rukuni, 1994). However, the success was not across the board. Over half of the marketed maize production from 1980 to 1988 originated from six well endowed districts out of a total of 53 (Stack, 1994, 261). And even within those districts the bulk of excess production originated from a minority of resource rich smallholder farmers (Stack, 1994, 262; see also Cliffe, 1988). The 1991/92 drought impacted severely on the smallholder sector, decimating cattle herds and forcing most communal farmers to rely on food hand-outs for their survival. Drought relief, grain loan schemes and handing out of seed packs remained a recurring phenomenon during the 1990s.

Resettlement did not provide a vibrant, alternative model of smallholder farming. Production levels have hardly exceeded those attained in communal areas. Model A resettlement schemes have typically copied the mixed farming model that is so commonly found in Zimbabwe's communal areas. Model B resettlement schemes, incorporating a cooperative farming approach, have failed to become productive (Moyo, 1995).

The commercial agricultural sector flourished in the period under review through increased export production of tobacco, as well as through the opening up of a new market for horticultural products like flowers, fruit and vegetables (Muir and Blackie, 1994). Around 85% of the total marketed agricultural production in the mid-1990s originated from large-scale farmers (Beynon et al., 1998, 82). Their sustained success can be attributed to favourable credit facilities provided by the Agricultural Finance Corporation and the National Farm Irrigation Fund; strong organisation of the sector through the Commercial Farmers Union and privately funded research institutions like ART Farm; and generally very good tobacco prices (Rukuni et al., 1998; Muir and Blackie, 1994).

The number and role of NGOs quickly increased after the liberalisation policies were effected. NGOs proved themselves capable of filling the gap left by a retreating state, but often failed to support farmer-led approaches to agricultural development on a country-wide scale. They often attracted capable staff from Agritex and other government agencies through favourable job packages (Vivian, 1994). The role of the Zimbabwe Farmers' Union in improving the fate of smallholder farmers remained small, in spite of many donor-funded capacity building programmes (Arnaiz *et al*, 1995). Finally, the on-going liberalisation of the economy and privatisation of marketing agencies led to the further marginalisation of remote communal areas, whereas favourably located areas, close to urban outlets and roads have flourished through increased opportunities of contract farming and ease of marketing.

ORGANIZATION OF THE BOOK

The book consists of four parts. The first part puts agricultural extension within the context of the smallholder sector in Zimbabwe. Different aspects of state interventions in both the past and the present are dealt with by comparing government attempts to improve and modernise livelihoods in communal areas with the day-to-day concerns of communal people themselves. The first chapter of the book looks into the origins and consequences of the agricultural demonstrator programme as developed in the 1920s by E.D. Alvord, the American missionary known as the father of agricultural extension in Zimbabwe. Bolding observes that the impact of the programme has been such that a number of continuities in development objectives, technical innovations and methods of extension have persisted to the present day. However, the majority of communal farmers ignored the extension messages and continued to pursue labour extensive methods of agriculture that pay tribute to the vagaries of the environment in communal areas. In Chapter 2, Mutimba extends this point by looking into the production constraints farmers are presently facing in three different communal areas situated in Natural Regions 2, 3 and 4, respectively. He shows how farmers, with the help of local knowledge and experience, actively adapt and re-work official extension recommendations to suit their farming conditions. Manzungu, in Chapter 3, tackles the controversy surrounding Agritex's attempts to transform rural farm families from subsistence into commercial agriculturists within the context of a smallholder irrigation scheme. By examining the way Agritex staff train Master Farmers, determine cropping programmes and tackle risk management in the scheme, it is shown that the state's perception on commercialisation differs from that of farmers. Agritex's efforts to impose 'scientific' farming as a way of commercial farming clashes with the pragmatic attitude of farmers who primarily seek to improve their livelihoods. Finally Andersson, in Chapter 4, challenges the common perception that people in communal areas are predominantly farmers. Labour migration has become an important aspect of social and economic life in communal areas, affecting the availability of labour for agricultural activities as well as the preparedness to invest in agriculture. Andersson argues that future agricultural interventions should take account of the existing multiplicity of rural livelihood strategies of which agriculture is only a part.

The second part of the book provides two examples of the transfer of technology at work. Hanyani-Mlambo, in Chapter 5, looks at the deforestation paradox in a resettlement area, where in spite of low population pressures and intensive extension efforts to conserve forests, deforestation continues to increase. The author analyses the failure of an exotic technology (re-forestation of gum trees) by assessing the motivations and practices of different strata of farmers and extension staff. In Chapter 6, van der Zaag observes that the success of the bench terrace is not simply a matter of a smooth transfer of conservation technology, but rather the result of mutual efforts of invention and intervention on the part of local farmers and extension officials. Whereas state intervenors favoured the use of the contour ridge under gradual slopes, a biography of the bench terrace teaches us that its development, appropriation and use were an outcome of farmers' interest in cultivating steep hills and extension agents' concern with devising a conservation technology that would suit the local conditions.

Part 3 looks at how extension interventions could benefit from the use of local knowledge and social networks. In Chapter 7, Vijfhuizen shows how agricultural produce in a smallholder irrigation scheme obtains its exchange value through cash sales, labour, barter, gifts, consumption and seed reproduction. For groundnuts and tomatoes the value agreed upon is influenced by social relationships between the actors involved rather than by the economic imperatives of the market. A convincing case is made for extension agencies to pay more attention to the role played by women farmers and the various ways in which crops attain a value when assessing the performance of smallholder irrigation schemes. Muchena looks at the different bases of 'scientific' and 'indigenous' knowledge in Chapter 8. She advocates the use of Shona proverbs by extension agents as a way to benefit from indigenous knowledge concerning natural resources.

Differences in perception often prohibit the establishment of effective linkages between researchers, extensionists and farmers. The fourth part of

---

Box 3: Dialogue between a Medical Doctor and a Rain Doctor

Medical Doctor: So you really believe that you can command the clouds? I think that can be done by God alone.

Rain Doctor: We both believe the very same thing. It is God that makes the rain, but I pray to him by means of these medicines, and, the rain coming, of course it is then mine. It was I who made it for the Bakwains [Kwena] for many years . . .; through my wisdom, too, their women became fat and shining. Ask them; they will tell you the same as I do.

Medical Doctor: But we are distinctly told in the parting words of our Saviour that we can pray to God acceptably in his name alone, and not by means of medicines.

Rain Doctor: Truly! but God told us differently. . . . God has given us one little thing, which you know nothing of. He has given us the knowledge of certain medicines by which we can make rain. *We* do not despise those things which you possess, though we are ignorant of them. We don't understand your book, yet we don't despise it. *You* ought not to despise our little knowledge, though you are ignorant of it.

Medical Doctor: I don't despise what I am ignorant of; I only think you are mistaken in saying that you have medicines which can influence the rain at all.

Rain Doctor: That's just the way people speak when they talk on a subject of which they have no knowledge. When first we opened our eyes, we found our forefathers making rain, and we follow in their footsteps. You, who send to Kuruman for corn, and irrigate your garden, may do without rain; we can not manage in that way . . .

Medical Doctor: I quite agree with you as to the value of the rain; but you can not charm the clouds by medicines. You wait till you see the clouds come, then you use your medicines, and take the credit which belongs to God only.

Rain Doctor: I use my medicines, and you employ yours; we are both doctors, and doctors are not deceivers. You give a patient medicine. Sometimes God is pleased to heal him by means of your medicine; sometimes not - he dies. When he is cured, you take the credit of what God does. I do the same. Sometimes God grants us rain, sometimes not. When he does, we take the credit of the charm. When a patient dies, you don't give up trust in your medicine, neither do I when rain fails. If you wish me to leave off my medicines, why continue your own?

[Livingstone, 1857, 23-25; cited in Comaroff and Comaroff, 1992, 243-4] the book deals with the nature of such encounters. Magadlela, in Chapter 9, presents a case study of a smallholder irrigation scheme where Agritex tries to impose a management model on farmers, who in turn manipulate existing and new management committees to further their own separate causes. Ultimately Agritex staff get caught between administrative and advisory roles, forcing the manager to close the scheme. The chapter raises important questions concerning the role of Agritex and the way power struggles affect the impact of extension interventions. Chapter 10, by Manzungu, deals with a case of information gathering by researchers using a 'quick and dirty' survey questionnaire. In this case, the irrigation farmers appear uncooperative, refusing to be treated as mere statistics. This raises questions on how researchers and other outsiders can establish effective rapport with farmers that satisfies the needs of both parties. Finally, Mutimba in Chapter 11, presents two case studies of university research projects that aim to improve farmer-research linkages through a process of participatory technology development. Mutimba demonstrates that the orientation of many researchers towards transfer of technology prohibits full realisation of the potential of farmer participatory research.

The concluding chapter distils the salient issues raised in the empirical chapters by outlining the contours of two separate worlds, the world of smallholder farmers on the one hand, and the perspectives of the national extension and research agencies on the other. The crux of the clash between these two world-views has been unveiled before by Livingstone nearly 150 years ago (Box 3). The chapter explores ways of bridging these worlds.

Part I
Past and Present of Agricultural Extension

Chapter 1

Alvord and the demonstration concept
Origins and consequences of the agricultural demonstrator scheme, 1920-44

ALEX BOLDING

INTRODUCTION

Many communal farmers in present-day Zimbabwe have observed a change in the services delivered by Agritex extension workers. If we may believe them, some extension workers do not visit farmers' fields as often as they used to, spend a lot of time writing reports in their offices and most important of all they do not not practically demonstrate the knowledge and skills they try to impart to farmers. According to many of the older farmers, Agritex extension agents stand no comparison with their predecessors, the demonstrators or *madhomeni* as they are known in Shona. One communal farmer in Makoni district noted: 'the profession of extension worker has become a white-collar job nowadays'.

This apparent change in agricultural extension is not only expressed by communal farmers, but also felt in some quarters of Agritex. During a workshop for Agritex extension workers and supervisors in Manicaland in 1995, a District Agricultural Extension Officer confronted his audience with a riddle:

> What is extension? Passing on of new ideas from research stations to farmers. Let me ask you something else: How many farmers attend your meetings? Aha, very few . . . Why? Because they know . . . The point is that we have run out of ideas. If you run out of ideas it is time to disappear. If you don't disappear, the farmers will.

In this chapter a close look is taken at the emergence of the agricultural demonstrator scheme, which has been the backbone of agricultural extension in Zimbabwe. A Native Commissioner, Keigwin, gave the first push towards the establishment of a government agency concerned with the industrial and agricultural development of the black majority of the population. His plan in 1920 led to a training scheme for African instructors who were to demonstrate improved agricultural practices in the Reserves. With the appointment of E.D. Alvord, an agricultural missionary, in the post of Agriculturist for Instruction of Natives in 1926, a more encompassing development scheme was embarked on. The elaboration of his philosophy of improving the livelihood of Africans led to the appointment of instructors, called demonstrators, in the fields of agriculture, community development, home industries, forestry, irrigation, livestock and conservation. During the 1930s the emphasis of the demonstrator policy was shifted from livelihood improvement to the prevention of destruction of natural resources (Anderson, 1984; Beinart, 1984; Drinkwater, 1989; McGregor, 1995). With the establishment of the Department of Native Agriculture [NAD] in 1944 and its white counterpart, the Department of Conservation and Extension [CONEX], in 1950 this shift from persuasive agricultural improvement towards enforced conservation was completed.

This chapter pays special attention to the methods of extension that were developed to persuade indigenous farmers of the validity and benefits of the propagated agricultural practices. Many of these methods and recommended practices still stand today as the standard within Agritex, the post-independence successor of NAD and CONEX. However, what was considered to be a success in the 1930s became a 'traditional' approach in the post-independence era (Pazvakavambwa, 1994, 113). To understand this change of heart in smallholder agriculture and agricultural extension in Zimbabwe's communal areas, I focus on certain continuities and discontinuities in extension and development practices.

First, I reconstruct the emergence of a philosophy that stressed segregated development of Africans. By means of introducing improved husbandry practices it was hoped to set free the agricultural potential of communal areas. The faith in the inherent capacity of communal areas to meet the desires of the black majority still holds today, as is most clearly implied by Agritex's mission statement of 1993.

Second, I will show that the concept of 'seeing is believing' and 'learning by doing' in agricultural extension, as elaborated in detail by Alvord in his demonstrator programme, has been dropped over time. During the 1940s this voluntary change approach was replaced by an authoritarian, rational instruction on the basis of 'scientific' findings in the face of increasing environmental concerns. It was Alvord himself, mourning the limited impact of his ideas, who succumbed to this change of approach and institutionalised it in the Department of Native Agriculture.

Third, the 'ingredients' of the agricultural success of the demonstrator programme are assessed. Research has shown that African agriculture was a relative success in the 1920s contrary to official views of the wasteful, ineffective nature of African agricultural practices (Palmer, 1977a, b; Ranger, 1985). How then could the Alvord 'cocktail' of crop rotation, mono-cropping, early winter ploughing and manuring result in such astonishing yield improvements? It is argued that the success of this 'cocktail' on demonstration plots was to a large extent a construction of its proponents. The 'forced' shift in agriculture from an extensive to an intensive basis and the fact that mainly resource rich African cultivators participated in the demonstration programme, may explain the marked yield differences achieved. However, recent research on inter-cropping, zero-tillage and agro-forestry suggests that under the present conditions in communal areas, the improvements of yesterday could well be the set-backs of today.

THE KEIGWIN PLAN: 1920-26

At the cradle of the agricultural demonstrator programme stands a famous cricketer, H.S. Keigwin. As a Native Commissioner he set out to develop a proposal for the development of 'Natives'. The Southern Rhodesia Native Affairs Committee of 1910, with Keigwin as its secretary, had opened the road for more Government attention towards education and industrial development of the black masses. Hitherto education opportunities for Africans had been limited to mission schools. The Education Ordinance of 1899 had stipulated native mission schools to devote at least two hours education to industrial training, which was meant to make Africans more effective workers in basic agricultural pursuits, and in the service of European employers (Atkinson, 1974, 146-7). Keigwin travelled extensively and consulted many high positioned officials in the British Colonial Office before he came up with a proposal for Industrial Development of Natives in 1920: 'the Keigwin plan.'[1]

Keigwin proposed the establishment of institutions to train 'progressive natives' in seven industries: hides and skins; food production; rope and mat making; basket and chair making; pottery and tiles; carpentry and wagon-work; and smithing.[2] He wanted to help the African 'to realise a better ideal of life', which Keigwin considered his obligation as member of a privileged, 'civilised' race. His ideal contrasted sharply with what most settlers had in mind regarding African development. Africans had been located on poor soils in (labour) Reserves; were considered a backward, lazy people by nature[3] and could only be used to the benefit of the nation by providing cheap labour for European farms, mines and industries. The main concern for most settlers was to get the Africans work for them. Keigwin was well aware of these sentiments and argued that by supporting the Africans to help themselves, they would by the same process help the State. By means of industrial development Africans would be taught to provide things for themselves and then they would of their own free will go out to look for work to earn money to 'purchase those things they will have learned to value'. To prove his point Keigwin drew heavily from experiences with black education in the United States of America. There a policy of lifting the economic status of black communities resulted in a process where, in the words of Booker T. Washington, Blacks had come to value education as means of dignifying labour and not as a means to escape labour (Washington, 1901).[4]

The general idea behind a native industrial school was to train the progressive Africans in the above-mentioned industries and send them back to the Reserves to work as instructors and set up small industries for the African market. With regard to food production, Keigwin noted that the common African methods of agriculture (e.g. shifting cultivation) could not be sustained:

> As the reserves get filled up, the soil is being exhausted faster than is practicable, while the extensive destruction of timber should be checked. Better methods of tillage, improvement in seed selection, more constant cultivation, proper rotation of crops and less wastage of corn in beer-making, are some of the subjects to be dealt with. **One of the first needs is for an instructor, who shall go round the reserves and demonstrate with a plough what should be done.**[5] [My emphasis, AB]

African agriculture had to be improved and intensified not only to sustain a larger population, but also to show that the Reserve lands were being put to good use, otherwise pressure from settlers to expropriate Africans of land and demands for coercive labour arrangements and increases of taxes would be impossible to contain, maintained Keigwin. The increased food production was to be achieved by instructors who would demonstrate the proper agricultural practices to their fellow Africans. Such a scheme had worked very successfully for the black population in America and also in the Union of South Africa.[6]

After passing through the Legislative Council in 1920, Keigwin's plan was approved and Keigwin himself became the first director for the Department of Native Development. He set up the first 'industrial farm' in Chinamora Reserve, where the soils were light and sandy as in most Reserves: Domboshawa Native Industrial School. The first eleven students were drawn from mission schools and helped to construct the school buildings. In 1922 a second Government school was opened in Tjolotjo.[7]

Keigwin wanted both schools to be mainly training grounds in improved agricultural methods. However, the students thought otherwise. Two strikes in Domboshawa, in 1921 and 1922, and a call of protest at the Superintendent of Natives' office in Bulawayo by Tjolotjo students, all expressing the desire for more literary instruction in English, made Keigwin realise that the Africans wanted to be taught things which would 'help them to earn better wages' (Lloyd, 1962, 9; see also Atkinson, 1974).[8]

In 1920, the Missionary Conference of Southern Rhodesia had supported Keigwin's plan, despite Keigwin's critical remarks that missionaries had in their educational endeavours been mainly focused on mental rather than material development of Africans. But by 1922, with both government schools remaining non-denominational and highly competitive of government subsidies on African education, the missionaries lashed out at the lack of Christian education at the two schools. They maintained that they would 'rather have an uneducated Native than education without Christianity' (Lloyd, 1962, 9). Keigwin replied by observing that in 10 Missionary stations agricultural instruction was being neglected.[9] In 1925, the Committee of Enquiry on Native Education, however, recommended that Christianity was to continue to form the basis of African Education (Atkinson, 1974, 171). The intimate relationship between agricultural improvement and Christian ideals was to remain part of the later demonstrator programme [see below].

With the first classes of African students in school, the question arose: What would be done with the educated Africans after completion of their training? Wilson (1923), in the first issue of the Native Department Annual, argued that the central issue of the 'Native Question' was not whether to educate the African, but how to make full economic use of the educated African in the nation's industrial development. His answer was to develop Native Reserves in such a way as to meet the economic, social and political wants of the Africans, meanwhile avoiding competition with the Europeans.[10] In his broad vision, Native Reserves would become attractive centres for progressive Africans, where booming activities like agriculture, crafts industries, house and road construction and schooling would take place under the guidance of Native Commissioners. Africans selected by Native Commissioners would go for training and return as enlightened practitioners and demonstrators.

In January 1924, Keigwin submitted a memorandum on the training of selected Africans to 'carry out instructional work in agriculture in Native Reserves under the control of their Native Commissioners'.[11] After an enquiry amongst all Native Commissioners, Keigwin drew up a policy on the training of native demonstrators. The reply from most Native Commissioners was positive and suitable candidates could be found.[12] Keigwin's policy was approved in July 1924[13] and at the end of the year the first 12 candidates for the two-year demonstrator course had been enrolled at both Domboshawa and Tjolotjo. The contours of the agricultural demonstrator programme started to take shape. Some salient features are described below.

Candidates were selected from districts where the people were 'most progressive, and where the Native Commissioner is most interested'. Main selection criteria were good character, accepted standing amongst fellow Africans, and some form of education, 'though not too bookish'. The training was to emphasise practical demonstration skills: trainees had to work on the common school plot as well as on their own demonstration plot. Prominent in the curriculum were preparation of the land, seed selection, cultivation and early ploughing on both native crops and commercial crops like cotton and wheat. After completion of the training the demonstrators were to be administratively treated as Native messengers attached to the Native Commissioner. Concerning their work in the field, Keigwin indicated he was still studying the Transkeian system of demonstrators, in operation since 1916. Keigwin himself envisaged demonstrators to have a demonstration plot of themselves, which would arouse the interest of other Africans in the area, who then submit (part of) their land for the demonstrator to work on according to the principles of improved agriculture. Demonstrators were to be judged by the results in those demonstration plots.[14]

However, Keigwin was never to see his work come to fruition. He resigned in August 1926, frustrated by a lack of cooperation on the part of the Southern Rhodesian administration, who had given in to Missionary pressure and disbanded the Department of Native Development (Alvord, 1958, 10). He had failed to get 'men of the right mind towards the work for the Natives. Our staff of whites [at Domboshawa, AB] has not been of the right sort.' He felt it was time to 'lapse into obscurity for a few years.'[15] However, he was very pleased with the appointment of somebody he considered his best possible successor: Emery Delmont Alvord.

THE LAUNCH OF THE DEMONSTRATOR PROGRAMME

In October 1926, Alvord, a distinguished football player, was appointed Agriculturist for the instruction of Natives within the (white) Department of Agriculture. His prime task, according to the Chief Native Commissioner in his annual report of 1932, was to 'develop native reserves so as to enable them to carry a larger population, and so avoid, as far as possible, the necessity for acquisition of more land for native occupation.' Alvord's immediate superior, Mr Mundy, the Chief Agriculturist had in a way created the post of Native Agriculturist in appreciation of Alvord's work as agricultural missionary in Mount Selinda, from 1919 to 1926. Mundy had defined Alvord's principal duties as organisation of better agricultural training at mission schools and organisation of native agricultural instruction in Reserves by means of the agricultural demonstrator programme.[16] Alvord himself saw the appointment as an opportunity to carry out his ideas about the development of Africans on a nation-wide scale, outside the limited confines of a Mission station.[17] Alvord's ideas were wide-ranging and comprehensive and taken together this formed a development ideology which pretended to be fore-seeing and covering all aspects of life. Alvord's practical attitude and perseverance enabled him to carry on, where an administrator like Keigwin had stranded.

During his period as agricultural missionary at Mount Selinda, Alvord had 'discovered' the merits of the demonstration concept in transforming African agriculture. Soon after his arrival, Alvord noted that the tenants on the mission farm at Mount Selinda could not maintain their 'wasteful, destructive' agricultural practice of 'shifting tillage', because of over-population. As a consequence the soils were wearing out and soil erosion became a problem. In November 1920 he started working on a six-acre school plot, practising improved methods of agriculture. Four acres were put under a crop rotation and two acres were well manured with kraal manure. The maize crop on the manured plot showed astonishing results in April 1921 and Alvord decided to hold a "before harvest" meeting at the edge of the plot for all mission tenants and explain the merits of manuring, cultivating and rotating crops. However, to Alvord's dismay, the African audience had quite different explanations for the success of the crop: 'Mtakati, Uroyi and muti' or else the 'magic of the whiteman' had done the trick. Alvord wrote on this experience in his autobiography:

> It was then that I made my prodigious discovery . . . that in spite of high qualifications and experience, a white man could not teach agriculture to the superstition steeped African who attributed high crop yields to "muti", witchcraft and favour of the ancestral spirits. I concluded that it was impossible to Christianize the Africans without, first of all intellectualizing their agricultural practices, so filled with superstition, ignorance, witchcraft and worship of the unknown. I must create in them wants and desires which would automatically lift them out of the sea of superstition and fear which engulfed them. I made the discovery that the African must see things demonstrated on his own level, within his reach, by Demonstrators of his own black colour and kinky wool, and before we could successfully preach to him the Gospel of Christ, we must first preach to him the "Gospel of the Plow". (Alvord, n.d.)

And so agricultural improvement by means of demonstration became a way, not only to increase agricultural production based on a change from an extensive, shifting land use system to an intensive, permanent land use system, but also to 'civilise' and 'Christianise' the African. In June 1921 the first African Demonstrator was appointed at Mount Selinda mission farm and by the end of 1921 several tenant farmers were persuaded to conduct crop rotation demonstration plots. During the 1921-22 drought these plots proved to be the only one's yielding crops in the wide surroundings of the Mission, attracting country-wide attention. In 1923 the average yield on the demonstration plots was 100 bushels per acre compared to a mere six bushels per acre on ordinary African plots. Alvord noted to his joy that 'natives started to believe that witchcraft and ancestral spirits had nothing to do with crop production.' (Alvord, n.d.)

The astonishing results that Alvord achieved drew many of the Rhodesian administrators and policy makers to Mount Selinda. It was during these visits that Alvord got to know Mundy, Keigwin, Taylor (Chief Native Commissioner (CNC) from 1923-28), and Jackson (CNC from 1929-31) and the committee members of the Phelps-Stokes Committee on Native Education (1924-25). This committee subsequently recommended the implementation of the agricultural demonstration programme. And Mundy, as we saw, appointed Alvord in his new job as Agriculturist for Natives in 1926.

The fact that by the end of 1926 the 12 demonstrators in training at Domboshawa and Tjolotjo school would complete their course, compelled Alvord to set up a policy on how these demonstrators were to operate in the Reserves. He decided to resort to the demonstrator scheme that had been reported to be so successful in the Transkei, Union of South Africa. In a letter to Transkei, Alvord formulated three alternative ways of utilising demonstrators and requested advice which would be best suited to African instruction:

> One is to place the demonstrator on a small area of land which he will retain and work as a demonstration farm or native garden, and on which will be established demonstration plots where he can illustrate to the natives of the locality the proper methods of crop culture, and, as his time permits visit and advise other natives living in the area on better farming methods.
>
> A second alternative would be for the demonstrator to have no land of his own but to select in his area a certain native willing to give a portion of his land to be worked by the demonstrator, who would work this land for 1 or 2 years alongside the ordinarily worked land of the resident native and would then pass on to another similarly selected native farmer elsewhere in the Reserve.
>
> A third plan is to locate the demonstrator in a reserve where he will merely go about as agricultural advisor, advising the various native farmers on better methods and showing them how to do the different operations properly, but not actually himself working any land as a demonstrator.[18]

The Agricultural director for Transkei recommended option No. 2 and informed Alvord of the demonstrator system that had been in operation for 11 years by then.[19] The general outline from Transkei was almost literally copied by Alvord in his own policy paper, published in February 1927.[20] His superior Mr Mundy, had expressed his preference for a government demonstration farm worked by a demonstrator (option 1), instead of having a demonstrator working on private land on behalf of the owner (option

No. 2).[21] Alvord emphasised that demonstrators were supposed to work on the demonstration plots together with the owner of the plot. '[T]hus the owners receive actual demonstration in methods and "learn to do by doing".'[22] Option No. 3 resembles very much the current mode of operation of Agritex agricultural extension workers. The modalities of Alvord's demonstrator programme were copied from the co-operative demonstration approach that had been developed in the USA during the first decade of the 20th century (Box 1).

Meanwhile Alvord noted in 1926 that 'agriculture had no place in the Domboshawa school curriculum.' The demonstrators in training knew nothing of crop rotations and the use of kraal manure, according to Alvord. He downright refused to send these men out to the Reserves, and ordered that their course be extended to mid-1927, with explicit focus on how to demonstrate improved farming to Africans in the Reserves near the two Government schools. Thus the first two 'demonstration centres' were opened up and run by the demonstrators in training. This intervention yielded Alvord his first conflict with the principal of Domboshawa school, who was more interested in developing the industrial crafts curriculum (Alvord, n.d.).

This was certainly not the last conflict Alvord met in his plight to improve African farming practices. A Member of the Legislative Assembly of Rhodesia had objected to Alvord's appointment, because the post had not been gazetted, Alvord was not a British national, and he was not sufficiently qualified. Alvord had to rely on his institutional network (Mundy and CNC Jackson and Taylor) to avert immediate dismissal.[23] After touring several Mission stations and criticising their agricultural instruction efforts, Alvord was also at loggerheads with numerous missionaries.[24] Finally the (white) Department of Agriculture was also not particularly supportive when it allocated all money set aside for Africans to Tobacco advisors (Alvord, n.d.). This virtually made Alvord office-bound during the first year in post, and can be interpreted as white administrative opposition to increased attention for African Development (cf. Page and Page, 1991, 6). All this seemed to change after Alvord's transfer to the Department of Native Affairs in April 1927, when Alvord could finally tour the country to mobilise support for his demonstrator scheme.[25]

Alvord was anxious to convince African cultivators, missionaries and administrators of his 'seeing is believing' approach. Under the guidance of demonstrators in training in 1925, an African cultivator (Mr Mvundhlana) in

---

Box 1 The development of an extension system based on demonstration in the USA (Sanders, 1966; Scott, 1970)

By the end of the nineteenth century American farmers could rely on the most extensive system of agricultural information in the world. Granges (farmers' lobby groups), farmers' institutes (local study clubs), land-grant colleges, experiment stations and various farm magazines were available in every single state. Still, the predominantly 'scientific' knowledge that was disseminated by these various organisations only made an isolated impact and was of little practical value in the field.

It was left to a practical man, Seaman Knapp, to find the ideal approach to bridge the gap between modern agricultural science and farmer practice. In 1902 Knapp was appointed in the Department of Agriculture as a special agent for the promotion of agriculture in the South of the USA. Southern agriculture had been characterised as stagnant and backward with poor agricultural methods and exhausted soils due to continuous cotton and corn growing. For his agricultural improvement programme Knapp selected practical farmers as agents and not college graduates. In most cases agents met an apathetic, suspicious response. They organised a meeting through local businessmen and farmer leaders, raised a guarantee fund and then selected some of the better farmers who would run a 10 acres demonstration plot. The demonstrator farmer signed a contract that he would follow all instructions given to him. Agents would personally visit the areas. Field meetings were held during the growing season and at harvest time instruction was given. Besides community demonstration farmers there was a second class of co-operators who would follow advice, but rarely received visits of agents. Yields of demonstrator farmers were often 10 times as high as ordinary yields. Knapp drafted his 10 commandments of modern farming (1904):

1. Prepare a suitable seed bed.
2. Ploughing during fall at proper depth (8-10 inches).
3. Selection, testing and storing of seed to ensure high germination.
4. Adequate plant spacing between plants and rows.
5. Frequent, shallow cultivation during the growing season.
6. Simple crop rotation: cotton, maize, cowpeas, followed by a winter cover crop.
7. Use of manures and fertilisers to rebuild soil fertility.
8. Optimum livestock production by means of pasture development and feeding.
9. Keeping of accurate records to monitor profits.
10. Achievement of self-sufficiency through production of all food needs for the family and its livestock.

Agents visited farmers, inspected demonstrations, launched other demonstrations and sometimes helped farmers out with routine farm jobs. In that way agents lived with their farmers and were truly itinerant teachers. Two distinctive farmer characteristics were exploited to the advantage of the programme: (1) farmers' sense of rivalry and competition; (2) farmers' dislike of experts and outsiders.

In 1906 Knapp linked up with Booker T. Washington and the first black agent was deployed from Washington's Tuskegee Institute. By 1912, 32 black agents were working with 3,500 black demonstrator farmers and 10 to 15,000 co-operators. There were no differences between white and black farmers regarding instruction given and results achieved. Gradually the work spread over the Northern States at the initiative of local farmers, businessmen and college lecturers. Ultimately each agriculturally oriented county (district) had its own county agent, who worked as a connecting link between the sources of information (land-grant colleges) and the people. The Smith-Lever Act of 1914 formalised this system of co-operative extension for the whole of the USA.

Tjolotjo had taken up the advice of winter ploughing and reaped a bumper harvest, whereas his neighbour (Mr Makothliso) reaped nothing.

> Whenever Europeans and Natives were shown the excellent crop on Mvundhlana's land, the poor crop growing adjacent to it on Makothliso's land, was pointed out as the "horrible example".[26]

As a consequence 14 African farmers asked for demonstrator assistance the next season, amongst whom Mr Makothliso. The latter's field promised to reap the best harvest. Alvord increasingly promoted his work by means of such stories of 'losers and winners' and was convinced that it stimulated competition as in the above case. The rationale was that when Africans had seen a fellow African do the trick, they would start to copy from and compete with this 'winner'. Thus a multiplying effect called 'extension' would occur.

Alvord made extensive use of the 'seeing is believing' principle in his public relations strategy promoting the demonstrator programme. He toured the country's mission stations from 1927 onwards showing slides on good farming with the help of a stereopticon lantern at night.[27] This drew big crowds and ultimately gained the missionaries' support for his agricultural instruction programme (Alvord, n.d.). Alvord also used photographs to report on the success of his demonstration techniques. Two of his publications in the leading journals for Rhodesian agricultural administrators, *Rhodesia Agricultural Journal* and *Native Affairs Department Annual*, are littered with pictures of proud, well-dressed demonstration plotholders holding a measuring gauge in front of their tall standing maize crop on the one hand and scantly dressed 'ordinary' African farmers standing in front of their poor, stunted maize crop on the other hand [see also book cover].

In one of these articles (Alvord, 1928), entitled 'the Great Hunger', Alvord outlined the superiority of the demonstration method in convincing Africans of the benefits of good farming practices according to scientific principles.[28] The story was based on the experiences of one of the first agricultural demonstrators deployed in the low-veld during a long dry spell in January 1928. The people of Chief Pandi had brewed beer to please their ancestors on several occasions, but no rain fell. Finally their 'witch-doctor' Mavudzi pointed out that the real reason for the great drought was 'the coming of the Native Farming Demonstrator'. During a 'before harvest' meeting at the edge of the demonstration plot, where Alvord himself was supposed to come and explain why the demonstration crop had done so well, the people split 'naturally' in two parties. On the one hand the *vatendi* who had taken up the teachings of missionaries and were well dressed and on the other hand the 'spirit-worshippers, almost naked, and draped in dirty pieces of limbo or soiled antelope skin aprons.' The latter had noted that the *vatendi* also had better crops than they had, which was attributed to the fact that the Agriculturist for Natives 'was angry with them for not becoming "believers", and had cast a spell upon their crops.' A heated debate ensued between the two groups: the "believers" trying to convince the others that it was only because they had taken up the good advice of the Agriculturist that they now had such good crops. By then Alvord arrived and promised to explain the secret of the demonstration plot mealies in a nearby ox-kraal. Here everybody saw that mulched kraal manure contained a lot of moisture. Alvord explained that he didn't make rain clouds in ox-kraals, but that the mulch stover prevented the sun from 'drinking up the moisture.' Taking his audience back to the maize field, Alvord went on to ask:

> You can see with your own eyes what these mealies are, and you can touch them with your hands. Do you now believe it is possible to grow mealies like this? (Alvord, 1928, 42).

Everybody did and was surprised to find moisture under the loose soil in the maize field. However, when Alvord asked for the reasons for this moisture to be there, he was accused of witchcraft. Alvord reacted furiously:

> No! You are wrong. There was no witchcraft here. A white man had nothing to do with this field. It was prepared and planted by your Native Farming Demonstrator, a man with black skin and woolly hair, just like your own. He did not make rain. He only followed the simple rules of good farming, which you in your superstitious ignorance have not learned. (Alvord, 1928, 42)

Alvord then proceeded explaining the good principles of farming. According to Alvord the 'following season there were scores of disillusioned natives begging the Demonstrator to do plots for them.'

Alvord's well orchestrated public relations campaign resulted in a government decision to support the scheme even further. A circular letter from CNC Jackson,[29] in April 1927, calling for more candidates for the demonstrator training, was met by a greater number of candidates than Domboshawa could absorb.[30] After their final examinations, by July 1927, the first 12 demonstrators had been sent out to selected Reserves. Selection criteria that Alvord used for placement of a demonstrator were progressive attitude of local African farmers, a supportive Native Commissioner and affinity of the Demonstrator with the local people. Alvord would then introduce the demonstrator to the community with the help of the local Native Commissioner. Alvord further paid frequent visits to the Demonstrators in the field to supervise their work. Despite initial successes as reported in 'the Great Hunger', in Chiweshe Reserve, demonstrator Philimon failed to make any impression. In June 1927, during an initial visit, Alvord had observed the eagerness of the local African farmers to advance in farming as well as their high standards of farming, which Alvord attributed to the example of the white farmers in the neighbouring Mazoe valley. The Chiweshe people:

> ... displayed greater enthusiasm for availing themselves of the services of the Native Farming Demonstrators than has been evident at any other Reserve where demonstrators are being located, and their chief concern was that they could not all of them have a plot under the direction of the demonstrator.[31]

However, first impressions can be deceptive. By November 1927, this enthusiasm had turned sour and the work of Philimon had amounted to 'little less than a failure.' The blame was with the local people:

> They have been on continual beer drinking for the past three months and the demonstrators can get no assistance in the way of oxen or help for other operations required. Under our demonstration scheme the work is entirely dependent upon the cooperation of kraal owners with the demonstrators. The people seem to have no confidence in us and are suspicious that there is a catch in the free advice we offer.[32]

In May 1927, Chief Nhema of Selukwe[33] Reserve had been more explicit on the 'catch' he suspected. During the introduction of the demonstrator programme to the people of Selukwe, Chief Nhema rose and

> . . . harangued his people, warning them not to believe or follow. This was only a scheme of the Government to test their land. If they found it was good, they would take it away and give it to the whitemen. (Alvord, n.d.,15)

Still Alvord could call the first year of actual demonstration work 'more successful than we had hoped for.'[34] The suspicions remained a feature of the programme throughout the first 10 years of its operation.[35]

Before an account is given of the expansion of the demonstrator programme, a closer look is taken at the agricultural practices that comprised the backbone of the demonstrator programme.

THE ALVORD PACKAGE REVIEWED

Alvord's package of crop rotations, heavy manuring, use of improved seed, weeding, ploughing and mono-cropping reflected the contemporary standards of highly productive agriculture in the West (Europe and USA). And highly productive it proved to be! The crop yields on demonstration plots on average exceeded ordinary African yields by a factor 8 and European yields by a factor 1.5, if we may believe the yield data supplied by Alvord and his demonstrators.[36]

These results seemed to pay tribute to the prevalent ideas of Western superiority and disregard of the merits of African methods of agriculture. Denouncing African agricultural methods had become a favourite hobby-horse for Rhodesian administrators and missionaries in the 1920s. Typologies of African methods of shifting tillage as 'destructive', 'wasteful' and 'inefficient' were rife (see Alvord, 1930; 1958; n.d.; CNC annual reports 1915-1930). The image of African agriculturists was even worse. They were regarded as an unwanting, lazy, beer drinking lot (Alvord, 1930). These conceptions successfully painted an image of 'decline' as far as African agriculture was concerned. However, Palmer (1977a) and Ranger (1985) have shown that African agriculture reached a peak of productivity in the late 1920s, which was a result of increased security and the emergence of new markets. This eager response of African farmers to new opportunities points to a highly dynamic and vibrant peasantry.

At the heart of this apparent contradiction lie differences in perceptions and interests regarding agricultural development of the African peasantry in the reserves. Whereas the African system capitalised on extensive land use and maximisation of returns per unit of labour, the demonstrator programme sought to maximise production per unit of land by means of intensification of production. Whereas African methods aimed at risk reduction and were based on an appreciation of local variations in soil quality and rainfall, Alvord and his European colleagues sought to develop a uniform standard of farming based on rationalisation of land use and vindication of drought hazards by applying 'scientific' methods. The latter involved a radical break away from African methods in favour of European styled techniques propagating continuous cultivation of the soil as practised in the moderate climates of the Northern hemisphere. African agriculture, it was believed, required rigorous modernisation. It was the idiom of segregation and growing political importance of land apportionment as the corner-stone of the settler state which necessitated changes in the prevalent African system of intensive wetland cultivation combined with extensive shifting cultivation of dry lands (Mtetwa, 1976; McGregor, 1991).

However, a second school of thought suggested a transformation of African agriculture by learning from African knowledge and experience accumulated over centuries of adaptation and adjustment to difficult environments (Floyd, 1960, 287-89). This school maintained that the slogans of modern agriculture had to be questioned and local systems evaluated on their merits before imposing alien concepts of farming (Allan, 1945, 1949; Worthington, 1938; Hailey, 1938; Floyd, 1960). An early proponent of this school of thought was Dr Shantz (1923) who hailed African systems of agriculture by pointing at the advantages of land fallowing; the fact that rotation of land ensured new rich soil; the system effectively dealt with soil fertility and plant diseases; and the fact that the practice of burning added a considerable quantity of mineral fertiliser in a short time.[37]

Taylor (1925) and Alvord (1929) both studied, to some extent, the merits of prevalent African agricultural methods, but were quick to discard them. Taylor (1925, 88-89) singled out five African practices in particular. He observed the planting of different crops on different patches of soil, inter-cropping, staggered planting, burning during clearance of new land, and ridging. However, as top administrator for Africans in Rhodesia, Taylor was interested in sealing off future African demands for more land by pointing at the unsustainability of prevalent African agricultural practices and promising that 'improved agriculture' could cope with the needs of the African people from the land that was specifically set aside for them: the Reserves. The potential of Alvord's package to enable Reserve land to carry a larger African population, became the main reason for support for the demonstrator scheme by Native Commissioners and top Rhodesian administrators.[38]

Alvord (1929) noted that Africans could tap from a 'remarkable variety of foods', by cultivating seven grass family crops, three legumes and twelve other crops as well as relying on a great variety of wild plants, fruits and animal foods in times of drought. Despite this high level of sophistication, Alvord ruled the African system as primitive and destructive. Alvord's mission was one of civilisation and providing better livelihoods through agricultural modernisation.[39]

From the first survey on agriculture in the reserves Alvord produced statistics indicating that the prevalent practices resulted in chronic food shortages (Alvord, 1929, 11). Despite the fact that these statistics were mere estimates and by no means accurate, they instilled a sense of urgency and objectivity which Alvord later frequently used to wield support for his demonstrator scheme (Alvord, 1930). In typical segregationist terms Alvord (1929, 16) concluded that the African population in the Reserves was 'a liability rather than an asset, but this can change . . .', once they were taught to till their lands better. Demonstrators had to teach plot holders the 'value of kraal manuring and good tillage' during the first two years and a 'system of permanent farming by means of a system of crop rotation' from the third year onwards. Permanent farming would concentrate cultivation on smaller areas and thus set free more lands for grazing purposes.[40] Thus Alvord seemed to be able to succeed where Major Mundy, the Chief Agriculturist and nestor of European agriculture in Rhodesia, had failed during many years of fruitless research: a sustainable land use system based on mixed farming, instead of the common settler practice of exhaustive maize mono-cropping.[41]

One may wonder how successful Alvord's package was in achieving this permanent system of agriculture. The important ingredients of the package, viz., manuring, crop rotation, ploughing and mono-cropping, are reviewed below, together with the response of African farmers, and scholarly critique.

Manuring

Application of manure was of crucial importance to maintain soil fertility, reduce erosion hazards by improving the soil structure and increase biological activity setting free plant nutrients for the crop roots. Alvord proposed an application of 10-15 tons kraal manure per acre once every four years (approx. 37 tons/ha) (Alvord, 1958, 8; Grant, 1976, 252).[42]

However, most Reserve cultivators did not have access to the required 12 to 16 heads of cattle per arable hectare to supply that manure (Page and Page, 1991, 10; see also Floyd, 1960, 303; McGregor, 1991, 92). Realising this, Alvord started a massive campaign for the construction of compost pits and refuse pits in the 1930s to add on to the limited amounts of kraal manure, despite its relatively low fertility value (Floyd, 1960, 304). The use of artificial fertilisers was rejected by Alvord because it would confirm African superstitions of "muti" (witchcraft)[43] and was too expensive. Furthermore Alvord stressed the benefits of intensive cultivation in terms of the increased grazing areas it would set free.[44] Later components of the demonstrator programme were specifically directed at improving the quality and size of the grazing area.[45] Floyd (1960, 303-4) has pointed out that neither of these suggestions provided lasting solutions for the decline in soil fertility: cattle do not add fertility, they merely transfer it from the grazing to the arable block.

Yield responses to manure vary according to soil type and prevailing moisture levels (McGregor, 1991, 90; Theisen, 1979). In dolerite soils and drier regions manuring does not translate into marked yield improvements. In areas where both soil fertility and moisture levels are limiting factors to crop growth, heavy manure applications increase the crops vulnerability to moisture stress to the extent of crop failure (McGregor, 1991, 90; Theisen, 1979; Grant, 1981; Nyamudeza, 1996, 9). Farmers describe this effect in terms of crop 'burning' or soil 'cooking' (Wilson, 1990 in McGregor, 1991, 90). Floyd (1960) therefore discarded the Alvord package as a high rainfall package, mainly suitable for infertile, well leached granite soils.

In light of the above African farmers had their reservations regarding the use of manure. The CNC reported in 1920 that Africans did not use manure, because it increased weed infestation, posed great labour demands for 'cleaning' and brought insect pests.[46] Faced with a shortage of kraal manure, even adopting Africans had to resort to alternative sources of fertility, modifying Alvord's package in the process. Theisen (1979) reports extensive application of anthill soil alone or in combination with kraal manure (see also Nyamapfene, 1989). McGregor (1991, 92) points at Africans' extensive knowledge of different qualities of leaf litters as fertility inputs. Chikukwa *et al.* (1996) report a mixture of various sources of fertility that are applied by present day communal and resettlement farmers (see also Scoones *et al.*, 1996, 113-120). The heterogeneous character of soils within one field (Grant, 1981, 170) provided another motive to modify blanket recommendations to suit local conditions, thus spreading risks (McGregor, 1991, 91).

Crop rotation

After the initial success of the well manured demonstration plots at Mt Selinda Mission during the 1922 drought, Alvord was looking for a more permanent system of agriculture that could be systematically promoted by means of demonstration. Many different crop rotation systems were experimented with, until a 4 course crop rotation system was found that could 'rapidly build up the fertility of the soil to a high state of productivity and . . . maintain that productivity indefinitely' (Alvord, 1958, 8). The adopted rotation, that became a 'standard' in agricultural extension up to the present day, was as follows: (1) maize with manure; (2) maize or other intertilled farinaceous crop; (3) groundnuts, beans or other legume crop; (4) rapoko or other close-growing millet crop.[47] These crops followed each other in systematic order on the same land, completing the cycle in 4 years.

In true missionary fashion Alvord laid down his 10 'commandments' for permanent agriculture (Alvord, 1958, 8; cf. Box 1):

1. Thorough stumping and clearing to ensure continuous easy tillage.
2. Winter ploughing to conserve moisture and decompose crop residue.
3. The application of manure once every 4 years to every land in the rotation.
4. A second ploughing just before planting time to aerate the soil.
5. Thorough seed-bed preparation to ensure uniform germination of seeds.
6. The planting of maize on the land where manure is applied each year.
7. Proper spacing and planting of all crops, row planted and broadcast.
8. The planting of a legume crop two years after manure is applied to revive nitrogen fixation and maintain a healthy soil by a necessary change of crop.
9. A heavy rooted, close growing crop after the legume, to smoothen all weeds and fill the soil with dense growth of fibrous roots.
10. Crops must not be mixed together in the same land.

Alvord claimed that with the help of this system 'even the poorest sand veld soil, depleted of most of its fertility, can be brought to a high state of productivity' (Alvord, 1928, 1106).[48]

However, even during his career as agriculturist, Alvord was challenged as to the permanence and feasibility of this system. During a hearing of the Natural Resources Board, in July 1942, Alvord was criticised by Dr Pole-Evans, a leading expert on agriculture from South Africa. Pole-Evans emphasised the necessity of including a perennial grass crop in the rotation to supply sufficient fodder to the hard needed cattle. Alvord replied that large grazing areas and liberal application of (composted) manure could meet the problem of fertility and avoid the difficult task of teaching the African to grow a crop for fodder purposes only. Dr Pole-Evans was not convinced: 'I am afraid there I can't agree with you.'[49] Later authors seem to support Pole-Evans, though the issue is still mired in controversy. Grant (1976, 252) points out that only the tobacco-maize-grass rotations practised by settler farmers from the 1950s onwards and the African systems of mould or ash culture with shifting cultivation were ecologically sound permanent systems of agriculture on sand veld soils. The African system included a simple rotation of a maize or rapoko crop followed by a legume crop, and long fallow periods for ecological recovery. However, this system required about 200 hectares per family, which was unfeasible in the face of the Land Apportionment Act. Regarding the Alvord rotation, Grant remarks in 1976 that it was an intermediate system that ultimately proved unfeasible: 'The rotation did not supply enough nitrogen, insufficient manure was used, results were disappointing, and the system has broken down in most areas.' (Grant, 1976, 252). Scoones *et al.* (1996, 122, 244) in their research on crop rotations practised in Chivi from 1988-1991, found that the most common rotation was 'maize-maize-maize.'

African preference for maize cultivation was already noted by the Chief Native Commissioner in 1920[50] and was frequently reported during the 1930s. In 1931, the superintendent for Natives in Fort Victoria accused Alvord of suffering from a 'maize complex': promoting maize cultivation at the expense of *rukweza* (rapoko[51]), which was in his opinion the main food crop for Africans and should be a compulsory part of any demonstrator's programme.[52] In his reply Alvord pleaded 'not guilty' and passed the buck onto the Africans:

> It is true, however, that we are very seriously handicapped in our work by the fact that Natives on every Reserve where Demonstrators are working have a very decided "Maize Complex". It is most difficult to persuade plot owners to undertake crop rotation demonstrations and equally difficult to induce them to plant any other crop than maize.[53]

Alvord went on to explain that it was the European missionaries, store keepers and farmers who had encouraged Africans to grow this crop. Farmers bought maize from Africans to feed their livestock and all employers throughout the country rationed maize meal to their African employees for reasons of easy availability, cheap pricing and easy preparation. Africans knew that 'maize is much easier and cheaper to grow than rapoko' and better returns to labour could be fetched with maize. And thus maize increasingly became the main food crop, whilst rapoko was cultivated mainly as a beer crop. Alvord finally discarded compulsory demonstration of rapoko, since 'continuous cropping of *rukweza* [rapoko, AB] rapidly depletes soil fertility'.[54]

Ploughing

Alvord lamented the African use of hoes to 'scratch the soil.' These 'primitive' implements were seen as indicative of the 'backwardness' of Africans as agriculturists. A staunch supporter of (winter) ploughing, Alvord often described his agricultural mission as the 'gospel of the plough'.

The spread of the plough amongst Africans in the Union of South Africa started in the early nineteenth century (Bundy, 1988, 44-46). By the 1870s African plough use had become wide-spread. Ploughs were bought from European traders and their use was hailed by officials as a sign of African agricultural improvement (Bundy, 1988, 71). In Rhodesia African plough use occurred later, the first ploughs being used by African cultivators close to mission stations. Early mission converts at Mount Selinda who went to work in South Africa upon their return brought or purchased a plough (Rennie, 1973). With the advance of European settlement came the European traders selling ploughs to Africans (Weinmann, 1975, 202). From then on the plough spread fast within the Reserves. By 1930 the use of the plough was 'almost general' (Alvord, 1930, 6) and by 1940 nearly every Reserve family owned a plough (Scoones et al, 1996, 27).

The plough 'stuck' very well with African cultivators. Its extensive use caused nothing less than a 'revolution' of the prevalent African agricultural system. It facilitated a shift from labour intensive wetland cultivation to labour extensive ploughing of (dry) top lands. As such it improved returns to labour and facilitated agricultural accumulation independent of lineage elders, who possessed most of the wetlands (McGregor, 1991, 78; Scoones et al., 1996, 27). Despite the fact that this new technology only rose to dominance between 1910 and 1930, it had been identified by Colonial officials as the traditional African shifting tillage system and had been condemned severely during the first decades of the twentieth century as destructive tillage (McGregor, 1991, `79).

As early as 1917 the Chief Native Commissioner noted that large scale African plough cultivation was not accompanied by efforts to improve the productive power of the soil.[55] Official alarm over destructive effects of extensive ploughing spread during the 1920s. By 1930 Alvord summed up prevalent opinions by referring to the rise of the plough as a 'mixed' blessing. Despite advances in the economy of labour, particularly of female labour,[56] there were higher yields attained at hand hoed fields:

> This misguided use of the plough does not improve Native farming,
> but only increases the acreage of poorly tilled lands (Alvord, 1930, 6).

The dangers associated with extensive ploughing were increased erosion, more rapid destruction of timber and the ploughing of more land than one was able to cultivate (Alvord, 1930, 11).

How was it possible that in such a short space of time African converts to the gospel of the plough turned from progressive heroes into destructive villains? How did this development tie in with the demonstrator programme? To evaluate these questions a closer look is taken at events in Chinamora Reserve at a stone's throw distance from Domboshawa Government school. This Reserve had enjoyed ample attention from demonstrators in training. However, in 1929 the school principal and two instructors expressed their disappointment:

> [T]hese men who have received instruction are not attempting to carry out the new methods. Our plot system has resulted in those men having for three years had for sale a much larger surplus crop than ever before, they have become accustomed to a larger income, but this has resulted from the work of the School demonstrators and not really from their own efforts. Now to continue they are breaking up more and more land which they are working almost entirely on the old methods . . . really large acreages are being put under the plough by a few men, . . . timber is being ruthlessly destroyed, and . . . the amount of grazing is diminishing at an appalling rate . . .[57]

On the basis of this observation the writers argued for a change of the demonstrator scheme, by setting up small demonstration farms run by demonstrators to serve as community centres, instead of having the demonstrator to work on the fields of plot holders. The CNC in his reply suggested that plot holders in Chinamora Reserve were 'temporarily suffering from an over-dose of demonstration' due to the zeal of the pupils to qualify as demonstrators. He also suggested that the lack of sufficient manure combined with a 'get-rich-quick spirit' had led the plot holders in Chinamora to 'fly to the alternative of extensive cultivation'.[58]

Alvord, who had suffered a strained relationship with the Domboshawa principal, realised that the moment of truth for his demonstrator scheme had arrived. He set out to defend it with vigour. He pointed out that plot holders in Chinamora had never received proper instruction, since the principle of 'learning by doing' had not been applied as on other Reserves. Furthermore 'these six men were progressive natives, with implements and wagons etc. before our demonstration work was started among them'. They had had a ready sale of grain and meal to the store supplying the school. They were never taught that permanent farming on small areas would set free more grazing land. Instead,

> ... production, rather than intensified farming, has been given greatest emphasis. These men have been encouraged to form a *Farmers* association and to become farmers.[59]

Alvord rejected proposed changes in the demonstrator programme, as 'the evils reported from Domboshawa are local only.' However, as the Domboshawa principal had already indicated in his letter, the case was not merely local. Ranger (1985, 61-70) reports on the emergence of 'reserve or plough entrepreneurs': progressive men opening up vast tracts of land to produce maize under plough cultivation. The Native Commissioner for Ndanga District complained in 1944:

> There have been two demonstrators stationed in this Reserve for some years and [they] have made no headway in producing 'Master Farmers'; their followers are still plot-holders, cultivating their one acre under improved methods and a large number of acres under old destructive methods.[60]

These entrepreneurs exploited the communal character of land tenure in Reserves, and were able to flourish till such time that most Reserves were centralised.[61] Ranger (1985) continues to suggest that during the world depression and consequent collapse of the maize price in the early 1930s, agricultural demonstrators assisted these plough entrepreneurs in finding alternative crops, like wheat, to sustain their levels of production. Davis and Döpcke (1987, 73) report for Gutu District that the African reaction to the Maize Control Act was to increase maize production even further or alternatively switch to wheat, which still fetched good prices. In any case the example set by Domboshawa entrepreneurs was not incidental, but widespread. Even those plot holders who had learned by doing, did not do as learned.

Whereas Alvord attributed the negative (erosive) effects of the plough to its 'misguided use' by Reserve entrepreneurs, the beneficial effects of (winter) ploughing were beyond dispute during the formative stages of the demonstrator programme. Mainwaring (1921) and Alvord (in two of his 'commandments') stressed increased moisture retention, decomposition of roots and weeds, aeration of the soil and the facilitation of convenient and timely tillage operations, as the main benefits.

However, a 'change of thought' started setting in, even before Alvord's retirement, with the publication of Faulkner's controversial book 'Ploughman's Folly' (Faulkner, 1945; see also Allan, 1945,[62] and Faulkner, 1948). Faulkner (1945, 9) claimed that 'the mouldboard plough . . . is the least satisfactory implement for the preparation of land for the production of crops.' The controversy has deepened over time resulting in two opposed camps amongst agricultural researchers and farmers in present day Zimbabwe.

The opponents to regular (winter) ploughing claim that mouldboard ploughing is 'not sustainable' (Chuma & Hagmann, 1995) and 'the main single culprit causing the loss of soil and water' (Oldrieve, 1995). Conventional ploughing allegedly leads to high levels of sheet erosion,[63] compaction of the soil below plough depth, burial of protective mulch cover, destruction of organic matter and beneficial micro-organisms through exposure to the sun, and drying out of the soil (Norton, 1984; Oldrieve, 1993 & 1995; Chuma & Hagmann, 1995). Instead the practice of zero or minimum tillage, combined with mulching of crop residues and/or ripping, is hailed as the truly sustainable tillage system (Lal, 1979, 1986; Oldrieve, 1993, 1995; Chuma & Hagmann, 1995). The latter comes close to the zero tillage that was practised in the indigenous African agricultural system (Page & Page, 1991, 10).

On the other hand protagonists of (winter) ploughing have stressed the unfeasibility and negative impacts associated with minimum tillage. Mulch is absent in most communal areas; the water holding capacity of the soil is not increased; the soil is more vulnerable to erosion; there are problems in weed control and it leads to poor crop establishment (see among others Mashavira et al., 1995).

Cultivation, weeding and mono-cropping

Alvord had learned at college (in the USA) that high yields could be obtained from one acre plots by planting single crops in rows with proper spacing and doing cultivation combined with regular weeding. It should come as no surprise that regarding the low yields of Africans, he observed that:

> [o]n the most fertile soils Native crops are much lower than they should be, because of poor tillage, planting in mixtures, too thick planting, lack of cultivation and overcrowding with weeds . . . This

results in under-sized, slender plants, which produce small grain heads
or none at all . . . the Native fails to see that this practice of over-
crowding plants on the soil is just as logical as expecting ten calves to
live and grow on the milk of one cow (Alvord, 1930, 6; see also
Alvord, 1929).

Alvord experimented extensively to determine proper crop spacing and
show the effects of regular cultivation and weeding on school demonstration
plots at Mount Selinda and both Government schools.[64] Every demonstrator
was equipped with a cultivator and supposed to continually go round the
demonstration plots to cultivate during the growing season. The objectives
of cultivation were the destruction of weeds; setting free of plant food and
conservation of soil moisture.[65]

In 1930 the CNC reported the use of planters and cultivators in three
districts and hailed this development as a further step towards adoption of
European methods in agriculture.[66] However, this was exceptional. During
the rise of the plough entrepreneurs, their lack of weeding was singled out
for attack. But again to the Africans extensive planting proved to give higher
returns to labour than regular weeding. To add to this, weeding of striga
(witchweed)[67] is not very effective. To control striga and other weeds Africans
burnt prior to planting (McGregor, 1991, 81).

Recent research has pointed out that whilst row planting, mono-cropping
and use of cultivators makes sense in moderate climates with a highly
mechanised agricultural sector, mixed or intercropping gives more benefits
in semi-arid to arid conditions. Some of the benefits of intercropping are a
greater yield stability; better use of space, light, moisture and nutrients; and
less severe pest and disease outbreaks due to availability of horizontal
resistance. Furthermore the intercropping of spreading plants like pumpkin
and cowpea smothers weeds and protects the soil against excessive rain and
heat (Page and Page, 1991, 11).[68] It seems Alvord was not completely unaware
of these benefits. For instance, in 1929 he suggested the inclusion of a
demonstration plot with maize intercropped with cowpea (nyemba) at
Domboshawa school.[69] However, this suggestion might have fallen prey to
standardisation efforts and the exclusive focus of the demonstrator
programme on increasing output per unit of land. In present day Zimbabwe
various forms of intercropping are a dominant farming strategy amongst
smallholders in semi-arid areas. Scoones et al. (1996, 105-107, 244) report
that 75% of all maize area is intercropped in Zimbabwe's southern districts.

In conclusion, it can be observed that Alvord's package was highly successful under some conditions only. As demonstrators were eager for success they made sure that demonstration plots were taken good care of resulting in the exceptionally high yields. However, the standardised package did not take into account variations in soil conditions, rainfall patterns and socio-economic position of Reserve farmers. Neither did it provide a sustainable solution to problems of fertility management.

Alvord was not blind to these deficiencies. However, with the growth of the demonstrator programme over time, Alvord felt a need to standardise the package at the expense of concessions to suit local conditions. Furthermore, increasing political pressure exerted by both settler farmers and Rhodesian administrators implied that the initial emphasis on maximising yield returns per unit of land became the cornerstone of the programme. The strict enforcement of the Land Apportionment Act over-ruled any consideration of the African strategy of maximising yields per unit of labour.

Africans reacted in their own fashion to the opportunities offered. Plough entrepreneurs opened up large tracts of land. Lack of cattle manure led to increased use of alternative sources of fertility. Initially favourable market conditions led to an increase in maize mono-cropping. And climatic vagaries implied a sustained African commitment to inter-cropping.

ALVORD'S MODERNIZATION AND COMMUNITY DEVELOPMENT VISION

Alvord's vision stretched beyond agricultural improvement or squeezing as many Africans into the Reserves. In his modernisation programme Alvord pursued the missionary ideal of satisfied, self-contained Christian communities with hard-working, rational individuals using agriculture as a base for the accumulation of personal wealth (see also Rennie, 1973, 522-527). Alvord incorporated the idea of segregation in his thinking, but not explicitly.[70] He merely romanticised the idea of self-contained village life, as he had experienced himself in his rural home in the United States of America (see Alvord, n.d., 1-20). Village industries and local (agri-)business would absorb the labour of those people who could not be accommodated in agriculture, either for reasons of land shortage or because of inability. Artisans and skilled workers would provide for the material needs of these communities. Alvord also anticipated the emergence of African leaders and associations that could articulate their own wants and desires and serve as an example to other Africans as well as providing services to their communities. These leaders would emerge, not on the basis of royal descent or mastery of the spiritual world, but on merit by developing skills and abilities (Alvord, n.d., 1958).

Alvord's demonstrator programme embodied a two-level thrust of modernising the household, as well as modernising entire Reserves.

Reserve modernisation started with the placement of an agricultural demonstrator in a central location in the Reserve. Once this demonstrator had succeeded in gaining the confidence of a number of 'progressive' farmers as plot holders, he would organise his best producers in a Native Farmers Association, which would engage itself in promoting improved farming practices. Sometimes these NFAs linked up to do collective buying and selling or purchase weighing scales and grinding mills in order to attain better prices. However in most cases their impact was limited and they soon collapsed.[71]

After the agricultural demonstrator had established himself, a community demonstrator would be appointed, who started building 'improved' houses, roads and toilet blocks for these initial plot holders and other progressive or respected men (often chiefs).[72] The community demonstrator soon became an essential ingredient in Alvord's centralisation policy. Centralisation of Reserves entailed a rudimentary soil classification survey on the basis of which the available land was split in a block of arable land and a block of grazing land, with a line of residential stands in-between. By means of centralisation it was hoped to reclaim degraded land and increase the carrying capacity of the Reserve. Whereas initially centralisation was presented as a conservation measure, over the years it became a measure to facilitate administrative control. Centralisation itself did nothing to stop degradation (see below). Rather it aggravated erosion through concentration of people and cattle movements (McGregor, 1991). It was only when it was followed by limitation of arable acreages, grazing improvement schemes (paddocking) and registration of arable land holdings to facilitate construction of contour ridges, that it could have any beneficial effects in terms of increasing the carrying capacity and combating sheet erosion. The biggest benefit to government was the consolidation of households and their farming operations, which was to prove essential in the implementation of the Land Husbandry Act in the 1950s (McGregor, 1991; 1995).

If there was a mission post close to or in the Reserve, normally a home demonstrator, also known as "Jeanes teacher", followed suit.[73] These would

---

Box 2: Leading Reserves

— **Chinamora Reserve** (end of 1920s). Chinamora was exposed to a large number of demonstrator trainees and was the first reserve to have a Native Farmers Association (NFA). Certain wealthy individuals capitalised on demonstrators working for them as 'farm managers' resulting in increased cash income by means of higher yields. By 1929 the 'improvement' drive of Alvord and his demonstrators had been transformed into an accumulation drive by a few leading cultivators, who expanded acreages and marketed their produce through a business consortium (NFA). The state tried to curb this undesirable development by enforcing centralisation (and resultant land re-distribution) in 1932/33. However, up to the present day Domboshawa area hosts a relatively wealthy peasant class that capitalises on its proximity to the Harare market, by focusing on horticultural production and private marketing (hawking).

— **Selukwe[78] Reserve** (end of 1920s, beginning 1930s). Selukwe was Alvord's leading example of modernisation that turned out to be a paper reality after the visit in 1942 of the NRB on tour (see below). It was the first Reserve to be centralised (1928-1930) with the help of some opportunistic headmen and the use of persuasive force (McGregor, 1991). The first community demonstrator was stationed in Selukwe in 1932. In many respects Selukwe was an experimenting ground for Alvord in the early days of the programme. McGregor (1991) describes how the programme had differential impacts in spatial and social terms.

— **Zimunya and Mutambara Reserves** (early 1930s). These reserves had already witnessed a boom in agricultural production triggered by the early establishment of irrigation furrows (in Zimunya initiated by an industrious Native Commissioner and in Mutambara by the mission in 1912). When Alvord came in with his programme to re-direct this intensification drive according to his own modalities, he and his demonstrators met stiff resistance, which proved lasting.[79] Alvord's experiment with Zimunya furrow results in a folly[80] (despite early efforts to set up a cooperative farmer managed irrigation scheme) and in Mutambara the Chief and his relatively wealthy followers did not accept state interference (Manzungu, 1995b). The resistance in both areas went beyond centralisation. Both schemes in the end proved technical nightmares and examples of the antagonistic effects produced by a 'tabula rasa' intervention approach by the state (Manzungu, 1995b; Roder, 1965).

— **Zimutu Reserve** (1930s). Despite a difficult start in Zimutu, which saw the first demonstrator clashing with vested interests of certain native messengers organised in the Southern Rhodesia Native Association,[81] Zimutu became Alvord's shining example in the mid and late 1930s. Alvord himself faced some stiff opposition in Victoria (Masvingo) from white settler farmers, who disliked his native production drive and resulting competition with settler farming interests, and from the superintendent for Natives, who accused Alvord of suffering from a maize complex (see above). However, a successful switch to wheat production saved the programme and culminated in the so-called 'voluntary de-stocking' exercise in the mid-1930s.[82] Zimutu was centralised in 1933 and was the venue for pasture improvement experiments in the late 1930s. Later Victoria Province and Zimutu in particular became examples of successful agricultural extension by means of the Master Farmer Associations of the late 1960s and 1970s (Kennan, 1980; Plowes, 1980).

— **Chiduku Reserve** (1942). Mainly through the industrious efforts of an authoritarian Native Commissioner, Chiduku Reserve witnessed the start of a more compulsory and systematic conservation drive using forced labour. It became an example for the NRB on tour and set the future *modus operandi* of the Native Land Husbandry Act. It marked Alvord's loss of initiative in rural state intervention and heralded the paternalistic, coercive drive of Native Commissioners and NRB technocrats.

— **Shiota[83] Reserve** (late 1930s and 1940s). Shiota is one of the Reserves that witnessed early involvement in the demonstrator programme. It went through its various ingredients (agricultural demonstrator in 1929, NFA in 1930, community demonstrator in 1936, and home demonstrators at the nearby Waddilove mission). Monumental was the start of the Master Farmer programme (see Box 3). These Master Farmers organised themselves in the Shiota Leaders Association, which organised an annual agricultural show and hosted field days.[84] Many of the later modalities of the Agritex training programme originated from Shiota. The tedious and thorough re-implementation of centralisation in 1943 facilitated strict control by the Land Development Officer.[85]

— **Nyanyadzi Irrigation Scheme** (1940s onwards). This scheme became the pride of Alvord and example of accelerated agricultural modernisation. The native industries policy (1943-1949) comprised a lime kiln, cart factory, sawmill and departmental marketing scheme.[86] However, the African industries proved non-viable and only the introduction of new cash crops led to sustained success of the scheme in the 1950s and early 1960s.

be recruited from devout female mission students or else from respected house wives. After a short training at Domboshawa school, they would concern themselves with health issues, nursing and household duties in European fashion.[74] Their impact was normally limited to the mission confines, but Native Commissioners were very sceptical of their work.[75]

After the establishment of an agricultural, community and home demonstrator, and implementation of centralisation, attention could divert to other areas of improvement depending on the type of demonstrator (soil erosion, forestry, livestock) that was deployed next, which often depended on the interest of the resident Native Commissioner (see Box 2 for some

leading Reserves). And so in some Reserves attention was paid to the
establishment of wood lots, in others to tackling the plight of livestock by
means of pasture improvement experiments and/or 'voluntary de-stocking'
or else a start was made with constructing contours and storm drains.[76] By
1948 there were 163 agricultural, 86 community, 25 erosion control, 34
forestry and 20 livestock demonstrators operating in the Reserves under the
guidance of 46 African supervisors and 50 (white) Land Development
Officers.[77]

Household modernisation was defined by an evolutionary process of
'improved' farming. First one became a *co-operator:* a person who is copying
some components of the crop improvement package. As the programme
further developed co-operators were identified by means of estimated average
crop yields.[87] The co-operators did not form a group. They rather formed a
category of starters who were distinguished by demonstrators on the basis
of their agricultural performance. They formed a reservoir of future plot
holder candidates.[88]

Next one would be selected as *plot holder.* Plot holders' activities were
carefully monitored by demonstrators. Before harvest meetings were held at
the fields of successful plot holders. Demonstrators were evaluated on the
basis of the performance of these plot holders and consequently
demonstrators carefully selected plot holders and pushed them on. Since the
improved practices required a lot of resources (oxen, plough, access to
markets), mostly well endowed Reserve farmers were selected as plot holders.
When one did not own two oxen for instance, one could never dream of
qualifying to become a Master Farmer.[89] Mind that, since the demonstrators
had an interest in well performing plot holders, yield estimates may have
been inflated.

Finally, after showing consistent determination in practising improved
agricultural methods, one could qualify as a *Master Farmer* (see box 3). In
order to qualify, one was required to build a square, permanent house, farm
sheds and possess oxen.[90] Thus only relatively wealthy cultivators managed
to attain this highly regarded 'status'. The limited opportunities within the
Reserves to improve one's living standard meant that established Master
Farmers looked beyond the Reserves to satisfy their aspirations.

The entire modernisation path thus led to increasing pressures on the settler
government to open up freehold land and business opportunities for Africans

(see Ranger 1985; Weinrich, 1973, 1975). Many Master Farmers in the end left the Reserve, in some cases together with their former demonstrator, to become *African Purchase Farmers*.[91] This was later institutionalised by government who required aspiring purchase farmers to have a Master Farmer certificate.[92] Others left the reserve in order to become *irrigation plot holders* (Weinrich, 1975).

The above model only applied to the male head of the household. For women, the modernisation effort implied a move away from agricultural activities to the confines of domestic duties. Home demonstrators played a crucial role in providing women of upwardly mobile Reserve cultivators with an example of a domesticated career. Processing of farm products, sanitary duties and house keeping became the mainstay of these women, though some became nurses or moral crusaders for the church. Their direct agricultural involvement was supposed to be confined to gardening near the homestead.[93]

The ideals of the caring mother and the hard working father fused in the model of the *Christian nuclear family*. Children were supposed to go to school and on Sundays the whole family would attend a church service.[94] To cater for the necessary cash for schooling, clothing and building of a permanent house and farm structures, part of the crop proceeds had to be sold. These ideals had serious practical implications in the state's agricultural intervention programmes. Hardly any women could qualify as Master Farmers, the only exceptions being widows. Similarly land registration and access to irrigation plots was limited to the male head of household or the single widow.[95]

COMPULSION AND CONTROL

The demonstrator programme received a considerable blow during the Great Depression of the early 1930s. The world recession saw prices for agricultural products plummet.[96] These gloomy events made the fears for African competition more pronounced amongst European settler farmers. In 1931, they directed their anger towards Alvord and his efforts to increase African maize production: '[m]any remarked that Alvord ought to be hung.' Whenever Alvord was seen in Fort Victoria 'he was shunned and treated as a pariah' (Alvord, 1958, 20). Some Native Commissioners and Superintendents used the opportunity to express their discontent with the programme. The assistant Native Commissioner for Goromonzi lamented the fact that demonstrators imbued Africans with a 'get rich quickly' attitude:

Box 3: The story of Mr Vambe, the first Master Farmer

Alvord reports in his autobiography the story of Mr Vambe Mutombgera, one of
the first plot holders in Shiota Reserve. The first agricultural demonstrator in Shiota
had been deployed in 1928 at the invitation of a small group of men. However the
demonstrator met with considerable resistance and 'even open hostility' from the
majority of Reserve dwellers, to the extent that 'he had to be guarded by his
plotholders when visiting some parts of the Reserve.'[97] Despite this initial set-
back Shiota was to witness the start of the Master Farmer programme.

'Back of this movement is the story of Vambe, a poor man, dressed in ragged
clothes and living in primitive pole and mud huts. His one wife was dressed in a
goat skin drape, naked from the waist up, and his children ran around in their
birthday suits. He was tilling a total of 32 acres of worn out, almost pure sand.
But, his cattle kraals were belly deep in well rotted manure. In 1929, he harvested
more bushels from his one acre demonstration plot than from his other 31 acres
combined. He immediately realised what a fool he was to waste his time, labour
and seed in ineffective farming methods.' (Alvord, n.d., 39). However, his wife
suspected witchcraft and prohibited her children to eat the maize thus grown,
until Vambe himself and the demonstrator had eaten first and emerged healthy
from the experience.[98] 'I [Alvord] persuaded him to add three more acres and
plant the four acres to a systematic four-course crop rotation, with outstanding
results. In 1932, the crops in this demonstration were so outstanding that news
of it spread for a hundred miles in all directions. Native farmers travelled by foot
and bicycles from distant Reserves to see it for themselves. That year, at the
'before harvest' meeting, he told the assembled 2,300 Natives that never again
would he scratch the soil "like a baboon looking for worms". He increased his
rotation to more than double the area, abandoned all primitive methods of tillage
and put his entire land under the systematic crop rotation. He also made his two
wives limit their labours to proper crop rotations on 2 acres each. By 1934, we
decided to officially honour him as a "Master Farmer", and, at the following before
harvest meeting, which was held alongside Vambe's land, he was promoted with
a certificate as a "Master Farmer". Thus began the scheme for "Master Farmer
Awards" to which, later, an ornate badge was added to a printed certificate. Today,
Vambe is a progressive and prosperous farmer; and an outstanding leader of his
people, who he dresses as well as any white man.' (Alvord, n.d., 39) The Land
Development Officer for Marandellas added in 1947 that Vambe had made so
much money 'that he was able to purchase four wives'.[99] Later Vambe bought a
230 acres Native Purchase farm for his son (Alvord, 1958, 26).

Alvord hoped to produce African leaders from this new 'class' of Master Farmers.
In Shiota the small group of recognised Master Farmers organised themselves in
the Shiota Leaders Association in 1937, that set out to organise the annual
agricultural show from then onwards. Prizes were given to the best performers in
agriculture, livestock, grinding and men's handwork.[100] If Master Farmers were
found to be practising agriculture in another way than the stipulated modern
practices, their badges and certificates were withdrawn.[101]

whereas the goal should be the raising of the level of agriculture throughout the Reserves, . . . I see a marked tendency for the demonstrators to become, in effect, the farm managers of the few enterprising and money seeking plot owners and it is a tendency that should be checked.[102]

The CNC wielded to these pressures from European farmers and members of his own staff by ruling that 'the saturation point had been reached with regard to the number of agricultural demonstrators and ordered that demonstrator training at Domboshawa should be discontinued' (Alvord, 1958, 27).[103] The result was that from 1935 to 1939 the number of agricultural demonstrators only increased by six.[104]

In order to save the white farming sector from imminent bankruptcy, the government of Rhodesia introduced the Maize Control Acts (1930, 1934) and Cattle Levy Acts (1931, 1934). By means of these Acts African cattle owners and maize growers were forced to subsidise white ranchers and farmers.[105] The reaction of African farmers was to withdraw from the demonstrator programme. Many plot holders complained bitterly: 'What is the use of adopting better methods and producing more crops when there is no market?' (Alvord, 1958, 26).[106] Alvord protested strongly against the Maize Control Act in his annual report for 1933 and pointed out the dangers to the country as a whole, if Africans were denied opportunities to sell excess grain for cash.[107] Alvord feared not only that his demonstrator programme would come to a halt, but also that his ideals of self-supporting rural communities would have to be shelved.[108] In an attempt to ensure African plotholders a certain cash income, Alvord paid increasing attention to the promotion of alternative (cash) crops like wheat and cotton as well as possibilities for establishing cooperative marketing ventures amongst Africans. However, these efforts only translated in tangible results in some well watered Reserves (wheat)[109] and in African irrigation schemes in the 1940s.[110] Cotton growing experiments started in 1934, and despite initial set-backs,[111] gradual expansion took place. By 1950 cotton was one of the biggest money spinners in Mhondoro Reserve (Alvord, 1958, 27).

After 1935 conservationist concerns started to take prevalence over agricultural production issues in the Reserves. As a consequence of erosion alarms raised in the USA (Dust Bowl) and South Africa in the 1930s (see Beinart, 1984; Beinart and Coates, 1995; Anderson, 1984; Drinkwater, 1989; McGregor, 1995), Alvord and Native Affairs officials became increasingly concerned with over-population, over-stocking and the impeding erosion menace in the Reserves.[112] Alvord himself fuelled these fears by providing comprehensive figures on the situation in the Reserves after his first visit to the USA in 1930 (see Alvord, 1930). During a second visit to the USA in 1935, Alvord undertook a study of the soil conservation work done amongst Blacks.[113] Upon his return he reported to the CNC that some Native Reserves were as badly eroded as some of the worst areas in the United States. He estimated that 1.5 million acres in the Reserves (16% of all arable land) had been badly damaged by sheet and gully erosion (Alvord, 1958, 29). Faced with such alarmist figures the CNC once more felt compelled to render support to Alvord and his programme. In 1936, the first Soil Conservation Officer was appointed,[114] assisted by three soil erosion demonstrators. This officer set out to construct contour ridges in recently centralised Reserves with the help of paid labour gangs.[115] In addition all agricultural demonstrators were called for a short training in soil conservation and pegging in 1936, and returned to their Reserves with ox-drawn terracers.[116]

Centralisation became the main thrust of Alvord's programme. It became conditional that a Reserve had been centralised before the services of a demonstrator would be rendered to it.[117] However, soil survey work preceding actual centralisation and allocation of individual land holdings proved tedious and time-consuming, even though the number of Land Inspectors under Alvord's command was increased from two to four in 1938.[118] Meanwhile requests from Native Commissioners to centralise Reserves kept on pouring in, convinced as they were that centralisation was the only remedy left to save the Reserves.[119] A request for centralisation of Sabi Reserve in 1938 was met with a gloomy reply by Alvord: 'With our present limited staff it is possible that we might be able to do Sabi Reserve in 12 to 15 years from now . . . .'[120] But it was not only a lack of staff that threw spanners in the centralisation drive. The NRB on tour in 1942 discovered that, despite Alvord's frequent reports of successful rehabilitation of Selukwe Reserve,[121] the reality of centralisation was devastating:

> The Selukwe Reserve was held up as an example of how a degenerated
> area had been rehabilitated and it is very disappointing to us now to
> find that it has reverted to desert conditions practically.[122]

The NRB found that the compulsory ridge construction scheme in Chiduku reserve under the authoritarian leadership of the local Native Commissioner offered more hope for a quick remedy.[123] The Natural Resources Act of

1941 placed the initiative of state intervention in Reserves firmly in the hands of Native Commissioners. Alvord succumbed to the increasing call for 'compulsion and control' in the quest for soil salvation in the Reserves, and concluded that: 'As far as its effect on the general Native is concerned we have wasted our time for 17 years in conducting agricultural demonstrator work in Native Reserves.' Alvord proceeded to call for a unified Department to implement de-stocking, compulsory crop rotation and control over tillage methods.[124] The Native Production and Trade Commission of 1944 effectively sealed the fate of the Africans in Reserves and processes of voluntary change by suggesting imposition of good husbandry conditions in Reserves.[125] Despite the fact that Alvord became the first director of the Department of Native Agriculture in 1944 and commanded considerably more staff in his final years before retiring in 1950, he had lost the initiative of African modernisation. He was forced to concentrate his improvement efforts on limited areas and did so increasingly on irrigation schemes, which offered the biggest potential for fast modernisation. And so Nyanyadzi witnessed the launch of the native industries development drive in 1943. However, despite the successful introduction of new cash crops in the irrigation scheme, the drive to develop rural industries that absorb excess labour and provide incentives for the growth of rural townships floundered. By 1948 the initiative was bankrupt.[126]

By the time of his retirement Alvord had become thoroughly frustrated in his efforts to demonstrate ways to raise productivity and conserve the soil:

> [W]e are faced with a stubborn, childish, conservative mass of people who are resistant to change . . . During the past 20 years, millions of acres of once good, arable lands have been ill-treated and mismanaged under improper tillage methods, in spite of the fact that for 20 years we have conducted demonstration plots throughout the country which have shown to large masses of people the results of good tillage methods. But their eyes are shut and their ears are tight (Alvord, 1948 18).

CONCLUSION

In conclusion one can observe that Alvord's demonstrator programme produced a number of continuities that bear relevance to agricultural extension as practised today.

The conviction, based on 'scientifically' calculated carrying capacities, that communal areas harbour agricultural potential which can be further developed to the benefit of its inhabitants, if only the right technology, approach and commitment can be found, is still dominant. This belief served during both the colonial and post-independence era (up to 1997) to curb demands by the indigenous population for more and better land. Despite all (donor) discourse on participatory, sustainable, and community development not much seems to have changed in the general development view, the roots of which were planted by the likes of Keigwin and Alvord.

Another continuity can be found in the extension methods that Agritex presently employs. Field days, annual agricultural shows and the Master Farmer training programme still form the core of Agritex' strategies towards improving agricultural production in communal areas. Furthermore the curriculum of Master Farmer training shows a striking resemblance to the contents of Alvord's improvement package. As the communal areas have experienced huge changes since 1950, the Master Farmer curriculum may be inadequate to address present-day challenges.

Furthermore the intimate relationship between the extension worker and a limited number of resource-rich smallholder farmers seems to find its origin in Alvord's demonstrator programme. To make the Alvord package work one required access to resources which only a select few would have.

Another legacy of the Alvord programme, demonstration of the benefits of improved farming practices in the field, became redundant after 1944 with the application of increasing amounts of force in agricultural interventions. Now that Agritex has re-established a voluntary change approach, it should consider the re-introduction of the demonstration concept in extension.

Despite some notable early successes, agricultural extension in Zimbabwe should squarely address the needs and demands of the majority of communal area farmers. If it fails to do so, the farmers will indeed ignore the extension service, as phrased by the District Agricultural Extension Officer's riddle. Without farmers, the service will become a relic of the past.

NOTES

1. National Archives of Zimbabwe [NAZ] file SRG4: Report by H. S. Keigwin, esquire, Native Commissioner, presented to the Legislative Council, 1920, Salisbury, Rhodesia.

2. As a Native Commissioner in Lomagundi Keigwin had observed that such activities could be undertaken by Africans. Underlying his vision of development were two assumptions. Firstly Keigwin assumed that it was better to develop the African among his 'own people in their villages', rather than developing and reproducing the African according to European lines and example. This implied, according to Keigwin, an emphasis on simple, primitive handicrafts close to the desires and abilities of a 'backward people'. In this process of segregated, primitive development competition with Europeans could be avoided. Secondly he believed in a concept of raising the mass ever so little rather than advance the few according to European lines. The latter would create the dangers apparently associated with 'over-education' of a few men who could then cause racial frictions by inciting the 'ignorant masses'.

3. See for instance Bevan (1924, 13) observing that 'If crops are good and their [African, AB] limited financial requirements can be met, they prefer to live in comparative idleness ...'. Moyo (1925, 47) considered that '... older men ... only look forward to three things, ... , namely, cattle, kaffir beer and polygamy. As long as he gets well with the three things mentioned above he does not worry: he is quite content ...'.

4. Keigwin refers to Booker T. Washington and his Tuskegee Institute in his report to the Legislative Council, 1920, Salisbury, Rhodesia. NAZ, SRG4.

5. NAZ, SRG4, Report by H.S. Keigwin, esquire, Native Commissioner, presented to the Legislative Council, 1920, Salisbury, Rhodesia.

6. Keigwin quotes extensively from a report of the General Education Board of America to show how such a scheme could be put to work. The description fits very closely with the system that was ultimately embarked on in Rhodesia.

7. In this chapter, the colonial names of Reserves are used. The present name of Tjolotjo is Tsholotsho.

8. Moyo (1925, 47), a black teacher at Tjolotjo school, later re-confirmed this image by stating that '... young men ... go to town and work ...' instead of putting up beautiful buildings and towns with their acquired carpentry skills.

9. NAZ S138/69, Report of missionary tour, to Chief Native Commissioner, from Keigwin, Director of Native Development, dd 14 October 1922.

10. In the build up of his argument, Wilson (1923, 87) refers to the industrial machine metaphor. Africans have to play a part in Rhodesia's industrial fabric, a la Ford Motor Car Industries, or else 'we shall not only break the machine, but we [the Rhodesian settlers] ourselves shall be buried under the debris.'

11. NAZ S138/206, Letter from Keigwin to the Chief Native Commissioner, dd 30 January 1924. Reply from Minister of Native Affairs to the Chief Native Commissioner, dd 13 February 1924.

12. NAZ S138/206 Circular on Development of Native Areas, from Chief Native Commissioner, dd 6 March 1924. Notable exceptions to the over-all positive response of Native Commissioners were: NC for Charter, who remarked that 'natives are successful without demonstrations and will not welcome them'; NC Buhera stating that 'natives are apathetic'; NC Bubi 'no benefit whatever will accrue to natives' and NC Sebungwe 'doubtful whether there is any use in the proposal'.

13. NAZ S138/206 Letter from the Secretary to the Premier, to the Chief Native Commissioner, dd 11 July 1924.

14. NAZ, S138/206, Training of Native demonstrators, report by Keigwin, to Chief Native Commissioner, dd 18 June 1924.

15. NAZ, AL6/1/1/14, Letter from Keigwin, to Alvord, dd 12 August 1926. After his resignation Keigwin became Director of Education in Sierra Leone (*NADA*, 1963, 122).

16. NAZ, AL6/1/1, Letter from Mundy to Alvord, dd 6 May 1926.

17. NAZ, AL6/1/1, Letter from Alvord to Mundy, dd 12 May 1926.

18. NAZ, S138/69, Undated draft letter to be sent to the Transkei, by E. D. Alvord to the Chief Native Commissioner. Attached to Native Agricultural Demonstrators, letter from Chief Native Commissioner to the Secretary of Native Affairs, Pretoria, dd 2 January 1927.

19. NAZ, S138/69, Letter from General Council Agricultural Director, J. W. D. Hughes, Umtata, Transkei, to the Chief Magistrate of the Transkeian Territories, Umtata, dd 26 January 1927.

20. NAZ, S138/69, Scheme for work of Native demonstrators, Letter from Alvord to Chief Agriculturist (Mundy), dd 25 February 1927.

21. Letter from Chief Agriculturist (Mundy) to the Chief Native Commissioner, dd 14 February 1927. The system preferred by Mundy was later adopted for Native Purchase Areas, where initially demonstrators had been chased away by farmers claiming to know everything about farming [see NAZ, S1542/D7, Letter from Alvord to Secretary for Native Affairs, dd 16 March 1939].

22. NAZ, S138/72, Letter from Alvord to the Director of Native Education, dd 13 May 1929.

23. Alvord replied with two lengthy letters to this MLA explaining that 'My position is that I was offered employment along lines for which my whole life training has been a preparation.'. Alvord dismissed the other adverse claims by outlining his ancestors came from Britain and that his qualifications were very high (a Bachelors and Masters in Agriculture from Washington State University). NAZ, AL6/1/1/23-24, Letters from Alvord to Mrs Tawse-Jollie, dd 11 and 14 April 1927.

24. NAZ, S138/72, Report from E.D. Alvord to Chief Native Commissioner, dd 16 June 1927. According to Alvord the missionaries' 'lack of interest' in his agricultural instruction programme was mainly due to a legacy of Keigwin's nondenominational stand ('a hangover of the antagonism they had had toward the Native Development Scheme and the Government Schools'). Alvord's criticism was directed at the fact that both Mutambara and Old Umtali missions only trained school boys in growing of vegetables under irrigation. He remarked that most school-leavers would not have access to irrigated gardens and thus would lose interest in agriculture, because they had not been trained on how to make money with ordinary Native field crops.

25. Alvord applied for the post of Director of Native Education in August 1927. To underline his abilities he had used his time in the office to write an extensive policy and curriculum for Agricultural courses to be taught at Native schools. He was however, turned down for the post. NAZ, AL 6/1/1, Letter from Alvord to Acting Director of Education, dd 2 August 1927.

26. NAZ, S138/206, Report on training of demonstrators at Tjolotjo, from Alvord to the Chief Agriculturist [Mundy], dd 13 November 1926.

27. These slides had been supplied to Alvord during his visit to the USA in 1926 by the agricultural extension department of the International Harvester Company (IHC). Alvord later requested for slides on black farming in America, and this greatly improved the effects on the Rhodesian Audience: 'Seeing people of their own colour on the screen improved them [Africans, AB] greatly, and brought the message home better than before. Because of these negro slides they give more attention than before and are not so apt to feel that it is all white men's wizardy and that they themselves cannot do such things.' NAZ, AL6/1/1/33, Letter from Alvord to G.J. Sammons, International Harvester Company, Chicago, USA, dd 17 February 1928.

28. The story was so powerful and convincing, that the BBC in October 1940 broadcast the story in a series on 'development of British Africa' for school children in the United Kingdom. NAZ, AL6/1/1, Letter from McGregor, BBC, London, to Alvord, dd 7 September 1940.

29. NAZ, S138/69, Training of native farming demonstrators, circular letter No. C 272/1927, from Chief Native Commissioner Jackson, to all Native Commissioners, dd 1 April 1927.

30. NAZ, S138/69, Letter from Alvord to the Principal of Tjolotjo school, dd 16 September 1927.

31. NAZ, S138/72, Report on Chiweshe and Chibi reserve trips from Alvord to the Chief Native Commissioner, dd 5 July 1927.

32. NAZ, S138/72, Report of progress of demonstration work, Chiweshe Reserve, from Alvord to the Chief Native Commissioner, dd 23 November 1927.

33. Selukwe's present name is Shurugwi.

34. Annual Report of Agriculturist for Instruction of Natives, to the Chief Native Commissioner, dd 14 January 1929.

35. For instance in 1933, the NC for Mount Darwin reports that in Madziwa Reserve people thought that: '... plots would be taken over by the Government as soon as they became established; that it was the first step in a scheme to restrict cultivation to small plots only and that the plot owners would have to hand over the yield to authorities, and that it was a scheme to increase taxation ...' NAZ, S235/511, Annual report for the year ending 31st December 1933, NC Mount Darwin.

36. Average yields in ton/ha over the period 1927-1949 were 2.2, 1.3 and 0.3 for demonstration plots, Europeans and ordinary Africans respectively. The European yield data were derived from Weinmann (1975). The yields from demonstration plots from annual reports by Alvord. Finally, the ordinary African yields were based on visual estimates by demonstrators and presented in Annual Reports published by Alvord.

37. Taylor (1925), CNC from 1923-1928 and supporter of the demonstration programme, in his review of Shantz' book systematically discarded these advantages. In Taylor's view the inherent infertility of Rhodesian soils rendered fallowing a waste of efforts. Virgin land still yielded less than a well cultivated, permanent garden. Modern, intensive agriculture could overcome the dual problems of loss of soil fertility and plant diseases. And finally natural processes of decay resulted in a bigger, less erosive and better supply of minerals than burning (Taylor, 1925).

38. A sense of the importance of this objective and the growing impetus of the Land Apportionment Act (1930) and associated squeezing of more Africans in Reserves can be obtained by assessing three annual reports of the CNC. In the report for 1928, the main aim of the demonstrator programme is to 'substitute intensive' cultivation of a smaller area in place of the extensive and harmful cultivation of a wide area'. In the annual report for 1932, it is to 'develop native reserves so as to enable them to carry a larger population and so avoid necessity of acquisition of more land for native occupation.' By 1933 it is bluntly 'to reduce the amount of land required to sustain a family.'

39. Alvord did not realise that shifting cultivation was 'less a device of barbarism than a concession to the character of the soil' (Hailey, 1938 in Floyd, 1960, 296).

40. NAZ, S138/72, Letter from Alvord to the Director, Native Education, dd 13 May 1929.

41. Major Mundy, in many respects, has been the nestor of European agriculture in Southern Rhodesia. During his career [1909-1943] as Chief Agriculturist and later Secretary of the Department of Agriculture, he was responsible for 41 official publications in the *Rhodesia Agricultural Journal* on various crop experiments and produced the first comprehensive book on agriculture in Rhodesia (Mundy, 1928). Thus he 'translated' many farming innovations from other 'pioneer' countries like USA, Australia and South Africa into locally tested recommendations. Mundy's prime concern was to develop a system of agriculture that would maintain soil fertility levels, reduce erosion hazards and be profitable to the white commercial farmer. Mundy recommended the introduction of a 4 course crop rotation involving (manured) maize, wheat and velvet beans, that would safeguard a permanent future for settler agriculture (Mundy and Walters, 1919). However, the lack of viable markets for other crops than maize and limited cattle holdings, inhibited wide-spread adoption of crop rotations and manuring. It should come as no surprise that Mundy's ideas made few in-roads in settler agriculture. The establishment of tobacco-maize-grass rotations with heavy applications of fertilisers in the 1950s proved to be the first profitable and sustainable system of European permanent farming on sandvelds in Rhodesia (Grant, 1976, 252). However, in Alvord and his programme for agricultural improvement Mundy found a ready response from the early 1920s onwards. Besides supplying Alvord with suggestions for crop rotations and manure trials on the mission farm at Mt Selinda in 1921 and 1923, Mundy himself paid several visits to Mt Selinda to support Alvord's work (Alvord, 1958; see also NAZ, S840/2/23, Letter from Mundy to Director of Education, dd 10 October 1923).

42. Later recommended rates rose to over 10 tons/ha annually (McGregor, 1991, 90).

43. NAZ, AL6/1/1, Letter from Alvord to Chief Agriculturist (Mundy), dd 28 August 1928.

44. NAZ, S138/72, Letter from Alvord to Director, Native Education, dd 13 May 1929.

45. Reference is made in particular to the experiments with improved fodder grasses and the effects of the centralisation programme (curbing and fixing the amount of arable land, thus setting free the grazing land).

46. NAZ, MF580, Annual report for the year 1920 by the Chief Native Commissioner.

47. In 1934 cotton was included in the rotation after the manured maize crop. MF 580, Annual report for the year 1934 by the Chief Native Commissioner.

48. Encouraged by Major Mundy, Alvord even wrote an article on sand veld farming in the *Rhodesia Agricultural Journal*, in order to convince its (largely) white settler readership. Mundy, realising that the results of his crop rotation trials had mainly fallen on deaf ears amongst settler farmers, must have been very pleased with Alvord's concluding remarks: 'My advice to every sand veld farmer is to screw up his courage, take the plunge, and, without loss of time, get started on a well considered and definite crop rotation scheme ...' (Alvord, 1928, 1110).

49. NAZ, S988, NRB interview with Mr E. D. Alvord, 14 July 1942, Rusape.

50. NAZ, MF580, Annual report for the year 1920 by the Chief Native Commissioner.

51. Rapoko or *rukweza* is finger millet.

52. NAZ, S138/72/2, Letter from Superintendent of Natives, Fort Victoria, to CNC, dd 9 May 1931.

53. NAZ, S138/72/2, Letter from Alvord to Director of Native Development, dd 8 June 1931.

54. NAZ, S138/72/2, Letter from Alvord to Director of Native Development, dd 8 June 1931.

55. NAZ, MF580, Annual report for the year 1917 by the Chief Native Commissioner.

56. Alvord mentions women in particular, since his gender view postulated that labour of women should be confined to domestic tasks. Earlier on Alvord had condemned hoeing by African women as being a kind of slavery.

57. NAZ, S138/72, Letter from GEP Broderick, Principal Domboshawa Government School, to the Director of Native Education, dd 10 April 1929.

58. S138/72, Letter from CNC to Director of Native Education, dd 30 April 1929.

59. S138/72, Letter from Alvord to Director of Native Education, dd 13 May 1929.

60. NAZ, S1619, NC Ndanga, quoted in Ranger (1985, 62).

61. Under the centralisation scheme, cultivators in the Reserves were allocated equitable land holdings.

62. As Allan (1945, 19) phrased it: 'It is a time of changing thoughts in our whole conception of the soil and man's relation to it.'

63. During tillage trials over 5 consecutive years (1989/90 to 1993/94) at Makoholi, near Masvingo, soil losses in tons/ha varied from 59.7 for conventional ploughing; 26.2 for hand hoeing and 5.9 for mulch ripping (Chuma & Hagmann, 1995, 44).

64. See S840/2/23, Letter from E.D. Alvord to L.M. Foggin, Education Department, dd 15 October 1921. S138/72, Report on 'Suggestions for agricultural instruction at government schools.', from Alvord to CNC, dd 23 July 1927.

65. S138/72, Circular letter from the Agriculturist for Natives to all Native demonstrators, dd 21 December 1927.

66. MF580, Annual Report for the year 1930 by the Chief Native Commissioner.

67. The shift from intensive wetland cultivation to extensive ploughing of dry land had been accompanied by a striga infestation (McGregor, 1991, 79).

68. High soil temperatures, caused by for instance row mono-cropping of maize, also encourages germination of striga (witchweed) (Parker, 1984).

69. S138/72, Letter from E. D. Alvord to the Director of Native Education, dd 13 May 1929.

70. An example of Alvord's rather independent thinking with regard to segregation is a proposal by him, in 1936, to start an African tenant farming scheme on European farm land. During his visit to the USA in 1935, Alvord had observed that black tenant farmers on white land were the most productive of all black farmers. Alvord put this proposal forward since he was faced with over-population and growing congestion in Reserves due to the strict implementation of the Land Apportionment Act to move all African tenant farmers out of European and Crown land. It is not known how the CNC reacted to this countervailing and courageous proposal. NAZ, S1542/A4/5, "Native Tenant Farming. A report on a survey of negro tenant farming in America: with recommendations for Rhodesia" written by E. D. Alvord for CNC, 27 January 1936.

71. For instance, in his annual report for 1932, the NC Marandellas reports that 'The farmers' associations in Shiota and Wedza reserves are almost defunct ... It is understood that the apathetic view of many of the members, which is now most apparent, was brought about when they found that membership did not necessarily ensure better markets or increased prices.' NAZ, S235/510. The NC for Lomagundi reported in his annual report for 1933 that the NFA in Zwimba Reserve confined its activities mainly to 'meetings discussing improved methods of agriculture and to receiving lectures by the Native demonstrator on agricultural and stock matters.' NAZ, S235/511.

72. The first Community demonstrator was deployed in Selukwe reserve on 1 October 1932, under the title of industrial demonstrator. NAZ, S1007/7, Letter from Alvord to NC Selukwe, dd 26 September 1932. The work of community demonstrators was to lay out 'model rural villages' during or after a centralisation survey had been done. 'All pole and mud huts were to be abolished and substituted with larger, well-ventilated circular huts or houses in Kimberly brick and burned brick' (Alvord, 1958, 25). Pole and mud huts were considered a health hazard.

73. The first home demonstrator was appointed in 1931. By 1935 there were 25 home demonstrators operating from Mission posts. NAZ, S1542/D7, Letter

from G. G. Rudd, organising instructress of domestic science, to CNC, 8 February 1935.

74. NAZ, S170/1161, Letter from Minister of Native Affairs, to Secretary to Premier's Office, 14 February 1933. Later so-called village homecraft schools were started to instill 'habits of discipline and cleanliness' into African women and their children. These efforts were directed at 'promoting civilised methods and habits amongst the rising generation.' NAZ, SRG3, Report of the Director of Native Education for the year 1945.

75. '... Jeanes teachers in [the] Reserve go round the kraals giving instructions to the Natives as to what they must do. They are told they must sweep their kraals and keep them clean, they are told to make special sanitary arrangements, to build their cattle kraals half a mile from their huts and so forth. The natives are disturbed as they do not know who these people are who go round giving them orders. In some cases they endeavour to carry them out, in others they do not, and so the matter stands.' NAZ, S170/1161, Letter from Minister of Native Affairs, to the Secretary to the Premier's Office, 14 February 1933.

76. After 1938 it became policy to construct contour ridges and storm drains on Reserves that had been centralised. NAZ, S988, Evidence of E. D. Alvord to the NRB, Salisbury, 19 November 1942.

77. NAZ, SRG3, Annual report of the Director of Native Agriculture for the Year 1948.

78. Selukwe is known as Shurugwi in present day Zimbabwe.

79. See NAZ, S160/IP1, Files on Mutambara irrigation scheme.

80. This furrow is presently known as Nyachowa furrow, and only partially operational (see Van der Zaag and Röling, 1996).

81. NAZ, S170/1161, Letter from Alvord to Director for Native Development, 4 December 1931. Letter from Superintendent of Natives, Fort Victoria, to CNC, 16 December 1931.

82. NAZ, S1050, Annual report for the year ending 31st December 1938, NC Victoria.

83. Presently named Chihota Communal Area.

84. NAZ, S2384/K5843/3236, Overview of demonstrator work in Chiota Reserve, DLDO Marandellas, 9 September 1947.

85. NAZ, S2384/K5843, Individual allocation of land, Shiota Reserve, Land Inspector, 24 February 1943.

86. See NAZ, S160/MS2, Native industries files, 1943-47.

87. The number of co-operators and plotholders each demonstrator had under his guidance were meticulously recorded each year. A drop in their number could

spell imminent discharge of the job. In the late 1940s the average yield for co-operators hovered around 6 bags per acre; for Master Farmers around 8 bags per acre and for plot holders around 10 bags per acre. NAZ, SRG3, Annual reports of the Director of Native Agriculture, 1947-1949.

88. Interviews with E. Chikazhe (former demonstrator in Murehwa, started in 1946), on 28 June 1997 at Mount Darwin; with Kondo (former demonstrator in Chiweshe, started in 1954) on 29 June 1997; and with B. Mupinda (former demonstrator in Makoni, started in 1950) on 9 March 1996, in Rusape.

89. *Ibid.*

90. 'If you can't get sufficient manure on your acres, it is very, very difficult to qualify as Master Farmer. Yeah, he [the plot holder] is handicapped by not having enough manure.' Interview with former demonstrator, E. Chikazhe, Mount Darwin, 28 June 1997.

91. Mr Kondo, former demonstrator in Mazowe, resigned in 1966 and took out a purchase farm in Chesa APA, taking 8 Master Farmers with him. Mr Chikazhe, former demonstrator in Murehwa, resigned in 1958, again taking several of his best farmers with him to Chesa APA. Interview with E. Chikazhe, Mount Darwin, 28 June 1997. Interview with Kondo, Mount Darwin, 29 June 1997. Of the 14 most experienced African supervisors on Alvord's pay-roll in 1947, seven owned a purchase farm and two had plots in irrigation schemes, after either resigning or retiring. Interview with M. Sigauke, former demonstrator in Chakohwa irrigation scheme, Birchenough Bridge, 6 March 1998.

92. Later still the same criterion was applied for future resettlement farmers after independence (see Riddell Commission report, 1981).

93. Interview with Mrs Sigauke, hospital matron and wife of A. Sigauke (former community demonstrator in Mtoko), Harare, 19 February 1998. See also Weinrich, 1973, 189-190.

94. Interview with M. Sigauke (former demonstrator in Chakohwa irrigation scheme, started in 1942), Birchenough Bridge, 6 March 1998.

95. NAZ, S2814/3585, Instructions on the allocation, occupation and preservation of plots on irrigation projects in Native Reserves, CNC, 5 October 1939.

96. The export maize price dropped in 1930 from 10 shillings to 8 shillings per bag and in 1931 it slumped even further down to 4 shillings per bag, whereas production costs averaged 8 shillings per bag. Cattle exports slumped to nil during the 1931-2 outbreak of foot-and-mouth disease (Phimister, 1988, 172).

97. NAZ, S2384/K5843/3236, Report from the Land Development Officer, Marandellas on Agricultural demonstration work, Chiota Reserve, Marandellas, dd 9 September 1947.

98. NAZ, S2384/K5843/3236, Report of Land Development Officer for Marandellas on Agricultural demonstration work Chiota Reserve, Marandellas, dd 9 September 1947.

99. NAZ, S2384/K5843/3236, Report of Land Development Officer for Marandellas on Agricultural demonstration work Chiota Reserve, Marandellas, dd 9 September 1947.

100. NAZ, S160/IP1, Letter from Secretary Shiota Leaders Association, to Alvord, via NC Marandellas, dd 4 July 1943. NAZ, S2384/K5843/3236, Letter from LDO Marandellas, to Alvord, dd 8 August 1945. NAZ, S2384/K5843/3236, Report from LDO Marandellas on Agricultural demonstration work, Chiota Reserve, Marandellas, dd 9 September 1947.

101. In February 1945 the certificates and badges of six Shiota Master Farmers were withdrawn by the Land Development Officer for reasons of not winter ploughing, not sticking to proper crop rotations, and growing maize on unstumped land 'under native methods'. NAZ, S2384/K5843/3236, Letter from DLDO Marandellas, to Assistant NC Wedza, dd 14 February 1945.

102. NAZ, S235/510, Annual report for the year ending 31st December 1932, NC Goromonzi.

103. After protests by Alvord, the CNC allowed the training to continue at a rate commensurate with annual loss of staff that was discharged, resigned or died (Alvord, 1958, 27).

104. By 1935 there had been 60 agricultural demonstrators (excluding 5 community demonstrators) deployed in Reserves at a rate of increase of about 5 per annum. By 1939 their number had increased to 66 agricultural demonstrators (and 19 community demonstrators). NAZ, MF 580, Annual Reports of the Agriculturist, Native Department, 1927-1940.

105. Under the Maize Control Acts, African producers were paid between 1s.6d. to 6s.6d. per bag from 1934-39, whilst white growers received an average price of over 8s. per bag. Cattle Levy Acts imposed a 2s.6d. levy on slaughter of cattle for domestic consumption in order to pay a bounty for white cattle exports. In 1934 an additional 3d. tax per head on all cattle was imposed (Phimister, 1988, 184).

106. The CNC commented in 1932 on the increasing antagonism against demonstrators as follows: 'Much of it [the antagonism] undoubtedly arises from the fact that his [demonstrator] work is largely concerned with demonstrating the production of maize: that so long as this cereal commanded a remunerative price, the Natives welcomed the acquisition of methods resulting in a higher yield: that, like other people, they have been bitterly disappointed

by the fall in prices, which has deprived them of an easy means of earning a lucrative living, which seemed within their grasp: That again like other less ignorant people, when things go wrong, they blame the Government, and vent their feelings on the nearest representative of the government. In this case the agricultural demonstrator.' NAZ, S170/1161, Letter from CNC to the Director for Native Development, dd 30 January 1932.

107. 'On many reserves this year natives were not able to sell grain for cash and had difficulty in paying tax and dipping fees. At most "kaffir truck" stores they are not able to sell grain for cash and must take salt or goods for it. This imposes upon them a hand-to-mouth existence under which they cannot progress. With no income they are low consumers to the detriment of the country as a whole. If they could sell a part of their excess grain for cash their capacity as consumers would be greatly increased and the whole country would benefit. It is simply a question of sound economics. If reserve natives are not assisted in the development of self-supporting rural communities it will be ruinous to the future interests of Rhodesia.' NAZ, MF 580, Annual report of the Agriculturist: Native Department, 1933.

108. In actual fact the effects of the Maize Control Acts were differential. In reserves close to a railway or town, peasants were able to attain favourable prices by selling direct to a Maize Control Board depot. Only in remote Reserves with no white farmers buying African maize as cattle fodder and no MCB depots, maize became an unattractive crop. Overall, African maize output rose sharply during the 1930s (Ranger, 1985; Phimister, 1988).

109. Wheat was traditionally produced by Africans in *dambos* on the central water shed and in the Eastern Highlands by means of indigenous irrigation furrows (see Roder, 1965). However, its deliberate promotion by demonstrators and the low prices of maize attainable in the 1930s, led to increased acreages under wheat (see for instance NAZ, S1050, Annual report for the year ending 31st December 1936, by the NC for Victoria).

110. Organised marketing of cash crops in the Save Valley irrigation projects took place as from 1943 (Alvord, 1958, 36). Many cash crops, like sunnhemp, beans, wheat and later even tomatoes, were introduced on these African irrigation schemes during the 1940s (Alvord, n.d.).

111. In 1934 11 demonstration plots in as many Reserves were devoted to cotton growing. The results were disappointing in terms of yields and even more so in terms of money paid to the growers by the Bindura Ginnery. NAZ, S1542/A4, Letter from Alvord to Bindura and District Co-operative Ginnery, dd 9 January 1935.

112. Initial fears of degradation were raised in South Africa during the 18th century and focused on destructive agricultural practices employed by settler farmers. It was feared that ultimately settler farmers were going to ruin all productive land and thus leave no hope of sustained colonial presence. During the 1920s and 1930s attention increasingly shifted to destructive agricultural practices followed by Africans. The erosive effects of squeezing as many Africans as possible in marginal Reserves, became manifest and served as an excuse for comprehensive, coercive state intervention in the lives and means of these Reserve Africans (Beinart, 1984).

113. Alvord also met the influential American conservationist, Bennet during that visit [see NAZ, Al6/1/1].

114. NAZ, S1542/A4/5, Letter from Alvord to CNC, 18 March 1936.

115. NAZ, S1542/D7, Letter from Alvord to NC Selukwe, 3 July 1939.

116. NAZ, S2401, Letter from Alvord to CNC, 19 November 1936.

117. NAZ, S1542/D7, Letter from Alvord to CNC, 22 February 1938.

118. Only 10 Reserves had been centralised by 1936, and by 1938, 2,407,500 acres of Reserve land had been covered. NAZ, MF580, Annual report of the Agriculturist for Natives, 1936 & 1938.

119. See for instance NAZ, S1542/S10, Letter from NC Concession to CNC, 11 May 1939; S1542/D7, Letter from NC Buhera, to NC The Range, 5 November 1937.

120. Sabi reserve was listed as number 21 on a list of 31 surveys where centralisation had been requested by NCs. NAZ, S1542/D7, Letter from Alvord to CNC, 4 January 1938.

121. 'On Selukwe reserve, where centralisation was first carried out cattle are in much better condition, land under cultivation has been reduced by half and crop yields are larger than before centralisation, grazing areas are growing up to good timber and considerable erosion has been checked.' NAZ, MF 580, Annual report of the Agriculturist for Natives, 1933.

122. NAZ, S988, Evidence of CNC Simmonds to the NRB, 19 November 1942.

123. NAZ, S988, Evidence of NC Stead to the NRB, Rusape, 14 July 1942.

124. NAZ, S235/483, Memo on Soil Salvation in Native Reserves, sent by Alvord to the Secretary for Native Affairs, 25 February 1943.

125. The Commission first of all thought of compulsory planned production. However, since the European farmers were anticipated to protest against such measures it was decided to resort to good husbandry practices, applied to both African and European producers. 'If that be not practicable, they should

nevertheless be imposed on the Natives ...' The recommendations of the Commission led to the enactment of the Good Husbandry Act of 1948, the predecessor of the Native Land Husbandry Act of 1951.

126. See NAZ, S160/MS2, Native Industries, 1943-47.

Chapter 2

Towards an understanding of technology needs of smallholder farmers

JEFF MUTIMBA

INTRODUCTION

The social structure of agricultural knowledge can be studied in a variety of ways. In this chapter I look at agricultural knowledge from the point of view that it is, to a large extent, generated by farmers. Thirty six focal farmers were selected in Murehwa, Mutoko and Marange communal areas (12 in each area respectively) for in-depth analysis of research questions at the farmer level. The questions were concerned with: farmers' production motivation; their production constraints and how they deal with them; sources of their knowledge; how they use 'official' recommendations; their concept of research and what they try out, how and why; their needs for research; and, how professionals deal with the farmers' informal experimentation.

The criteria used to select these three communal areas were:

— agro-ecological zone;
— type of farming system;
— presence of research and other development agencies.

These criteria reflect my initial propositions about opportunities and innovativeness. The selected areas provided an instructive contrast, and the broad range required to examine agricultural knowledge. In this chapter, however, I do not systematically compare agricultural knowledge and production constraints along these selection criteria.[1]

Murehwa lies in Natural Region (NR) II and receives reliable rainfall in normal years. The area has been a focus of extensive research activities. The Department of Research and Specialist Services (DRSS) has had sites with on-farm trials in the area for 15 years. Non-governmental organisations

---

[1] Dr Enos Shumba made me aware of this limitation of the present chapter.

(NGOs), the Department of Agricultural Technical and Extension Services (Agritex) and other government organisations are active in the area. Infrastructure (roads, telephones, marketing) is well developed. The farming system is mixed subsistence-commercial with a strong focus on maize and vegetables as the principal cash crops. Many of the households are female-headed because of high male out-migration to urban areas — mainly to Harare which is only 100 kilometres away.

The village chosen in Mutoko lies in NR III and often suffers from mid-season droughts around January. The farming system is mixed subsistence-commercial with a strong focus on maize, mangoes and vegetables (especially tomatoes) as the principal cash crops. Attention on Mutoko district in general is only now beginning to grow. Coopibo, an NGO, is active promoting crop production demonstrations in collaboration with Agritex, and DRSS is beginning to conduct on-farm trials in the district. The Agricultural and Rural Development Authority (ARDA), a parastatal, has been active assisting farmers with the marketing of their mangoes but is starting to pull out because the scheme has not worked well. The area is 160 kilometres from Harare. Only 20 kilometres of this distance is gravel — the rest is tarred and busy. It is a highway leading to Mozambique and Malawi.

Marange has had very few agricultural development programmes apart from the general extension services. The Communal Area falls under NR IV and suffers from low and erratic rainfall. The farming system is much more subsistence-oriented; a larger proportion of production is for home consumption, and livestock are important within the diversified farming system. DRSS has had no on-farm trials in the area and the nearest research station (Grasslands) is some 300 kilometres away. Enda-Zimbabwe (an NGO) is trying to facilitate on-farm sorghum and millet variety trials in the area. The village that was selected is 80 kilometres from Mutare. Of this distance, 50 kilometres is rough gravel road and only a few buses pass through the area.

One village was selected in each of the three communal areas. The villages selected were Nyamburi (Murehwa), Nyamakope (Mutoko) and Makomwe (Marange). From each village, 12 focal farmers were selected through purposeful sampling to include some female-headed households, master farmers, master farmer trainees, ordinary farmers and different levels of resource endowment. Tables 1 to 3 provide some selected characteristics of the sampled farmers in the three villages.

Table 1: Case study farmers by gender

	Marange	Murehwa	Mutoko
Widows	2	3	5
Female farmers with absentee husbands	2	4	0
Husband and wife both present and farming	8	5	7

Table 2: Type of farmer by extension training

	Marange	Murehwa	Mutoko
Master farmers	3	4	0
Master farmer trainees	7	2	1
Ordinary farmers	2	6	11

Table 3: Mean access to selected resources per farm household

	Marange (n=12) mean	s.d.	Murehwa n=12) mean	s.d.	Mutoko (n=12) mean	s.d.
Labour (no. adults)	4.2	2.4	3.1	1.3	2.1	1.1
Ploughs (no.)	1.2	0.4	0.8	0.6	1.1	0.5
Ox-carts (no.)	0.6	0.5	0.5	0.5	0.5	0.5
Arable land (ha)	3.1	1.5	1.5	0.6	2.0	0.9
Garden (ha)	0.2	0.1	0.4	0.7	0.4	0.3
Cattle (no.)	5.0	3.1	5.7	5.8	7.4	4.9
Goats (no.)	4.2	5.0	1.7	4.3	3.2	4.8
Poultry (no.)	17.3	13.6	23.4	17.6	16.9	14.9

Dialogue was established and maintained through repeated regular visits, observations and discussions, with the 36 farmers over a period of one year (September 1994-August 1995). Where both husband and wife were present, discussions were conducted with the two or with whoever was present during a specific visit. On a few occasions, discussions were also held with older children (17 years and above) when both parents were not at home.

The field study culminated in a report-back workshop in each of the three areas where the findings were presented to the farmers to enable them to confirm or add further insights to the findings. The results were presented to them on flipcharts. Farmers then debated these findings in small groups of four to six (sometimes of women and men separately and other times mixed) and then presented their findings in plenary sessions where further debate took place and conclusions reached.

FARMERS' PRODUCTION MOTIVATION

My assumption in looking at production motivation was that researchers needed to understand why farmers produce certain commodities in order to appreciate why they follow certain farming strategies, some of which may not be recommended by scientists. My second assumption was that farmer innovativeness was linked to their motivation for production.

Farmers are motivated by a variety of reasons to produce a crop or crops. The reasons revolve around food and cash. When opportunities are limiting, like in Marange, farmers are more concerned with food security. In Marange for example, ten of the 12 farmers grew pearl millet, seven grew finger millet and five grew sorghum among other crops. In Murehwa, on the other hand, none of the 12 farmers grew pearl millet, six farmers grew finger millet and only one grew sorghum.

Food security seems to be of primary concern to all farmers. For example, although only about half the farmers grew finger millet, most farmers, especially the older ones, mentioned that they grew the crop once every two to three years to replenish their stocks. Finger millet can last several years in storage. The importance of this crop as a food security crop was also seen in the 1992/93 season when almost all the farmers grew it to replenish their stocks which had been used up during the previous year's drought. Farmers also mentioned that finger millet had a special traditional value in that the beer is used during traditional ceremonies. The thin porridge *(usvusvu)* is also used to nurse the sick.

All the 36 farmers grew maize, which occupied most of their lands, and all but one farmer (in Murehwa) grew groundnuts. About groundnuts, one farmer in Murehwa said, *'nzungu ioiri yemhuri'* (groundnuts is the lubricating oil of the family). These two crops are grown both for food and cash. Twenty five farmers grew two or more varieties of maize. Some of these varieties are grown in only small amounts for their food qualities. For example farmers who grew SR52 in Murehwa said it made nice-looking white and tasty *sadza* (staple food of thick porridge). They said it was also nice when boiled or roasted green. Those varieties grown for commercial purposes were chosen for their drought tolerance and plumpness of the seed (which meant good grades at the Grain Marketing Board — GMB).

Seventeen farmers grew at least two varieties of groundnuts. They liked Natal Common because of its high oil content. They said because of the oil content it was easy to roast (a process which is done before grinding into peanut butter) without burning — so it makes nice peanut butter which does not smell *chiutsi* (smoke). Because of the oil content they found it was easy to grind. They also said it got good grades at GMB. Farmers also grew other varieties like Valencia which they said had less oil content and were nice to eat green. *Hadzifinhi* (you can eat a lot before you feel you don't want more), they said. Because they are larger in size, farmers said these varieties were easy to handle and to shell when eating them green. For these two reasons, these varieties sold well on the informal market when green. One farmer in Murehwa also added that because they are large, it takes only a few nuts to fill a selling container.

When opportunities permit, farmers grow high value cash crops. Farmers in Murehwa and Mutoko are engaged in market gardening and they sell most of their produce in Harare. ARDA has even been trying to assist farmers to have access to the export market for their mangoes. However, this has not yet been successful with farmers by-passing ARDA's services. ARDA was contemplating withdrawing and moving to another communal area (Domboshawa) closer to Harare. The lesson to be drawn here is that, agricultural change is not necessarily a state directed process.

Farmers' food and cash requirements take many forms. To meet these needs, they keep different types of livestock and grow a wide range of crops (Table 4).

Table 4: Mean number of different types of crops and livestock per farm household

	Marange (n=12) mean	s.d.	Murehwa (n=12) mean	s.d.	Mutoko (n=12) mean	s.d.
Field crops	9.4	2.4	7.7	2.5	8.4	1.5
Garden vegetables	3.2	2.1	5.9	2.7	4.0	2.1
Types of fruit trees	5.4	4.1	7.7	4.8	5.2	2.1
Types of livestock	2.8	0.6	2.0	1.0	2.7	1.4
Total different crops and livestock	20.8	8.1	23.3	8.6	20.3	4.9

The livestock involved include cattle, poultry, donkeys, pigs, and rabbits. The crops include those that are eaten as snacks like sweet sorghum, cucumbers (local and non-commercial varieties), pumpkins, *mbambara* nuts, non-commercial varieties of watermelons, green mealies; those that provide staple food like maize; those that provide variety like rice; those that ensure food security like finger millet and pearl millet; those that provide relish like groundnuts, cowpeas, pumpkins and garden vegetables; those that form an important component of breakfast like sweet potatoes; those that provide beer like finger millet and pearl millet. Any excess to home requirements is sold for cash but there are also crops that are grown specifically for sale like maize, groundnuts, garden vegetables (e.g. tomatoes and onions in Murehwa and Mutoko) and fruit trees (mangoes in Murehwa and Mutoko).

PRODUCTION CONSTRAINTS AND HOW FARMERS DEAL WITH THEM

The purpose of looking at production constraints was to learn to what extent farmers were able to identify their production problems and what measures they employed to deal with the problems. My assumption was that in an effective agricultural knowledge system, scientists could benefit from farmers' knowledge of their priority production problems and the solutions they try, as well as problems they have no solutions for. These experiences of farmers would then form the basis for further research.

The biggest single production constraint that was observed during the season was moisture as a result of low and erratic rainfall throughout the three areas. Farmers had to replant several times either because there was poor, or no germination or crops were completely scotched after germination. The worst affected area was Marange where two of the 12 farmers reaped nothing from their fields and had already started receiving food handouts by harvest time.

The second most important problem was lack of draught power for ploughing. Six of the 36 farmers had no draught power of their own whilst another farmer only got three cattle during the season as dowry from a daughter who got married. Sixteen farmers had inadequate draught power with less than four mature oxen each. There are local and informal arrangements for sharing the available draught power. Those who do not have draught power help (or work for) those who have in return for draught power. The main problem here is that those who provide the draught power do so at their own time and convenience. Most of the time, the ploughing for those who do not have cattle is therefore done late. Those with inadequate numbers team up their cattle and plough their lands together.

Labour is another big problem especially during peak periods like planting and harvesting. Farmers resort to 'queuing' some of the tasks. This problem affects how certain tasks are performed like winter ploughing and fertilizer application discussed separately below.

Red spider mite was a big problem on tomatoes throughout the area. It appeared on almost all tomato fields. Some farmers used chemical control without satisfactory results, whilst others believed it was a varietal problem. They thought some varieties were more susceptible than others. Sweet potato weevil was another serious problem on sweet potatoes which causes high losses due to spoilage of the tubers. Farmers practised some rotation to try and reduce the incidence of the pest but it did not seem to help much. There was a small outbreak of army worm in Murehwa and Mutoko. This is not a regular pest and farmers have no control measures for it. Government, through Agritex, normally assists farmers whenever there is an outbreak. However, their reaction is usually slow so that considerable damage is done before they arrive with the chemicals. Farmers in Mutoko had problems with vermin (particularly baboons and monkeys). They guard their fields every day when crops are at their vulnerable stages.

Farmers also had problems with stalkborer on their maize and sorghum; white grabs *(hweva)* on their maize; ground ants *(mhamhasi)* and aphids on their groundnuts and cowpeas; bean fly maggot on their cowpeas; quelea birds on their pearl millet; webs *(dandemutande)* on their finger millet; and grasshoppers on their leaf vegetables. Whilst few farmers used chemical control for stalkborer on maize, none tried to control any of the other pests except scaring the birds away from their pearl millet fields. (Whilst farmers observed ground ants as a pest, the real pest that caused damage to the groundnuts was *Hilda patruleris*. Ground ants was really a secondary pest.)

During report-back meetings, farmers also added sickness and death in the families and villages, laziness, and lack of care on their part, as other constraints affecting their production. These were only brought in group meetings but were not raised during visits to individual farmers.

SOURCES OF FARMER KNOWLEDGE

This window of analysis was used in order to obtain some insights into how the farmers acquire, share and use knowledge and information related to technology development and the relative importance of the different sources used.

The most obvious source of information was Agritex. All the 36 farmers knew the Agritex staff in their areas. Some had more frequent contact with

Agritex staff than others but at least all knew where to go if they needed advice. However some lacked confidence to join the extension groups. They either felt these were for the better-off or they simply felt the group pressure to conform to group authority was just too demanding. They would be required to pay joining fees, to travel to meetings, to be exemplary in whatever they did and so forth.

'Other farmers' are an important source of information. As farmers move about their daily business, they listen to what others say, they see what others do and they learn from them. However, the farmers themselves do not readily recognise this as an important source of information, that is, they do not mention it explicitly to the visiting researcher. In response to a general question on where they get their knowledge from, none of the 36 farmers mentioned 'other farmers'. But discussions on specific activities indicated that they benefited from other farmers in many ways. They either had seen, learnt about and obtained a new seed of 'madumbe' (local type of yams) from another farmer or seen and learnt of a new maize variety at an agricultural show. I will come back to this point later.

Seed houses, particularly Seed Co-op, seem to be relatively well known in the three areas. Nearly all farmers knew about Seed Co-op and, in Marange, Seed Co-op had actually addressed pre-planting meetings which coincided with food distribution days organised by the local leadership. Fertilizer companies are another source of information. For example, Zimbabwe Fertilizer Corporation (ZFC) was mentioned in all areas. However, knowledge from the private sector sales persons tends to be confined to the utilisation of their products.

Other sources of knowledge tend to be confined to certain areas. For example ARDA was active in Mutoko and Murehwa promoting the marketing of tomatoes and mangoes. They were also distributing grafted seedlings of improved varieties of mangoes. The Zimbabwe Farmers' Union (ZFU) had facilitated a number of training courses in Murehwa and Mutoko. GMB in Murehwa held pre-marketing discussions with farmers talking about grades and grading of grain crops. The Agricultural Finance Corporation (AFC) was active in Murehwa and gave supportive lectures to farmers in the course of ensuring that farmers produce enough to repay their loans. One young farmer in Mutoko also got some of his information from farming magazines. Farmers had also heard about other sources of information which were active around their areas but had not dealt with them directly. For example, farmers in Mutoko knew of Coopibo which was supporting demonstrations in neighbouring villages and farmers in Murehwa had also heard of DRSS which was running on-farm trials in neighbouring villages.

HOW FARMERS USE 'OFFICIAL' RECOMMENDATIONS AND ADAPT TECHNOLOGIES

The purpose of looking through this window was to try and get an indication of the relevance of official recommendations, to identify opportunities for improvement, and to gather evidence of farmers' capacity to experiment.

In order to have a clearer understanding of how farmers used recommendations, it was necessary to focus on one technology theme. I decided to choose those technologies related to maize production. To grow maize successfully, there are certain things that the extension service recommends. Eleven of the key recommendations were selected for scrutiny. These are: crops recommended per natural region; winter ploughing; proper plough setting; ridging; monocropping; rotation; application of fertilizer at planting; using fertilizer cups; planting with the first rains; using planting wire; and, record keeping.

Crops recommended

In Mutoko and Marange where the rainfall is low and erratic, the extension service recommends planting of small grain crops like pearl millet and finger millet because these are drought tolerant. In the high rainfall areas of Murehwa, the extension service is also trying to push soyabean production because of the demand for it in the market, its value as a rotation crop because of its ability to fix nitrogen, and its potential as a protein source in the family diet.

Despite the recent successive droughts and the low inherent fertility in the soil, farmers still grow maize and always put sixty percent, or more, of their land to maize. Farmers gave many reasons for this preference. Maize has become a staple food in these areas. It can be eaten in various forms: as thick porridge (*sadza*) for main meals; as thin porridge for breakfast; as boiled or roasted green mealies; as roasted dry grain (a form of popcorn); as boiled dry grain either alone (*mangai*) or with cowpeas, groundnuts or *mbambara* nuts (*mutakura*), as *munxuchu*, or as *mbwirembwire*. Maize is less labour intensive than small grain crops. Weeding, harvesting and processing are labour intensive management operations of small grains. In addition, and more importantly, maize gives higher yield in normal years. It is easier to process into food than small grains. A farmer in Marange added, '*kana murume akasara ari ega pamba, anoenda kuchigayo, akabika sadza, akadya*' (even if the husband remains at home alone, he can go to the grinding mill, cook food and eat). Dehulling (which has to be done before grinding) small grains is a slow and cumbersome process and is done by women.

Maize is not attacked by birds whilst standing in the field the way small grains (particularly pearl millet) are. There is always a ready market for any excess to home requirements. Over the years, the publicity given to maize at national level has tended to favour maize production. For many years, there has been favourable pre-planting prices. Small grains are less talked about at national level and there is no market for them. Farmers therefore only grow a little of the small grains for use at home. Maize stover has become an important cattle feed source during winter when grazing is scarce. Small grains stover does not make good cattle feed — in fact, cattle do not even eat it. The shelled maize cobs are also an important source of fuel in Murehwa where fuel wood is now scarce.

Winter ploughing

The extension service recommends 'winter ploughing' for all crops — but even more so for maize. The term 'winter ploughing' is really a misnomer as what they really mean is 'ploughing during the harvesting time or immediately soon after'. This is between March and April. Several reasons are given for this recommendation: winter ploughing conserves moisture by breaking the soil capillary activity; farmers can then plant with the first rains (only a little of which will be required to link up with the residual moisture already conserved) the following season giving the crops an early start; trash and weeds are buried and have time to rot before the next crop; it breaks the cycle of several pests; it is done when the soil is still moist so less draught power is required; it is done whilst the cattle are still in good condition to pull the plough; it reduces soil erosion in that much of next rain soaks into the soil easily before there is any run-off; and, it is done when the weather is still mild so that it is more comfortable to work in the fields during this time. If weeds germinate during winter, the extension service recommends a second (shallow) ploughing at the time of planting to kill the weeds.

Farmers are aware of all these advantages and they were able to recite them to me. However only one (master) farmer winter ploughed the whole of his land and another three winter ploughed only parts of their lands. Farmers had several reasons for this trend which is quite common. They said that rain tails off too early in the season so that by the time they harvest their crops, the soil is already too dry to plough. Those without draught power said they could not borrow cattle at harvesting time and then again at planting time. Some simply wanted to avoid double job by winter ploughing then ploughing again at planting time. Farmers also mentioned that they would be too busy with harvesting crops. Among other things, if a farmer delays with the harvesting, the crops will be destroyed by cattle which are let loose around this time. Cattle are herded during summer and let loose soon after (the majority of) farmers have harvested.

One farmer said he wanted the weeds to grow so he can graze his cattle. Although only one farmer gave this as his reason for not winter ploughing, it appears that farmers do in fact take this into consideration. This was later confirmed by a village chairman in Mutoko who said, although he did not have cattle himself, he deliberately discouraged his people from winter ploughing to allow cattle to graze in the fields (Grazing management planners call this 'fifth paddock'.)

Another farmer said winter ploughing accelerates erosion, from both wind and water. He said a winter ploughed land is bare for a long time and during the dry period. It therefore suffers from wind erosion. He also argued that the second shallow ploughing pulverises the soil thereby making it even more vulnerable to erosion. (I was forced to recall some of the arguments I had heard at a minimum tillage field day on a commercial farm in Mutepatepa, north-east of Harare, the previous year.)

During the report-back meetings, farmers added other reasons for not winter ploughing. These were: laziness; procrastination; *kushaya shungu* (lack of strong will); and, lack of conviction that winter ploughing has in fact all these advantages.

Proper plough setting

Extension recommends specific ways of setting ox-drawn ploughs using the hitch assembly to get the correct depth and width of cut. This is done with the wheel completely removed, as they say the wheel is just for keeping the plough properly balanced. None of the farmers who had ploughs used the hitch assembly for setting the plough. In fact, none of the ploughs still had the hitch assembly. Farmers literally removed these and put them away —

usually in a tool shed or under their granaries. Farmers believe that these are too heavy for both the cattle and the person handling the plough. Farmers admitted that this was a belief handed down to them from years back and they have stuck to it. They could not say whether they actually experienced this themselves. One farmer pointed out during a report-back meeting in Marange that may be this practice started when people were still ploughing partially stumped lands. He said under those circumstances the hitch assembly would hook on to the stumps now and again causing pain to both cattle and the person holding the plough. This would result in frequent stoppages to untangle the assembly from the stumps thereby slowing down the work.

Farmers just use the wheel to set the depth and the person handling the plough determines the width of cut by pushing the plough towards, or away from, the last furrow — a process which is quite difficult and heavy.

Ridging

Extensive work has been done on ridging as a moisture conservation measure in low rainfall areas. Farmers have not adopted this practice because they said this meant more work making the ridges and planting. Most farmers drop the seed behind the plough (so that they plough and plant at the same time). They would have to use hand hoes if they had to plant on ridges. To make ridges properly, one needs a ridger or at least a cultivator. None had ridgers and only nine farmers had cultivators (none in Marange where ridges would be most beneficial, three in Murehwa and six in Mutoko).

Monocropping

Extension recommends monocropping or a single crop in time and space. Only 16 farmers had a large part of their maize in pure stands — but, even then, portions of their maize fields were intercropped with other crops. All the other maize fields had various combinations of pumpkins (local varieties), *mbambara* nuts, groundnuts, sunflower, sweet sorghum, cowpeas, rice, watermelons, finger millet and even fruit trees.

Farmers argued that land was short and they could not afford to allocate separate pieces of land to some minor crops. They also argued that some of the crops benefited from being intercropped. For example, rice growers said if intercropped with maize, rice is protected from too much sunlight, hailstorm, wind and birds. They said the planting is synchronised in such a way that maize is removed first and the rice is left to mature properly. If the maize is not removed the rice continues to grow tall and lodges easily. Two women farmers (one in Marange and another in Murehwa) said sunflower protects groundnuts from damage by craws. The Marange farmer was dramatic about it. She said, 'gunguwo haridi kudya risikaoni zvinobve yoyo — ahi ndinourawa' (a craw does not want to eat without seeing things coming from a distance — it is scared that it might be killed). So the sunflower provides cover which the craw does not want to go under.

The following crop combinations were observed: maize, pumpkins and sweet sorghum; maize and rice; maize and sunflower; maize and beans; maize and groundnuts; maize and *mbambara* nuts; maize and rapoko; maize, cowpeas and water melons. Whilst most of these associations were planned, there was an interesting case of how a woman farmer in Marange reacted to poor germination as a result of poor rainfall. When maize did not germinate well, she filled in the gaps with finger millet. When the finger millet did not germinate well, she filled in with pearl millet. So she ended up with maize, finger millet, pearl millet, sweet sorghum and pumpkins all in the same field. The last two were part of her original plan. She argued that if she had filled in with maize, the replanted maize would not have done well because of competition from the maize that germinated first. She could not plough under those maize plants she could 'see' so as to start afresh in anticipation of better germination which she could 'not see'. She got something from all the five crops.

Similar combinations were also observed of the other crops. Farmers said they knew which crops did well together. 'Zvinobva nemidzi yembeu. Dzimwe mbeu dzinamakaro' (It depends on the rooting system of the crops. Some crops are heavy feeders), remarked one farmer in Marange. They mentioned some crops which could not do well together, like maize and sorghum. They observed that commercial varieties of pumpkins did not do well under maize. They also mentioned that timing of planting of the different crops was important. For example when groundnuts and cowpeas are mixed, cowpeas should only be planted when the groundnuts are pegging so that the groundnuts are harvested first. If cowpeas is planted earlier it will smother the groundnuts. In any crop association, one is the major crop and the others they called *roja* (from lodger). (The dictionary definition of 'lodger' is 'one who pays for room(s) in someone else's house'.)

Rotation

Extension recommends crop rotation for a variety of reasons including pest and disease control, erosion control, proper utilisation of nutrients and as a fertility improvement technique. The intercropping discussed above makes it impractical for farmers to follow a specific crop rotation. In fact, that is one of the reasons why intercropping is discouraged.

The only crop they do not grow in the same land twice in succession is groundnuts. They say they do not do well. The other guiding principle is that maize must always occupy the best land available. Space for the other crops is negotiated between the families. The fact that maize occupies more land all the time makes it difficult to avoid growing it on the same land for several years. The only time when the farmers are compelled to rotate maize is when they have observed witchweed. They then rotate it with crops which are tolerant to the weed like sunflower or pearl millet.

Time of applying compound fertilizer

All farmers in Murehwa and Mutoko got fertilizer from the Government as part of its drought recovery programme. Most of the farmers also made their own arrangements to get more fertilizer — including borrowing from AFC. Farmers in Marange received only maize and pearl millet seed. They did not receive fertilizer. However, five farmers obtained some through their own means.

Extension recommends the application of compound fertilizer at, or before, planting of maize. Farmers do accept the need to use fertilizers and, apart from cash constraints, farmers do use fertilizer. However, out of the 29 farmers who had fertilizer, only one (master) farmer in Murehwa applied it at planting. Twenty four applied it after germination — when the maize had between two and four leaves. This they said was their normal practice. They punch or dig a hole beside each plant, apply the fertilizer and cover.

Farmers give three reasons for this practice. Firstly, they say if they apply fertilizer at planting, they may be applying it in places where nothing might germinate. In this case they would be wasting expensive fertilizer. If they apply it after germination, they will be applying it on plants they can see. The second reason they give is, in the event of rainfall disappearing for extended periods after planting, maize planted with fertilizer will burn. When they apply after germination, they only do so when there is adequate moisture in the soil. Four farmers, all in Marange, ended up not using the fertilizer at all because of low moisture content in the soil. The third reason is, planting is a crucial activity which has to be done over a limited period of time. Applying fertilizer at this time increases the amount of work during this peak period.

They will be racing against time. Soon after germination, there is usually little else happening and therefore less demand for labour. They can therefore devote time to applying fertilizer — which in itself is a slow process.

Using fertilizer cups

The official recommendation is that fertilizer should be applied using specially calibrated fertilizer cups. Some farmers have found this method to be too slow. Of the 25 farmers who used fertilizer, only two used cups. Another nine used tea spoons. The others just used their hands — so many plants (or planting stations) per hand-full. When I checked this with the fertilizer cups, I found that there was no difference in the amount of fertilizer applied per plant. Through practice, they got the quantities right.

Planting with the first rains

Extension recommends planting with the first rains. Whilst farmers accept this recommendation, what happened during this study was more a reaction to the season than planned planting programmes. Farmers had to plant and replant several times because either the germination was poor because of inadequate moisture or crops simply dried out soon after germination because the rain disappeared for extended periods of time. So farmers planted each time there was some soaking rain. Whereas the planting season is normally between end of October to beginning of December, in this case, farmers continued planting up to end of January because the rainfall was erratic.

Farmers had indeed planned to plant with the first rains but they always stagger their plantings. Some crops, and some portions of their lands, are planted with the first rains whilst others are planted later in order to spread the risk. In other words, they plant a little at a time and then 'watch the season'. They may even dry plant. One farmer in Murehwa dry planted one previously winter ploughed *gandiwa* (a *gandiwa*, also called *rozhi* in Marange, is the cultivable land in between two contour ridges) and the rest was planted with the rains. The dry planted maize germinated very well but later suffered from drought but survived. Those portions of the land which have a tendency of water-logging are usually planted first.

In deciding when to plant, farmers also look at natural signs. One women farmer in Marange said, *'Ndinotanga kusima rukweza rwangu kutanga kunongoita mimvee kubuda mazhizha matsva. Rukweza runoda kuti rumbopiswa ngezuva kuti ruzomere zvakanaka'* (I start planting my finger millet when the sausage trees start growing new leaves. Finger millet has to be heated in the sun first so that it germinates well). Another farmer, also in Marange said he looked at the position of the moon. He said when the moon starts coming out towards early morning (I understood this to be around 4 am), he can plant during that period if there is rain. He is then sure the rain is going to persist for some time. He said if the rain comes when the moon is in a different position, he doesn't plant never mind how much rain there may be. He has established a correlation between the position of the moon and the persistence of rain over some period.

The decision to plant is also taken after some intensive negotiation within the families. The same farmer who watches the moon in Marange has one *rozhi* which he insists should not be planted before he is happy. He feels this *rozhi* is key to the survival of the family and should not be 'gambled' with. The rest of the land can be planted if the wife thinks it is time to do so. His neighbour also complained, when the groundnuts were wilting, that he had argued with his wife that they should not have planted on that particular day — but she had insisted.

For the farmers with a commercial gardening orientation in Murehwa, planting is done at other times as well as looking for opportunities to make money. Four farmers planted maize in their gardens in August and September and were selling it green from November. Another farmer was still planting maize up to May. However, much of it was suffering from severe streak infection.

Using a planting wire

Extension recommends planting with a planting wire and using a hand hoe to get the rows straight and evenly spaced as is done at research stations. One farmer in Murehwa used a planting wire; one in Mutoko has made himself a wooden row marker to mark the rows; one farmer in Marange dropped the seed in cattle hoof-prints; another farmer in Marange just used her head to see where the rows should go; the rest dropped the seed behind the plough. In both cases, the overall number of seeds planted per hectare tended to be close to that expected by the extension service. However, the germination from those who dropped the seed behind the plough tended to be poor in many cases. There was little control over planting depth. Planting with a wire is a labour-intensive operation for farmers.

Keeping of records

For years Agritex has recommended the keeping of records by farmers. Together with ZFU, they have produced sets of record books and distributed them to farmers. It is a requirement for a master farmer trainee to keep records and to show these on the day of examination. None of the farmers in Marange kept written records. All the six master farmers and trainees in Murehwa kept some records of planting dates and sales. In Mutoko, the one master farmer trainee kept written records of planting dates and sales of garden vegetables. In reply to a question on record keeping a farmer in Mutoko once remarked, 'I know everything I did here. Do you want me to put it down on paper for you?' To the farmer, it seemed records were a requirement by outsiders. Otherwise he did not see the need to write anything as he knew it off hand.

Farmers, including those who did not keep written records, seemed to be concerned more about remembering dates of planting and, indeed, they could go several years back and plan on that experience. Otherwise, for the rest of their farm operations, farmers do not see why they should make the effort to write.

Discussion

Whilst the eleven practices discussed above are among the main extension recommendations for maize production, these are not very different from those promoted by Alvord [the first 'Agriculturalist for the Instruction of Natives' (Alvord, 1958)] in the 1920s. By 1923, Alvord had established a system of extension (though limited in extent) based on demonstrations that followed a four year rotation viz:

First year	maize with manure
Second year	maize or sorghum
Third year	groundnuts, beans or other legume
Fourth year	rapoko or pearl millet as a cover crop.

The instruction was based on what Alvord termed 'ten rules for permanent agriculture'. These were:
— stumping and clearing to ensure easy tillage;
— winter ploughing to conserve moisture and decompose residues;
— application of manure every four years;
— a second ploughing just before planting to aerate the soil;

— thorough seedbed preparation to ensure uniform germination;
— proper spacing and planting for all crops;
— planting maize on the land where manure is applied each year;
— planting a legume crop two years after manure is applied;
— a heavy rooted, close growing crop after the legume;
— crops not to be planted mixed together in the same land.

Seventy years later, the options have not increased much except for the introduction of fertilizers and new crop varieties. As can be seen from the discussion above, most of the recommendations are not appropriate for smallholder farmers. Researchers have not responded to how farmers adapt these recommendations. The extensionists therefore continue to 'push' these standard recommendations without regard to the farmer circumstances, concerns and limitations.

Whereas some scientists have indeed begun to challenge the recommendations, unfortunately, they do so through international organisations and publications like the World Bank (Norton, 1984) and the British Overseas Development Administration (Page *et al.*, 1985). Local scientists raise their reservations through annual reports where they document their on-station work (Elwell, 1989) or through specially sponsored workshops (Oldrieve, 1995). The extension workers in Marange or Mutoko have no access to annual reports and they do not attend workshops with researchers. Therefore, they never get to know about these questions being raised or whether there are any alternatives to what they are recommending. Given this general lack of new options in terms of technologies for communal farmers, there is need to re-examine the role of national research and extension systems. In Chapter 11 of this volume, the role of the University in research will be further examined.

WHAT FARMERS TRY OUT AND THEIR CONCEPT OF RESEARCH

In addition to farmers' experiences with official recommendations, I also wanted to get some insights into farmer-initiated experimentation to provide a basis for establishing a common understanding about the concept of research. From the farmers' complex needs discussed above, it can be seen that their needs for research are also complex. To meet those needs, farmers are engaged in different levels of experimentation.

Each of the sampled farm households had between nine (Marange) and 44 (Murehwa) crop and livestock enterprises over a period of one year — most of which research and extension scientists know very little about as their training and mandates are on cash crops only (see Table 4 above). If one considers the different varieties and breeds farmers deal with and the different management practices they apply to them, one can easily see that the technologies that the farmers are processing at any one time, run into hundreds.

Farmers also conduct soil fertility experiments. Two farmers who were adding anthill soil to their maize lands explained that there were differences in the nutrient content of the anthill soils — adding that they could also tell which anthill soil was richer and which one was poorer in nutrients just by looking at the colour of the soil. The decision to apply anthill soil on a particular patch of land depends on the performance of the previous crop. Although they do not write or make any growth measurements, they certainly make observations on the productivity of their entire lands. These observations also determine where the next maize crop is going to be planted the following season.

There are different levels of innovativeness between farmers. Some farmers try more things, and sometimes more complex things, than other farmers. But what drives farmers to innovate? There are several prime movers.

May be the most important prime mover is need. It could be need for survival, need for cash, need for variety in the food items and so forth. The wide range of technologies that farmers play around with as shown in Table 4 shows that the farmers' needs are diverse.

Some farmers experiment purely out of *curiosity*. Two farmers (one in Murehwa and another in Marange) were growing one *musawu* tree each simply to see if it would grow in their area. *Musawu* tree is commonly found in the low altitude and hot areas of the Zambezi Valley. The fruit is normally sold in the market place (*musika*) in Harare and sometimes in Mutare. The two farmers bought some fruit to eat and then grew the trees from seed to see if they would grow. The Marange farmer also planted a few short rows (10 rows x 20 metres long) of groundnuts in his garden in June. They dried up when they were beginning to flower because his well (from which he was watering) dried up. '*Mudzimai akaite nharo achiti 'muri kutambise mbeu yangu' — ini ndaida kuono kuti nzungu hadzindyiwi here chirimo. Hino mvura ndiyo yakandishaishira . . . Izvi handi zvemadhumeniba izvi. Vanhu vakaseka — asi handityi kusekwa'* (My wife argued saying 'you are playing with my seed' — I wanted to see if groundnuts could not be eaten in winter. But the water let me down . . . These ideas are not from the extension workers. People laughed — but I am not afraid of their laughing at me).

Some farmers are simply moved by *ingenuity*. Two farmers in Murehwa are able to get their avocado trees to produce fruit within three years of planting. One ties a wire around the tree trunk when the tree is about two years old. Another puts lots of old metal tins in the planting hole mixed with compost, puts a slab (stone or concrete) on top, covers with soil and then plants his seedling. He says when the roots get to the slab the tree starts flowering. One horticulturist could not believe this was possible and suggested the farmers might have been using early maturing varieties. There are however no known varieties that flower within three years. The normal time is eight or more years if the trees are not grafted. The scientific method of getting avocado trees to produce fruit within three years is by grafting. Another horticulturist thought this had to do with the stress the trees were being subjected to, that induced flowering. The farmer who used metal tins actually uses them on all fruit trees. He says this improves fruit quality.

On being asked what made them want to experiment this much, four farmers used words like *'shungu'* (a desire to achieve); *'mbiri'* (a desire for fame); *'unyanzvi'* (a desire to surpass others, or desire to become expert); and *'kashavi kokunakirwa'* (in-born interest in seeing things grow).

Exposure of the farmers is also an important factor in experimentation. Several farmers were trying crops they saw for the first time at agricultural shows; one farmer was trying a pumpkin variety she got from South Africa; another farmer was trying a bean variety she had seen in Harare and was attracted just by the appearance (the same farmer was also trying sesame she got from another district — Mudzi); another farmer was trying a pumpkin variety brought to her by her son from the United States; two farmers (one in Murehwa and another in Mutoko) were trying yams they had brought from Manicaland; another was growing a variety of paw paw brought from South Africa by an Indian friend. One very successful market gardener in Mutoko said that when he was working in a hotel as a waiter, he observed that all the food stuffs that were being used in the hotel came from the land. He decided to grow those crops himself and quit his hotel job. He established contacts with traders who played a central role in the types of vegetable crops he grew based on their knowledge of the market. Therefore, farmers have their own informal networks.

Rainfall affects the degree to which farmers can innovate. The range of innovations was generally wider in Murehwa than it was in Marange and Mutoko.

Access to the market has a lot of influence. The most obvious factors that have affected farming in the study areas in Murehwa and Mutoko relate to the incorporation of the areas into a wider network of economic exchange. The road linking Zimbabwe with Mozambique and Malawi has facilitated transport and haulage to Harare which, with its growing population, provides an expanding market for horticultural products (fruit and vegetables). Private transporters, including buses, are quite willing to do business in the two areas because of lower risk to their vehicles. This helps to explain why market gardening has become increasingly profitable for farmers in the two study areas. One farmer was even making a living purely from fruit trees.

The improvement in road infrastructure in Murehwa and Mutoko has not only facilitated the marketing of farm produce, but also meant that inputs became easily accessible. Most inputs are available from local shops whilst they are also easily accessible in Harare. Agro-chemical company representatives are also active in the areas.

This observation suggests that, although the pattern of development is shaped by social actors from below, it has to be complemented by other external initiatives. In Marange, where the road infrastructure is poor, farmers are unable to benefit from wider networks of economic exchange to the same degree as farmers in Murehwa and Mutoko. This is despite the fact that Marange is only 80 kilometres from a major city, Mutare, which could provide a lucrative market for horticultural products. There is therefore a level of interdependence and interrelatedness between local initiatives and the wider social processes.

Table 5 suggests that there is no difference between the number of things master farmers and master farmer trainees experiment with. It also seems to suggest that they experiment with slightly more things than ordinary farmers. Their exposure to extension ideas might have influence on their desire to experiment.

Table 5: Mean number of crops and livestock produced by master farmers, master farmer trainees and ordinary farmers

	mean	s.d.
Master farmers (n=7)	22.3	6.9
Master farmer trainees (n=10)	24.6	9.2
Ordinary farmers (n=19)	19.4	5.4

Table 6 suggests that when the husband is away working, there are opportunities for the farmers to experiment with more things. Several factors could be responsible for this. There is income in the family which can be used to try new things. Exposure might be another factor where the husband is away and, now and again, both travel to areas outside their own and come into contact with other ideas and technologies.

Table 6: Mean number of crops and livestock produced by type of house-headship

	mean	s.d.
Widows (n=10)	19.9	4.9
Female farmers with absentee husbands (n=6)	27.8	9.3
Husband and wife both present and farming (n=20)	20.3	6.9

The farmers' research is very much an individual initiative. The research process is informal but each research activity is purposeful. Information on the experiments is shared informally — there are no special meetings either to 'launch' the trials nor to 'announce' or to 'publicise' the results. Other farmers get to know about the experiments, and subsequently the results, informally. In other words farmer-to-farmer movement of technology takes place as they go about their daily lives — they do not plan special missions to look for it, but they keep their ears and eyes open for anything interesting.

When farmers come across anything that looks interesting to them, they try it first on a small scale. Their criteria for selection differs according to the specific technology they are trying and the use they want it for. It could be anything from simple appearance (the bean trial), or testing for shade tolerance (commercial varieties of pumpkins), to ease of roasting without burning (high oil content groundnut varieties).

Farmers operate seed production and distribution systems. I was surprised to see that Hickory King, a maize variety I thought had disappeared decades ago, was still around and being passed from farmer to farmer. They do the same with many other varieties of crops like sweet potatoes, millet, pumpkins, water melons, sweet sorghum and others.

Because of the informal nature of the technological development and dissemination process, farmers had difficulties accepting that they had anything they could teach us when we introduced our study saying 'we had come to learn from them'. Their information was not available in a systematised manner. Our study was too formal and systematic— we were strangers with a specific mission and we were going to be there for a specified period of time. One woman in Mutoko even remarked, 'We have no experience in teaching — tell us how each day's lesson should begin.' Another woman also remarked 'Unfortunately you have selected only women and old men — you will learn nothing.'

FARMERS' NEEDS FOR RESEARCH

The question of what types of research are needed by farmers was the most difficult to deal with mainly because farmers' concept of research differed from our understanding of it. Moreover, farmers often do not have information about research activities outside their locality, and lack an overview. They had not worked with formal researchers directly before and they did not know how they could collaborate with them or what they could expect from them. Therefore, each time the question of research needs was asked directly, they would mention problems like: local unavailability of inputs; poor transport for both outputs and inputs; they wanted fences for their grazing schemes; they wanted tractors to plough their lands; or, they wanted Cold Storage Company (CSC) cattle loans to be extended to them as well. But these problems I perceived as lying outside the scope of agricultural research.

I therefore decided that the best way of identifying their needs for research was by observing and taking note of those problems that did not seem to have many effective, or easy, alternative solutions either from farmers themselves or from scientists. Through this approach, the following needs for research were identified: land preparation in the face of growing shortage of draught power; soil fertility in the face of rising costs of fertilizers; drought tolerant maize varieties in the face of the huge preference for maize even in drier areas; sweet potato weevils in the face of the growing importance of sweet potatoes as a substitute for bread which was getting out of reach for many rural households; and red spider mite on tomatoes.

Researchers are doing some work to address some of these problems. Work on drought tolerant maize varieties is going on. In response to farmers' lack of access to ridgers for making ridges to enhance moisture conservation, researchers are now looking at encouraging communal ridgers (one ridger to be shared by a few farmers) and semi-permanent ridges (so that farmers do not have to ridge every year). However, farmers are experiencing an increase in termites on the semi-permanent ridges.

In discussing the problem of the sweet potato weevil, a senior horticulturalist said they had varieties which were not susceptible to this pest. He said farmers were free to go to Grasslands Research Station, near Marondera, to buy the cuttings in February of every year. He said this is what the large commercial farmers do. However, to start with, the research station is in the middle of a commercial area and several kilometres from the main road. One has to have personal transport to get there. Therefore, the station is out of reach for communal farmers. Secondly, the cuttings are obtainable at the end of the rainy season (in February) and farmers have to plant and water the seed cuttings over winter to increase the seed cuttings for the next summer. Not many communal farmers would have water to support 'seed multiplication' over winter. For these two reasons, communal farmers remain unable to benefit from research on sweet potatoes.

HOW PROFESSIONALS DEAL WITH INFORMAL EXPERIMENTATION

Previously, scientists have argued that compound fertilizers should be applied at or before planting so that plant roots can grow to the fertilizer. They have argued that, since these fertilizers release the nutrients slowly, applying it after germination was ineffective as the roots would grow away from the fertilizer. Farmers have argued that there was no difference between applying it at planting or after germination. As a result, researchers have done some work on this and have since confirmed the farmers' observations but have not published the results. Therefore, the official recommendation is still that the fertilizer should be applied at planting as the farmers' practice has not become part of the official literature yet.

The use of the row marker, which is generally a farmers' invention, has been accepted as an alternative recommendation to using planting wire; the use of anthill soil has been accepted as a soil improvement technique and is recommended. However, there has been no formal research work to legitimise these two pieces of technology. Extensionists are however, happy to make general recommendations from what they have seen the farmers doing.

In general however, the extension service's attitude ranges from indifference to outright condemnation of farmer experimentation and practices.

Extensionists stick to official recommendations like: winter ploughing; monocropping; planting of small grains in low rainfall areas; planting with the first rains; and then watch helplessly as farmers skip winter ploughing; practice intercropping; plant maize even in low rainfall areas; and stagger their planting and even plant out-of-season crops. One farmer in Marange has remained a trainee for 20 years because he does not want to abandon intercropping. 'The old man is untidy', the local Extension Worker once remarked with a smile when we were talking about the farmer's long stay as a trainee.

Researchers too have not paid much attention to farmers' informal experimentation. The case of farmers inducing early flowering of avocado trees with unconventional methods was mentioned above. Horticulturalists remained sceptical, and the extension service did not pick the method either for further examination nor recognise them for general recommendation saying they could not recommend practices that had not been proven by research. The practices needed to be tested by researchers first to establish their efficacy before farmers could be 'taught' to use them.

A senior researcher dismissed intercropping as practised by farmers as too random to be taken seriously as intercropping. He said this was not intercropping but 'scattering of seeds'.

In Mutoko, a farmer who was hosting a ridging trial for moisture conservation on maize noticed that there was an improvement in the moisture status of the soil. He also noticed that the ridging operation resulted in wider row spacing than when he planted on flat land. He then suggested to the researcher to plant a row of rice in between the maize rows. The researcher refused saying that this was not an intercropping trial and added that, if the farmer wanted, he could do so on another plot. The farmer felt discouraged and did not try his idea.

CONCLUSIONS

From the case studies above, it can be concluded that the specific agricultural practices, as informed by particular knowledge systems, are shaped by both social actors from below and other related external initiatives like roads, markets and exposure. The road through Murehwa and Mutoko has led to new practices, if compared with those found in Marange; horticultural crops grown in Murehwa and Mutoko fit into specific urban consumption patterns; and, farmers in both areas were experimenting with crops they had come across in areas outside their own. The personal networks of the social actors involved played an essential role in the range of crops grown. As they move to, and from, Harare, they share information on trends in the market. Farmers make important decisions based on such information. To this extent, therefore, an individual farmer's choice of crops thus appears to be socially regulated.

It follows therefore, that it is too simple to perceive the development of agricultural knowledge systems as separate, discrete, events· resulting from the innovative behaviour of particular individuals. Innovations should rather be looked upon as interlinked events which are part of a social environment. A rather varied range of factors like exposure, individual motivation, road networks, can all be considered as socially regulated forces that have shaped specific knowledge systems.

Direct state intervention *per se* does not necessarily lead to expected change. Farmers in Murehwa and Mutoko have continued to by-pass ARDA which was set up to help them. Many of the technologies developed and being pushed by the state research and extension systems have largely been ignored, or substantially adapted, by farmers in the three study areas. Communication between the different actors in the technology development process has been poor leading to the generation of technologies which are not entirely appropriate for communal farmers.

The chapter has also shown that farmers are not just 'users' or 'consumers' of knowledge generated by others, but rather that they are essential contributors to technological change — they are technology generators. Farmers experiment and their experiments indicate their problems. Other researchers (Richards, 1985; Röling, 1994; Rhoades and Bebbington, 1995) have come out with similar findings. Richards (1991) further argues that the archaeological and historical record show a long string of important agricultural technology breakthroughs made by farmers in traditional societies, although their rapidity and diffusion might have been slower than innovations in modern agricultural science.

Farmer experimentation, therefore, represents an untapped resource. The challenge now is how agricultural science could engage the farmers' knowledge, and their willingness and ability to pursue it.

Chapter 3

Of science and livelihood strategies
Two sides of the commercialisation debate in smallholder irrigation schemes

EMMANUEL MANZUNGU

Agritex, the government department charged with the responsibility for extension advice[1] throughout the country, has as one of its mission statements, to 'maintain a process of *transforming rural farm families*[2] from subsistence into commercial agriculture hence ensuring healthy farm families that have a sound base for economic growth' (Agritex, 1993, emphasis in the original; see appendix A). This is guided by the department's philosophy which says that 'agricultural (extension) activities should be based on (the importance of) exchange and sharing of knowledge, skills and ideas between the farmer and the extension agent thereby enabling the farmers to make decisions on those issues that affect their well-being' (*Ibid.*). This chapter looks at the commercialisation debate hinted at in the mission statement. The debate is examined in the context of smallholder irrigation.

The debate, as it is identified here, is between government officials on the one hand and farmers on the other. For a start the debate has been one sided. That is to say it was mainly the views of government (officials) that were sought and presented and not that of farmers. In the majority of cases commentators were preoccupied with finding out whether government investment could be justified through commercialisation measures such as the growing of 'cash' crops. A proliferation of economic analyses of irrigation projects ensued. Officials, before and after independence, have tended to prescribe certain crop production practices towards the attainment of the goal of maximising government investments. It was and is still envisaged that only through commercialisation can this be achieved. Commercialisation, it is asserted, should be based on adopting farming practices that have some scientific underpinning. As such extension workers rely on results from experiment stations that are handed to them from provincial crop specialists. In cases where results from the experiment stations are not available, information is obtained from textbooks. The information is made available through training sessions where written materials are also availed. This summarises Agritex's policy of the 'adoption of proven agricultural practices leading to increased, sustained and profitable production' (Agritex, 1990).

I shall argue that this type of commercial farming that is supposedly based on a scientific paradigm of agriculture is basically at loggerheads with farmer opinions because it is not universally applicable to different farmers in different circumstances. The difference between farmers and extension workers is that farmers approach the whole subject not just as a discussion matter but as part of their livelihood strategies in which their experiences play an important part. As such they wrestle with many difficult situations and try to weigh out the constraints and opportunities of different scenarios. Farmers are therefore less inclined to invoke scientific validity for their actions as extension workers are apt to do.

In the end the issue boils down to a question of the definition of commercial farming or commercialisation. On the basis of empirical material I conclude that farmers commercialise some of their farming activities in line with their circumstances. The fundamental difference is that while government officials can talk commercial farming, farmers have to practise it. It makes sense therefore to approach the subject of commercialisation by taking into account farmers' views as well. This is the line taken in this chapter.

Before going into detail, an outline of the environment of smallholder irrigation schemes is presented as an illustration of how the practices of the different actors take account of the given circumstances. The environment is presented in the form of constraints facing smallholder schemes. Since the official view to the commercialisation debate is well documented in literature, only the main points are highlighted in this chapter. Thereafter cases studies that illustrate the two sides of the debate are presented. The material is from Chibuwe and Fuve Panganai irrigation schemes in which the author was conducting research from 1993 to 1996. Chibuwe is in Manicaland province while Fuve Panganai is in the Masvingo province.

The case studies focus on the practices of both Agritex officials and farmers who are daily engaged with some aspect(s) of the debate. It is surmised that an examination of the goings-on in smallholder irrigation at the 'ground level' is likely to bring out how the subsistence-commercial agriculture is acted out. In this chapter I limit myself to two scenarios that are of crucial importance in the debate: the training that is offered to farmers so as to transform farms into 'commercial' agriculture and crop selection whose direct relevance to the debate cannot be overemphasized. A discussion of the material is undertaken after which a conclusion is presented.

SOME CONSTRAINTS FACING SMALLHOLDER SCHEMES

The disadvantages facing the generality of communal areas apply to smallholder irrigation schemes as well. The majority of the smallholder irrigation schemes (68 per cent by area) are located in the less endowed regions of IV and V (FAO, 1990a, 28). The colonial history of the country caused some projects to be implemented on sites largely unsuited to irrigation, a fact that compromised the economic viability of such schemes (Rukuni, 1988). Many schemes have water supply problems (Makadho, 1994). Land is also limiting as only about 0.1 per cent of the total communal area population have access to irrigated plots (Manzungu, 1994). Consequently the issue of production of food or cash crops in the schemes, which touches on the economic plot size, have been a subject of some discussion (see Rukuni, 1984; Nyoni, 1990; Meinzen-Dick, 1993; Agritex, 1993; Jansen, 1993). It is important to note that there is no consensus on the issue. But there is a paradox in that while cash crops can bring in more money, the markets for them are less readily available.

Mehretu (1994) noted that communal areas are generally located in remote areas inaccessible to urban, industrial and infrastructural developments. This has implications on input acquisition as well as the marketing of crops. The question of what type of commercialisation is to be pursued is important because the type of farming pursued in large-scale commercial farming areas is clearly not applicable to the smallholder farming sector.

THE CASE FOR COMMERCIALISATION: THE OFFICIAL VIEW

Rukuni (1988) notes that Alvord (see Bolding in this volume for the profile of the man and the extent of his work) in the late 1920s to the early 1930s promoted smallholder irrigation schemes for famine relief, i.e. for subsistence needs. But claims of famine relief were largely conjectural as no figures were produced in support of the assertions (Rukuni, 1988). A visit to the

United States of America in 1935 made Alvord turn around the way the schemes operated (Rukuni, 1988; Manzungu, 1995b). As a result the first seeds of doing away with what was seen as subsistence farming and embracing commercial farming in the sub-sector were sown. Farmers were not allowed to engage in rainfed farming and were expected to depend only on irrigation. They were also not allowed to engage in any other income generating activities. In time the 'seeds' of commercial farming germinated into a situation where

> farmers were expected to produce crops such as wheat and beans for sale, even though they were not part of the local diet. Farmers were (then) expected to purchase maize and sorghum from cash proceeds. Government agronomists designed cropping patterns based on wheat and beans and sunhemp. These measures proved unpopular with farmers because of numerous reasons, including the lack of markets for 'compulsory' crops especially, in years of good rains when bumper crops depressed local prices (Rukuni, 1988, 202–203; cf. Roder, 1965; Reynolds, 1969).

In time a variety of measures were put in place to make sure 'commercial' farming was undertaken. Because of technical and economic imperatives it was essential to ensure that plotholders cultivate their plots in the manner approved (Hughes, 1974, 213-214). The main devices relied on were the introduction of lease agreements, appointment of scheme managers to supervise the farmers, specification of precise areas to be planted by each plotholder, enforcing planting dates, specifying types of seed to be used, and quantities and types of fertilizers to be applied (*Ibid*).

This trend has continued after independence. Bourdillon and Madzudzo (1994) observed that many of the schemes they studied had similar rigid cropping programmes, a position which they found unfortunate. Although Jansen (1993) did not allude specifically to cropping programmes, her reference to the desirability of growing 'high value' crops as compared to food crops which were least efficient in the use of irrigation infrastructure, endorsed 'commercial' farming (cf. Nzima, 1990; Peacock, 1995) on the basis of government investment ticket.

Another angle which preoccupied and has preoccupied practitioners in irrigation was what plot size was appropriate for each plotholder. Because of different objectives schemes tended to have varied plot sizes. Thus, plots in supplementary schemes meant to supplement rainfed production were as small as 0.1 ha, while full-time irrigated plots averaged 1-1.5 ha. But the economic plot size debate meant to balance food and cash crops has not been settled (see Agritex, 1993).

In sympathy with the investment concerns, there is a huge commitment of human resources in irrigation schemes to the tune of one extension worker for every 50 farmers compared to rainfed communal agriculture where the extension worker to farmer ratio is 1 to 800 (Auret, 1990). Smallholder irrigation enjoys other forms of financial support which does not apply to rainfed agriculture. But this has come at a price. With costs of developing irrigation estimated at Z$70,000–Z$110,000 per hectare (Rukuni, 1994b, Vol. 2, 398) and maintenance costs at Z$6,000 per hectare per year (Peacock, 1995, A-3) it is considered not economically sound to engage in subsistence farming which is tantamount to using water and other resources 'inefficiently'. Commercial farming is advanced as the natural and economic proposition to take. Smallholder irrigation is a good candidate to implement 'commercial' farming programmes because there is already a measure of control; the sharing of irrigation facilities and water requires certain rules and regulations to be in place. This is in contrast to rainfed agriculture where farmers tend to be autonomous in their farming practices. Moreover, schemes are considered the locus of transfer of technology through intensification of agriculture (Derude, 1983). 'Commercial' farming then becomes the obvious choice.

There are a number of ways in which Agritex promotes 'commercial' farming. Central to this exercise is the drawing up of crop budgets, which are meant to show farmers what profit levels can be attained. This is often coupled with training such as the intensive Master Farmer Training programme, where farmers undergo training culminating in oral and written examinations.

THE COMMERCIALISATION DEBATE IN ACTION

The three cases presented below were chosen to show different aspects of the commercialisation debate. In the first case, attention is paid to how extension personnel teach 'commercial' farming to farmers. This is done by examining two training sessions of the Master Farmer Training programme. The aim was to uncover some of the conceptual ideas that underpin that training. The next cases look at how farmers (do not) apply their training to their field situations.

CASE 1: MASTER FARMER TRAINING: EQUIPPING THE FARMERS?

The Master Farmer Training Programme, which started with Alvord, an American Missionary-cum-agriculturalist, has been in existence for about 60 years and has hardly changed in its outlook (Page and Page, 1991). The Training Branch of Agritex undertakes reviews of the Master Farmer Training curriculum from time to time in order to come up with more appropriate packages. This review however, does not seem to be translating into the improvement of the programme as acknowledged by some of the Agritex officials (see Magwa, 1985; Tapererwa and Chivizhe, 1994; Madondo, 1995). But the programme is still an important component of Agritex's training programme.

Record keeping in Chibuwe

The training session reported here took place at Chibuwe irrigation scheme on 24 August 1994 in the afternoon. This was a Tuesday which coincided with the Advanced Master Farmer Training session. Thursdays were reserved for the Ordinary level training.[3]

On this particular day the session got to a late start because a group of visitors from the Ministry of Agriculture had come to the scheme to conduct a crop census (see Manzungu in this volume for what transpired there). This delayed both the extension worker and some of the farmers who had been chosen by the Irrigation Supervisor as respondents. In fact some of the farmers did not turn up at all. Before the extension worker arrived on his motor cycle the author talked to the first woman to turn up for training. This woman, who was in her sixties, said that she participated in the programme because she wanted to learn how to farm better. The second woman, who incidentally was the last student for the afternoon, arrived a few minutes after the extension worker. The session then got under way. The extension worker explained that this was a revision session that dealt with record keeping.

The main task of the afternoon was the computation of the net farm profit (see appendix B). This was to some extent a revision as this involved transferring various variable and overhead costs that were previously calculated. These were then compared to the money coming in. As an introduction to the subject the extension worker explained that the net farm profit represented the total surplus for all enterprises on the farm which included livestock and crops. This was an easy exercise, he explained, since the figures they were going to use had already been calculated. With this brief remark the students were told to open pages 57 and 58 of the Master Farmer Record book. The first part, which involved transferring variable costs was easy. The extension worker asked the women to transfer figures from the relevant previous pages, say from page 40 to the appropriate slot. When it came to getting the total variable costs the second woman had some problems. The extension worker let the first woman help her out. The student-help-student exercise worked out well. However the exercise was a mechanical shifting of figures. This became apparent later on.

Soon afterwards it became clear that the session could hardly be described as a revision as the women found it heavy going. A few lines down it was the extension worker who had a slight problem. He could not remember what the abbreviation O.L. stood for. He was open enough to admit that he had forgotten it. The author suggested the answer of oil and lubrication and (hopefully) explained the context in which this would be used. The extension worker however had an easy way out of it: 'Just put any figure which you think is suitable.'

It was obvious that the section on estimation of overheads was a problem to the extension worker as well. In all the filling in of this section he did not mention any figure but let the women farmers write down their own estimated figures. This was because the items contained in the section were not applicable to the situations of the farmers who were being trained. This also applied to the extension worker who groped for appropriate examples to give to his students. For example regarding fuel costs, the extension worker advised the farmers to consider the cost of paraffin and wood that they used for domestic purposes as part of the fuel costs! One other problem was that estimation of costs was difficult as these estimates spanned over 12 months.

The next hurdle for the extension worker was the term depreciation. After offering his explanation he realised that he had not done it to the satisfaction of the audience. But the training had to go on so he asked his audience to write down some estimates. The author again made another contribution towards the concept of depreciation. Since depreciation is a concept that is very particular to economics it is reproduced here from Gittinger (1982, 467). The reproduction is made to allow the reader to appreciate what the women were wrestling with.[4]

Depreciation . . . the anticipated reduction in the value of an ASSET over time that is brought about through physical use or obsolescence. In accounting, depreciation refers to the process of allocating a portion of the original cost of a FIXED asset to each ACCOUNTING PERIOD so that the value is gradually used up ("written off") during the course of the asset's estimated "useful life". . . . There are two principal types of depreciation methods: "straight-line" depreciation, which allocates the cost of the fixed asset in equal amounts for each accounting period, and "accelerated" depreciation, which allocates a larger proportion of the original cost to earlier accounting periods and a smaller proportion to the later periods. In DISCOUNTED CASH FLOW ANALYSIS depreciation is not treated as a cost. Instead, the cost of an asset is shown in the year it is incurred, and BENEFITS are shown in the year they are realised *(emphasis in the original)*.

After the depreciation woes the session was suspended by the extension worker advising that it would continue in the next session. As far as the author could judge, the suspension was because he was having difficulties with the whole exercise.[5]

As the empirical evidence has shown, a lot of difficulties on the part of both the farmers and the extension worker were encountered. For example in relation to oil and lubrication there was a mix up because paraffin and wood, which were given as examples are actually household items and not farm items and as such do not qualify to be included in the computation of the net farm profit. Depreciation engendered the same mix up. For example the extension worker did not explain that this was a paper transaction and not a cash item. The method of estimation was not explained either. This mix up cannot simply be laid at the door of the particular extension worker. A larger picture ought to be drawn. This larger picture should touch on how the extension workers are equipped to train farmers.

Extension workers, as all Agritex officials, are given in-service training in farm management alongside other subjects. The training is based on what may be considered 'textbook commercial' training. This seems to be the reason why the extension worker experienced difficulties in providing relevant monetary examples. In the end the 'commercial' farming knowledge that the extension worker was seeking to 'extend', which was not based on the experiences of the farmers or himself, was not effectively imparted. The lack of relevance of some aspects of Master Farming training may be the reason why graduate master farmers stop practising what they would have learnt as observed by Magwa (1995).

The plough lesson in Fuve Panganai

Another training session of the Master Farmer Training programme presented here was at Fuve Panganai irrigation scheme on Tuesday 17 October 1995. The extension worker, who was only three months in post, had indicated the previous day that his training sessions started at 6 am because he did not want to interfere too much with farmers' programmes such as irrigation duties. He explained that the training session would last 30 minutes. He also explained that currently he was concentrating on the Ordinary programme as he needed first to assess the books of those who failed the Advanced programme before training more farmers. His general approach was different from the extension worker whom he replaced who did not have any Master Farmer Training students. He also had earned the respect of farmers for his willingness to provide practical solutions to farmers' problems. For example, since he arrived he had successfully arranged the acquisition of fertilisers from Masvingo town. As a result the maize crop was the best that had been grown since irrigation was started, which was also affirmed by the Agritex District Office.

To some degree his practical orientation was revealed in the way he conducted the Master Farmer lesson. Before he proceeded with the lesson of the day he went through the lesson of last week. Last week's lesson had been about keeping track of replacement of plough parts. The main parts under discussion were *rushaya, vhiri, dishi* and *muromo*. He got off the discussion by asking farmers to estimate the time it took for different parts to wear out before replacement. Farmers gave answers that were based on time e.g. two seasons or three seasons, with some answers simply stating that they needed replacement when the parts had worn out.

The extension worker explained that this was necessary so that farmers would know when to replace the various parts. As a round up to the revision he glanced from the training manual and gave the 'correct' answers (Table 1). He commented to one farmer who had missed the lesson under discussion that if he had been present he would be knowing 'the correct answers'. The answers had the sole qualification that replacement was dependent on the soil moisture status of the field.

Table 1: Replacement time table for different plough parts

Part	When to replace
Muromo (shear)	2 ha
Vhiri (land wheel)	10 ha
Rushaya (landside)	15 ha
Dhishi (mouldboard)	50 ha

What was striking about the answers given by farmers and the official ones was that farmers 'measured' replacement in terms of seasons or time while official answers were based on area ploughed. It appeared farmers, in their answers, considered their plots and then thought back how long it took for the different plough parts to wear out. On the other hand the official answers were on the basis of experiments conducted at the Institute of Agricultural Engineering. Instead of trying to link the results of these experiments with the farmers' situation the option that was followed was to reproduce the experimental results. One farm machinery specialist when asked about the conduct of the lesson wondered whether the lesson was really that necessary given the fact that farmers in their own experiences knew when to replace the different plough parts.

When the farm machinery lesson ended the extension worker then proceeded with the lesson of the day which was about castration. The approach to the lesson was very similar to the revision exercise with farmers being asked to provide some answers before the official answers were given. After the lesson, in answer to how he picked his lessons, he indicated that he followed a calendar which he had been issued with.

CASE 2: CAUGHT UP BETWEEN AGRONOMIC AND ECONOMIC SENSE

Battles over cropping programmes

This case presents an illustration of conflicts between farmers and state officials concerning crop choice in Fuve Panganai. Each farmer in the scheme has two 0.5 ha plots that are located in different sections because of the block system. In summer one plot is planted to cotton and the other to groundnuts. Wheat and maize are grown in winter on the two plots.

As said crop selection lies at the heart of the commercialisation debate. In the 1993/94 season, as had been the case since the scheme started in 1990, farmers grew groundnuts and cotton in summer and wheat and maize in winter (Manzungu, 1995a). Most of the groundnuts were destroyed by rodents. As a result farmers were not keen to grow groundnuts the following summer. Besides, they said, the crop was labour intensive. Furthermore some farmers argued that their fields were not really suitable for groundnuts, which at the end of it all meant smaller harvests. All the same they could not just go ahead and plant cotton, which many favoured, without consulting Agritex. Talks were initiated with the extension workers who in turn sent the request to the district office in Zaka. The answer was negative. However farmers went ahead and planted cotton instead of groundnuts. Some did not plant the entire field; they grew a portion to groundnuts. Many farmers grew two plots of cotton and made a lot of money with some earning as much as Z$15,000 after input deductions. Apparently without the knowledge of the money made by farmers, complaints were raised by Agritex over this non-compliance to cropping programmes. Nothing much was done though. Spurred on by the possibility of growing crops of their choice, farmers also avoided growing wheat as 'it was a high input crop whose price was not attractive'. Instead of committing the 0.5 ha to wheat the majority wanted to grow more maize so that they could 'beat hunger'. This hunger was caused by three factors. Firstly, the 0.5 ha of maize was too small for some families. The second reason was that because the staple maize crop was a winter one it tended to be sold green which jeopardised food reserves. Related to that was that farmers were forced to harvest early (and sell green) so that they could prepare lands for cotton. One other factor was that the cash receipts from green mealies talked about favourably on the field day on 6 October 1994 by Agritex officials was not achieved by a majority of the farmers as the projected price of Z$1 per cob plummeted to Z$0.30 because of an oversupply of green mealies.

Meanwhile Agritex continued to insist that rotations had to be followed. It was not made clear though what was the object of the rotations. One officer, who mentioned control of pests and diseases, could not provide details of the pests and diseases.

After noticing that farmers were getting 'out of hand' Agritex was more determined in the 1995 winter season to keep the situation under control. However, by the time the District officials in Zaka realised what was happening farmers in block C had already planted maize before the stipulated time in June. Some farmers, such as the chairman of block C, planted as early as April. Wheat was also shunned by practically all the farmers. Since they could do very little about block C the officials tried to nip the problem in the bud in block D which contained the newer farmers.

One official from the district office was heard on block D threatening farmers, 'When I come back next week I want to find wheat growing here. You want me to lose my job? People from Harare asked why farmers were not adhering to the cropping programme.' Since many farmers had not yet planted in block D and also because it was their first year of irrigation, many farmers complied and grew wheat, but not every one. In block B the threats did not quite bear the same result because farmers had been longer in the scheme and could not easily be threatened.

When maize plants in block C were attacked by maize streak virus the Agritex officials were jubilant and reminded farmers: 'We told you so.' All the same, farmers were not repentant. Some farmers put in a lot of fertilizers to make the maize plants 'grow out of the problem'. One farmer summed up the feeling in block C by saying that although they had experienced problems with maize streak virus it was worth the risk. Besides, he said the only problem was that Jerera selling point (a growth point near Zaka) had not stocked the maize variety SC 613 which was maize streak virus resistant. The extension agent on the other hand knew about SC 613 but could not say why he had not advised farmers to buy the maize streak resistant variety. It could be that the message recommending this particular variety had not yet come through the normal Agritex channels.

Farmers' discourse on commercialisation aspects

Farmers were also engaged in another commercialisation debate that excluded Agritex as illustrated below. The case shows that farmers do their own 'experiments' as individuals and settle for those options that are suitable to them.

Zheve is the former chairman of block B. He resigned from the position because of problems related to water distribution in the block. On 12 October 1995 he was getting maize through the government-run grain loan scheme because his 'timing in the irrigated plot had failed him'. He now wanted some grain to carry him until his maize crop was ready for harvesting. When the subject of crop choice came up he mentioned cotton as his main cash crop which was marketed through Triangle or Cotton Company of Zimbabwe ginneries. He explained that credit inputs were also available from both. Farmers tended to go for whoever was first to bring inputs. Both companies seemed not to get timing of the supply of inputs right though. He, however, personally preferred marketing his crop to the Cotton Company of Zimbabwe because it paid advancement payments for each bale of cotton delivered and 'long after you have forgotten about it they give you bonus payment'.

Farmers Mugovera and Chiwomo had opposite views to him. They did not want to get involved in the grain loan scheme because they had 'planned well'. They thought that Agritex was bent on destroying farmers because of its rigid cropping programme. They did not want to grow wheat always. They wanted to grow it when the price was right. Next year they would grow it but not this year. Also they did not want to grow the whole 0.5 ha under wheat. They insisted that Agritex farmed on paper and not on the ground; 'if Agritex was given a plot they would also stop growing wheat!' These straight rows of maize could not be achieved if Agritex were farming, they continued. And the maize plants would dry up as the extension workers would be skirting the field not wanting to get muddy! As for themselves they preferred Triangle instead of the Cotton Company of Zimbabwe. They were not impressed with the advances from the Cotton Company of Zimbabwe; these were in bits and pieces which were not good for them. They did not want that, they wanted their money in bulk because then they would do something big with it, for example pay school fees. Triangle also gave good grades and had less deductions. The only problem was that Triangle workers were apt to make mistakes, e.g. they would credit cotton bales to farmers other than the rightful owners. All this information they had gathered because they had sent their cotton to both companies so as to experiment.

When the 1995 summer season came the 'free cropping' came to a halt. An Agritex delegation from the province came in early November and told farmers that under no circumstances would they be allowed to change the cropping programme. It appears eviction threats were used because one farmer in block C, who had already planted cotton, was observed uprooting it. —

The case shows that farmers are not just consumers of knowledge but are also involved in the generation of information and knowledge. It is also important to realise that the farmers did not hold the same views. If farmers are not a homogenous group the question becomes how scientific information produced on the basis of experiments in one location relates to other varying situations.

CASE 3: ON RISK MANAGEMENT

In the introduction it was remarked that while government officials can talk farming, farmers have to practise it. This point is illustrated here with reference to risk management in crop production. It will be clear that farmers are subject to risks that an extensionist, however well meaning, does not face. This reinforces the importance of casting the commercialisation debate from farmers' perspectives.

Towards the summer of 1994 the Irrigation Supervisor and the extension workers in Chibuwe held meetings with farmers on the need for them to devote a portion of their plots to green mealies by which they could 'make money'. An example was given of a farmer in block A who had made quite some cash from selling green mealies the previous year. Farmers were advised to prepare inputs for six *bunds* (border strips) for that purpose. Planting of the green mealie crop was to be done on 15 September. The rest of the maize would be planted in late October to early November.

Farmers' reaction was not easy to tell. It could be that they were more concerned with one other message that was coming from the Irrigation Supervisor's office through water bailiffs. Both the Supervisor and the extension workers were agreed on the fact that any farmer without fertilizer would not be allowed to plant. The water bailiffs who undertake such police duties were fully appraised. It was during a meeting in block D with the extension workers that farmers' reactions came to be known. The extension worker for the block called for a meeting to advise farmers on the issue. He explained that it was necessary so that farmers could make money. He assured the farmers that the market was available. He encouraged the farmers to take this opportunity so that they could be like the *vayungu* (Whites)[6] such as those on block A.

The 'good' message of the extension worker met with no enthusiasm. Some farmers sat there expressionless while a few voiced their concern. They said that they were not going to plant the six *bunds* because they knew that there would be no water at crucial times. Rather they would plant with the main rains when they would be assured of enough water. A discussion ensued with the extension worker assuring the farmers that there would be enough water. It was clear that he was not convincing anyone.

The extension worker was clearly exasperated. He assured the author that things were going to be alright.[7] When the 15 September deadline for planting arrived there was a visible reduction in the number of people present in block D. It appeared they were afraid to be asked why they were not planting. Of those that were there the following were typical reactions.

Mbuya **(Granny) Muchambeyi** said that her beans were still in the field and there was no point of uprooting them before they matured. She also did not have money (money was available after bean sales) to buy seed and fertilizers. Besides in block D water was a perpetual problem, for example she had only irrigated her bean crop four times. The water supply situation was unlikely to be reliable for the maize crop either. Therefore she would not sow maize before the rains.

Mrs Maidei Muchawaya mentioned the same problem but added that fertilizer was not available in the local shops. Since there was a ruling in force of no fertilizer no planting she would not plant the maize.

Mrs Nkumbuyani's worker said that the landlord was not sowing green mealies because she did not have the required fertilizers as insisted by Agritex.

Mr Mapepa said it was risky because the water bailiff was not diligent on his job. Sometimes he came late and at other times never. He was afraid that the maize would suffer because of the poor water supply. If the water bailiff were diligent enough and would ensure that water was distributed as equitably as possible, there was good reason to plant the maize as required.

Meanwhile in block A many farmers planted the green mealie crop on the date or close to it. This was because the block has the most reliable water supply in the scheme. Farmers had enough irrigation water for sometime. The rains fell sometime in October and people in block D planted their maize crop.

On 18 November 1994 irrigation ceased in the scheme because there was no water in the Save River. Irrigation was only resumed in late January. The maize crop in block A suffered terribly as it was then at a moisture sensitive stage. The crop in block D was younger and was not affected much. By February 1995 it was showing who had been wiser. The green colour in block A disappeared to give way to a grey and finally brown colour. In block D a good maize crop was eventually harvested while block A maize was practically a write-off.

For the first time I heard block D farmers being referred to as good farmers. They were even included in the list to receive tomato seed from a canning factory from Mutare. In block A the situation did not improve at all. The Supervisor forbade farmers to plant another crop as it would disturb the next winter crop. As it turned out the final irrigation for the year was on 29 March 1995. It was not possible to plant the winter crop.

The Secretary of the Irrigation Management Committee was very much annoyed by the whole affair. On the impending hunger because of a poor crop he was bitter because it was Agritex who advised about planting a green mealie crop which failed. Furthermore Agritex had disallowed the farmer to plant another crop, say vegetables, to get cash because of cropping programmes. 'We don't want Agritex any more. It is good for nothing. The scheme can be better without them,' he said.

DISCUSSION

It is submitted that the empirical material presented can best be understood if the notions of science and livelihood strategies alluded to in the introduction are brought into the debate. This is because these, to a very large degree, structure and influence the commercialisation debate in smallholder irrigation schemes.

The notions are important in that they assist in finding out the *raison d'être* of government officials' actions and those of farmers. In other words: Why is a certain form of 'commercial' farming insisted upon and why do farmers often resist the efforts of the extension personnel?

Firstly, there is every reason to believe that extension personnel are sincere in what they do, be it training or advising farmers which crops to grow. Their role in irrigation schemes is informed by the fact that they are custodians of government investment (which is substantial). Thus they feel responsible to ensure that government investment is used to the highest possible efficiency through commercial and not subsistence farming. For that to happen past and present officials feel that farmers need to have the necessary knowledge. That knowledge, we have seen, is supposed to be based on proven scientific facts which is then passed on through, for instance, training programmes.

By proven is meant recommendations given to farmers which emanate from research stations where scientific experiments are conducted so as to verify the 'truth' about farming. At the national level this is done by the Department of Research and Specialist Services (DR&SS) which is a sister department in the same ministry as Agritex. The scientific knowledge, which is meant to improve farming, is spread through the Transfer-of-Technology model where knowledge from research stations passes from scientists to extensionists and finally to farmers. Commentators like Chambers and Ghildyal (1985) among others, have criticised such a conceptualisation of knowledge. One of the criticisms is that the conditions faced by farmers are very different from those faced by researchers. Also, the meaning of 'proven' is an issue, as extrapolations can easily be mistaken for proof. I can also add here that in smallholder irrigation DR&SS is not involved to any meaningful extent. Thus the basis of the extrapolations is not really available. It would appear then that there is cause to doubt the inclination of the extension messages towards scientific positivism, the logic that seems to suggest that the cause of subsistence farming in communal areas is due to a lack of scientific or technical knowledge. Such reasoning fails to locate the subsistence-commercial debate in the context of problems of land, water scarcity, distance from markets, capital etc. But this is not to deny the importance of knowledge in farming.

Many farmers weighed out the options available to them and in some circumstances decided there was no point in investing into speculative commercial farming as farmers in block D in Chibuwe demonstrated. Cash crops, however, were still grown when the circumstances were more favourable, e.g. when water was available. On the same point we saw a very active debate about commercial farming activities involving cotton in Fuve Panganai. It appears that farmers took serious account of the risk factor involved in farming while extensionists tended to rely on models that were at variance with farmers' reality. In some instances these threatened farmers' livelihoods. The glut of maize in Fuve Panganai is a case in point.

The foregoing should not be misconstrued to mean that there is no role for scientific knowledge in agriculture. Rather what is being cautioned against is the denial of other forms of knowledge and the proper location or use of that knowledge. It is equally important to underline that progress is not going to be on the basis of proficient use of scientific knowledge alone. As has been said by many writers, science represents one view of looking at reality but not the only one. Some writers have argued that scientific knowledge in some cases may turn out to be spectator knowledge (Huizer, 1995, 15 citing Maslow, 1966), a knowledge that is neutral, quantifiable, objectified, detached and non-involved with the immediate practical reality. Roep and de Bruin (1994) have also argued that scientification of agriculture is not necessarily desirable because of the tendency towards standardisation. There are however other ways of knowing apart from scientific knowledge. Science, then, should be used together with other knowledge systems (Zweers, 1995; Watson-Verran and Turnball, 1995; Muchena, this volume).

It is significant that without or with little 'science', farmers in Chibuwe and Fuve Panganai displayed a good grasp of 'commercial' farming which however was different from that propounded by Agritex. This meant appraising the constraints and taking the necessary measures to eliminate or reduce them.

The Master Farmer training programme needs further comment. It was not easy for an extension worker, who saw the limitations of the programme, to change it just like that. The programme thus continues regardless. Bourdillon and Madzudzo (1994, 12-13) have described the programme in this way:

> In this country, education is seen largely as a process of passing examinations for improved paper qualifications, and not concerned with understanding real life. There is a danger that the Master Farmers' programme falls into this category. Possibly the prestige of a Master Farmer's Certificate distracts both farmers and officers from the main purpose of extension: improving their farming skills. It might improve the focus of both trainers and farmers if the only reward were to come from farming production.

The fact that Master Farmer trainees have to take an examination means that the room for manoeuvre to make the programme as relevant as could be to farmers' circumstances was limited. This brings in the crucial question; if what farmers are doing is successful commercially, what then is the transformation that was desired?

CONCLUSION

The point of departure between Agritex and farmers seems to be the definition of 'commercial' farming or what constitutes commercial farming. Agritex relies on scientific models of cropping programmes, crop rotations etc., and equates what is considered to be scientific farming with commercial farming. On the other hand farmers have no fixed philosophical position but look for practical solutions that can improve their livelihoods. In this 'game' farmers do not consider themselves permanent commercial farmers. Depending on circumstances, they can be commercial farmers but also turn subsistence when the situation so demands. Risk management was demonstrated to be an essential component of the enterprise. Finally it may pay if some of the entrenched biases towards scientific farming models and against farmers' livelihood strategies are removed so as to realise the Agritex philosophy which says that 'agricultural activities should be based on exchange and sharing of knowledge, skills and ideas between the farmer and the extension agent thereby enabling the farmers to make decisions on those issues that affect their well-being' (Agritex, 1993).

NOTES

1. Agritex has other responsibilities such as planning which are of less concern in this chapter.
2. This seems to be guided by Agritex's history. Since its emergence in 1981 as an amalgamation of the Department of Conservation and Extension (CONEX) formerly operating in White commercial areas and the Department of Agricultural Development (DEVAG) which operated in Black rural areas, Agritex concentrates its services towards the communal farming sector and offers its services to commercial farmers on request. In common parlance communal and resettlement areas are referred to as the rural areas. Mehretu (1994) however observes that there is also another rural sector, the commercial farming sector. In this chapter the popular meaning of rural areas is referred to.
3. The Master Farmer Training Programme has two levels of training. Farmers have to complete ordinary level first before they go on to the Advanced level. For more details see Bolding in this volume.
4. Ironically, the women did not know how they fared in the written test since their scripts were mislaid in one of the offices.
5. This was confirmed by a senior officer at the provincial office who said that in the annual tests undertaken by extension workers and supervisors to gauge their abilities the subject of farm management had the worst failure rates.
6. In this case to be like Whites refers to be rich or to have a lot of money. Apparently this is an expression which has become part of the local language. For example if people say 'that person is white' they mean he/she is rich and has a good standard of living. For the people Whites are rich and are never poor; in fact it is true to say that practically all of them have never seen a poor White. *Vayungu* (plural, singular *murungu*) is the Ndau version of *varungu/varumbi* (singular *murungu/murumbi*).
7. Later on I heard him saying that he had confronted one of the farmers who had been vociferous about the issue.

APPENDIX A

AGRITEX PHILOSOPHY AND MISSION STATEMENT

PHILOSOPHY

Agricultural extension activities should be based on the importance of exchange and sharing of knowledge, skills and ideas between the farmer and extension agent **thereby**, enabling the farmers to make decisions on those issues that affect their well-being.

MISSION STATEMENT

Agritex will serve the needs of different farmer clientele by generating, providing and promoting agricultural programmes which enhance competitive and economically viable productivity on a sustainable basis and in so doing the following areas will receive special attention:

- The provision of services and opportunities that *develop the human resource base* such that the full agricultural potential of men and women be they adults or youths is realised and exploited.

- *Develop and promote environmentally sound technologies and management practices* 'that aim at conserving renewable natural resources for use by future generations.

- Adopting an *integrated extension approach* by recognising the existence and roles that can be played by other agencies in agricultural development.

- *The development and wise management of the water resources* would contribute significantly to the transformation of drought prone areas of Zimbabwe into productive and habitable lands *thus*, enhancing food security and self-sufficiency at both national and household levels.

- Maintain a process of *transforming rural farm families* from subsistence into commercial agriculture *hence*, ensuring healthy farm families that have a sound base for economic growth.

- The realisation that all the above activities have to be carried out only if each member of the department upholds and maintains *high professional ethics and standards* and make these guide their day to day conduct.

(**Source:** Agritex, 1993)

APPENDIX B

EXTRACT OF THE MASTER FARMER TRAINING RECORD

TOTAL FARM

Money going out (variable costs)	$	c
Grain crop		
Other crop		
Livestock		
Total Variable Costs		

Overheads (estimate)	$	c
Contract ploughing costs		
O.L, fuel maintenance, repairs		
Interest on loans and depreciation		
Other		
Total overheads		

ACCOUNT

Money coming in		
Grain crop		
Other crop		
Livestock		
Gross income		
Less: Total variable costs		
Less: Combined gross margin		
Less: Overheads (estimate)		
Net Farm profit		
Note: No labour costs for your own labour have been charged. The "net farm profit" can be looked at as your reward for your labour		

Source: Master Farmer Training Book

Chapter 4

Contextualising communal agriculture

Observations on labour migration and farming in Save Communal Land

JENS A. ANDERSSON[1]

The previous chapters in this volume discussed the historical background of the department of Agricultural Technical and Extension Services (Agritex), and the development of its intervention strategies. Communal agriculture[2] was looked at from an interventionist perspective. This chapter takes a different perspective, neither concentrating on specific intervening institutions, nor their intervention practices. Instead, it sets out to discuss some aspects of the changing socio-economic context of communal agriculture, and subsequently, the context in which agricultural intervention takes place. The approach taken here is sociological, focusing upon how communal area households made, and make a living. More specifically, the importance of agriculture to rural livelihood will be discussed, building upon observations of one ward in Save Communal Land in Buhera district. In so doing, this chapter aims to raise questions on the interventionist perspective itself, notably its inherent pre-occupation with agricultural development.

Communal areas were created as 'Native Reserves' during the early colonial period. As part of an overall racial segregation policy, these areas were intended for African occupation only, and were designed as labour reserves; the 'Reserves' were to supply the labour for the expanding, white dominated, capitalist economy. Colonialism thus effectively turned African communities into migrant labour societies. It is this history of labour migration, which was circulatory in nature, that remains of great importance to an understanding of the rural economy in the communal areas today. The argument put forward in this chapter is that in order to formulate agricultural intervention policies better suited to the specific needs of communal farmers, one has to understand the importance of labour migration in rural livelihood strategies.

To argue for the importance of circulatory labour migration in contemporary livelihoods in the communal areas, implies also a different perspective on the role of agriculture in sustaining rural livelihood. Consequently, intervention policies aimed at improving agricultural production have to be looked at in this perspective. As will be discussed in the first part of this chapter, interventions are often based on certain stereotypes of the nature of communal agriculture that need not be valid for all communal areas in Zimbabwe. Buhera district is such an area. Here, labour migration has become such an important aspect of social and economic life, that communal agriculture cannot be understood adequately without considering its links to the urban economy of Harare, where most migrant labourers from Buhera district go. This multiplicity of rural livelihood strategies, of which agriculture is only a part, should inform future intervention strategies, better suited to the needs of (specific categories of) communal farmers and their families.

Buhera district is not typical of communal areas in Zimbabwe as a whole. For instance, the area never experienced large-scale land alienation during the colonial period.[3] Unlike many other areas, the Land Apportionment Act designated the whole area a communal area (then called 'Sabi Native Reserve'.[4] Nevertheless, the case of Buhera district, and that of the ward studied in more detail, can have us look differently at the situation of communal agriculture in general, since labour migration is a common phenomenon in rural Zimbabwe as a whole.

COLONIALISM AND COMMUNAL AGRICULTURE: RACIAL SEGREGATION AND RESTRICTIVE DEVELOPMENT

Any understanding of contemporary communal agriculture has its roots in Zimbabwe's colonial history, characterised by extensive state intervention in the rural economy. Colonial state intervention created the contemporary communal areas (then called Native Reserves) as part of an overall policy of racial segregation. It suppressed peasant[5] initiative in these areas whenever African producers' ability to respond to market opportunities threatened European agriculture, using discriminatory legislation such as the Maize Control Acts and Cattle Levy Acts in the 1930s. Simultaneously, white settler agriculture got massive government support in the form of financial assistance, agricultural extension services, infrastructural development, etc.

LABOUR MIGRATION AND FARMING IN SAVE COMMUNAL LAND

To be sure, the same colonial state that concentrated African producers on marginal lands with low developed infrastructure also implemented policies aimed at improving agricultural practices of African farmers. Often such policies originated from conservationist concerns (Beinart, 1984; Bolding, this volume), but at the same time they were aimed at enabling the growing number of people (pushed into) these areas to produce enough to sustain their livelihood.[6] State intervention in communal agriculture thus could be contradictory in nature; simultaneously suppressing and stimulating African agricultural production. Some thoughts of a colonial administrator stationed at Buhera in 1952 may illustrate this:

> Statistics may show that the Natives' contribution to the grain production of the country is colossal. The Natives are exhorted [by the government] to grow more food. This would be justified if all the Natives followed modern methods of cultivation and observed even the rudimentary principles of soil conservation but alas! the majority do not do so. The results of high prices for grain and exhortations to grow more food have been that the Natives have ripped the heart out of most Reserves (. . .) This ruination of the country's natural resources should not be allowed to continue. (. . .) The content that legislation will curb the ignorant native in his destructive ways is mere wishful thinking. The only solution is to drop the price of grain and reduce the profits allowed to the grain trader. This should discourage the Native from cultivating to excess and the grain trader from making excessive profits.[7]

Notwithstanding the self-contradictory policies concerning communal agriculture, interventions in African agriculture remained within the restrictive policy framework of racial segregation and the government's concern for the demands of the white settler economy. Communal agriculture was largely a neglected sector and peasant initiative within it, suppressed. Yet, the perspective underpinning colonial state intervention from the 1920s onwards has generally viewed Africans as agricultural producers.[8] This view was maintained despite the state's own policies that continued to concentrate people on marginal lands and undercut the 'peasant-option' of making a living from the soil.

Colonial state intervention did not, however, reduce communal agriculture to merely subsistence production. Although market production was to such a degree undercut that rural households had to seek wage labour employment to complement farm income and meet tax demands, at the same time agricultural output in the communal areas increased (Ranger, 1985, 140-142; 1978, 121). Aquina (1963, 26) has described how peasant households compensated for the labour shortages that resulted from this incorporation of labour migration into rural livelihood strategies. Through pooling resources (cattle, ploughs, labour), the mobilisation of kinship and local village structures, and the reallocation of agricultural tasks between men and women, households were able to sustain their agricultural production. Furthermore, income from labour migration was often used to support agricultural activities (Cheater, 1981, 356; Callear, 1985). While causing increased opposition, colonial intervention thus not always meant stagnation and decline in communal agriculture.

In post-colonial Zimbabwe, the colonial neglect and suppression of initiative in communal agriculture informed a concerted effort to redress the situation. Academics and interventionists alike now specifically focused on rural development and peasant agriculture in the communal areas. Agricultural intervention has become geared towards raising production in the communal areas.

Inherent to this concern for the improvement of communal agriculture are two general assumptions as regards the nature of peasant agriculture in the communal areas. As both often underpin agricultural intervention policies and relate to the colonial perspective that viewed Africans as agricultural producers, they are worth considering in the context of this book. The first assumption is that there exists an under-utilised potential in communal agriculture that can, and will, be exploited when agro-technical advice, finance and market facilities, etc. are provided. The second is that such efforts to enhance communal agriculture assume that the so-called 'peasant-option' — i.e. meeting cash and subsistence needs with agricultural production — has remained the preferred option for rural households in the communal areas.

To be sure, making a living as a peasant producer was and still is the preferred option for many people in the communal areas. However, the observations from Buhera district presented here suggest that this preference for the 'peasant-option' may have less universal applicability than is often assumed.

CONTINUITY OF INTERVENTION PRACTICES IN A CHANGING ENVIRONMENT

Agriculture in areas such as the Buhera district is, similar to many other communal areas in Zimbabwe, characterised by poor soils, soil erosion and recurrent water shortage.[9] Observations made in the district during the colonial period are still indicative of many of its contemporary problems:

> I have never yet seen a normal crop in the Reserve and although acreage cultivated increases, there has never been sufficient food produced to carry the population from season to season for the past three years. [Assistant Native Commissioner Buhera sub-district to Native Commissioner Charter, Nov. 1918][10]
>
> The first few miles led over a jumble of eroded waste-land, criss-crossed by a tangle of sleigh tracks (. . .). It was the time of the year when the crops should have stood high and ripening in the fields, but most of what I saw were undersized, withered stalks, crippled by a premature drought. The man at my side began to talk about hunger and starvation [Anthropologist J.F. Holleman in 1947 (1958, 145)].

The continuity in propagated 'solutions' to these problems is striking. For instance, it seems as if agricultural extension practices in the 1940s have been informed by one of Agritex's mission statements of 1994:

> In view of the drought, this area having suffered considerably, the prospects of a reasonable harvest yield is poor. In spite of this set back, the demonstrator is imparting all his knowledge to those who have suffered with the view that in the event of another rainless season a fairly good job can be made from a bad one by proper tillage methods and crop rotation in respect of the nature of the soil and any rain that may occur [Land Development Officer Buhera district, monthly report May 1947].[11]
>
> The development and wise management of water resources would contribute significantly to the transformation of drought prone areas of Zimbabwe into productive and habitable lands thus enhancing food security and self-sufficiency at both national and household levels [Agritex, 1994].

The construction of dams and boreholes constitutes another example of continuity in intervention policies. Intervening agencies in contemporary Buhera (local government institutions, Agritex and foreign NGOs) seem as busy with their construction as the colonial government some fifty years ago:

> All available teams are busy on wells and at least four should be completed within two weeks. Work on dams at (. . .) is still in the preliminary stages and has not been restarted this season. Water in the existing dams is sufficient for the next few months [Land Development Officer Buhera district, monthly report June 1948].

Yet, in the past such intervention practices often served the racial segregation policies of the colonial government, as a plan to move some 1,770 families into the Buhera district in 1950, illustrates:

> The rate of absorption of population is dependant entirely on the provision, firstly, of water by boreholes and wells — the speed by the Irrigation Department can be induced to supply this, for, without water (and, of course, good roads) they are unable to conserve water by the construction of dams, which are essential for stock [Native Commissioner Buhera district to Provincial Native Commissioner, August 1950].[12]

Hence, from an agricultural interventionist point of view it seems as if the situation in Buhera district has not changed much over the past fifty years. The same problems still exist and they are tackled with similar intervention strategies.

Taking a wider perspective, as this chapter aims to do, the situation of Buhera district has, however, altered considerably. These changes do not simply relate to the political changes, ending racial segregation, but to the social economic context in which agricultural production takes place — i.e. the growing importance of urban income as a source of rural livelihood.

RURAL LIVELIHOOD AND THE GROWING IMPORTANCE OF LABOUR MIGRATION

It is not hard to understand that in a drought prone agricultural environment, making a living from the soil is a difficult option to pursue. During the colonial period rural households in most wards of the Buhera district, including Murambinda ward, have therefore also undertaken economic initiatives outside agriculture. Pushed by cash needs (for ploughs, clothing, tax, etc.) circulatory migration from the area became a major feature of life.[13] Up to the present this is reflected in imbalances in the age structure of the population. Table 1 shows that the proportion of people in the age category 20-44 in Murambinda ward, as well as in Buhera district as a whole, is substantially smaller than the national figures (whereas in the age group 0-14 it is the other way round).

Table 1: Specific age groups as percentage of the total population, 1992

Age group	Zimbabwe	Buhera district	Murambinda ward[14]
0-14	45 %	52 %	52 %
15-19	12 %	13 %	14 %
20-24	10 %	7 %	7 %
25-44	21 %	15 %	14 %
45-64	9 %	9 %	9 %
65+	3 %	4 %	4 %
Total population	**10 412 548**	**203 739**	**6 660**

Computed from CSO, 1994a, 1994b

The dominant pattern of labour migration from the area is circulatory labour migration of economically active men. Indications for this kind of migration are imbalances in the sex-ratio in population statistics. In Buhera district in 1992, the number of men per 100 women was 84.7 as compared to 95.4 in Zimbabwe as a whole. For the economically active population these figures diverge even more.[15]

Temporary out-migration already started in the early colonial period but the dominant pattern of labour migration in those days differed from today's. Labour migration was much more a way of meeting immediate cash needs than an indispensable element in rural survival. Out-migration was set by the agricultural calendar, with most men seeking employment towards the end of the agricultural season, when labour demand in local agriculture decreased.[16] Furthermore, migration was closely related to a specific period in a person's life-cycle. Those who moved out were generally young men trying to start an independent life. They had to use wage income to accumulate enough money for bride wealth (roora) payments, to buy cloths, and agricultural inputs such as ploughs, yokes and cattle. But once having accumulated sufficient cash, they would return and take up life as a peasant farmer.

In general, staying at home and making a living from the soil was still a viable and preferred option in Buhera district.[17] Most men would remain

home with their wives to practise the common system of shifting cultivation
(see Holleman, 1958) — i.e. ploughing the same piece of land for several
years whereafter a new piece was cleared and cultivated. The recurrent food
scarcities in the vulnerable agricultural environment of the area were usually
tackled successfully through exchange of cattle for grain.[18]

In the 1940s and early 1950s however, a different migration pattern became
common. The expansion of urban employment opportunities in Harare (and
the wages that could be earned), made urban life attractive to young men
from Buhera. As one of these migrants recalls:

> It was quite easy to get a job then [early 1950s]. I went to Harare,
> because I saw people here who worked there. They had nice clothes
> and looked smart (...). They earned a lot of money and were respected
> at home.

Many of those who went to Harare returned after relatively short periods.
Their careers still followed the dominant pattern of the early colonial period;
temporary migration to start an independent life as a peasant producer, or
to meet immediate cash needs. But another category of migrants, and this
group became bigger, now would stay away in town for many years and
build up an urban existence, only to return home to take up farming when
old or pensioned. This type of urban migrants clearly had other ambitions
than starting a peasant career. I will call this new type of migrants 'semi-
urbanised'.

While in urban employment, these migrants did maintain close links with
their rural home. Migrants would regularly visit their home areas, send money,
preferred to marry at home, and their wives would generally stay at the rural
home cultivating the land. Moreover, when in town, migrants generally relied
on each other in arranging both employment and accommodation:[19]

> My boss said to me that he needed more workers and told me to
> bring my 'brothers'. So, I informed J., he did not want to come. He
> preferred to farm. But my younger brother K. liked to come to Harare.
> It was on a Sunday that he went with me, and we stayed together in
> town.

Thus, rural social organisation formed the basis for rural peoples' initiatives
outside agriculture.

That labour migration to towns gradually became of more importance to
rural livelihood also relates to population increase in the district. Natural

growth and forced immigrations of people who were pushed out of their lands elsewhere, put existing agricultural practices under pressure and increased land shortage. In the agriculturally more favourable north-western part of the district such a tendency could already be observed in the 1930s. As the Assistant Native Commissioner Buhera wrote in 1937:

> I have to report on the growing congestion of population in the north west portion of this Reserve. (. . .) During the past three years approximately 350 natives with their families have come here from other districts and nearly all these have settled in the part of the Reserve of which I am now speaking. The result is that, because of the way the land is being farmed, there is little, if any spare land either for pasture or agricultural purposes.
>
> . . . I have on every opportunity drawn the attention of the natives to this situation which exists and stressed the necessity for more economical methods of farming in order that the land may support the number of people who must continue to live there.[20]

Yet, increased land pressure does not necessarily increase labour migration from an area. People may respond differently to such forces, as seems indeed to have been the case in north-western Buhera. In this part of the district, an intensification of land use practices probably prevented a growth of labour migration, as nowadays the number of absentee men is generally less than in the south-eastern area of the district. Agriculture has remained more significant to rural livelihood in this part of Buhera district as compared to the south-eastern part where as a result of recurrent droughts people relied more on cattle for their survival.

In the Murambinda ward (see map) people perceive land shortages as a problem that became manifest in the late 1960s or later.[21] However, by that time the new pattern of labour migration — i.e. that of semi-urbanised migrant workers — was already established in many wards of the district, including Murambinda ward. Though important, and nowadays probably the major force behind labour migration to towns, growing land scarcity was not the sole factor causing an increasing number of men to seek an urban career instead of making a living from the soil. Peasant initiative outside agriculture did not simply develop out of a growing land shortage.

The Murambinda area in Buhera district

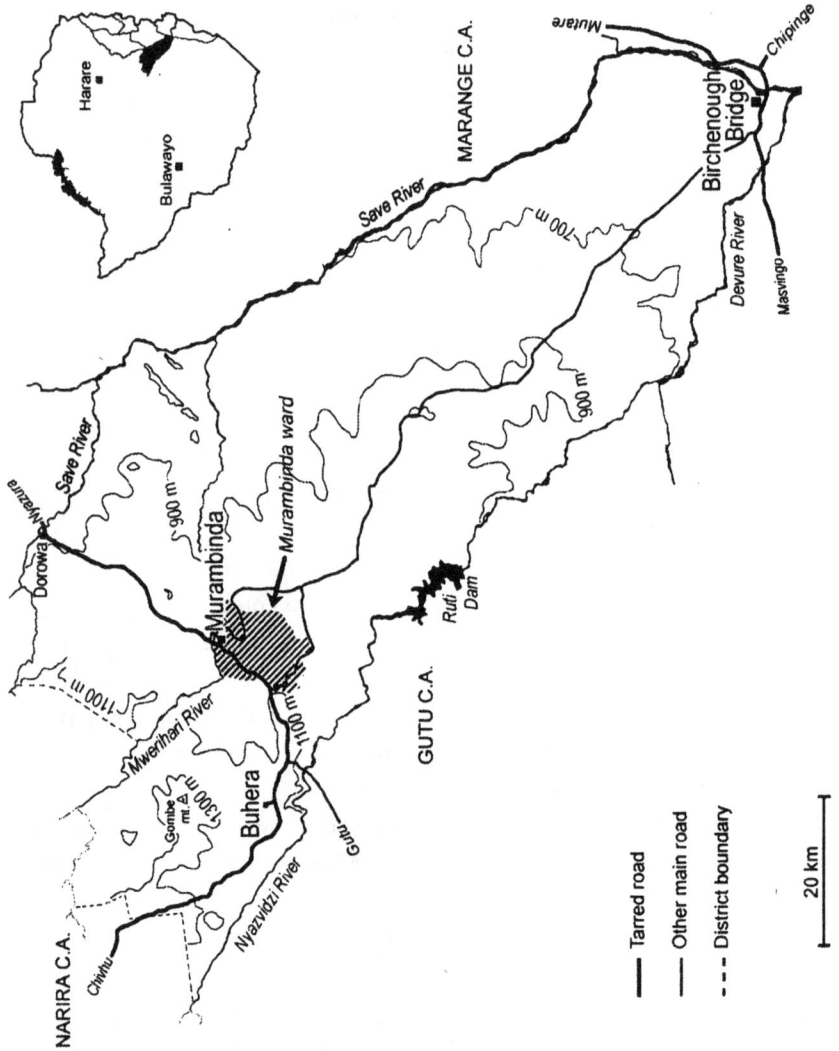

Population growth since the 1960s[22] has further increased land shortage, making the 'peasant option' increasingly difficult to pursue. It is therefore understandable that in contemporary Buhera district it has become common for school leavers to seek a career in the urban sector of the economy, instead of aiming at a life as peasant producer. The urban orientation of young people in the district should also be understood against the background of the ever-increasing investment in education during the colonial period. Remittances from labour migration were not only used to meet the tax requirements of the colonial government or the immediate needs of the rural household, but also invested in the education of children. Education of children was, understandably, aimed at increasing the opportunities as a (semi-urbanised) migrant worker in the urban economy. Hence, one may conclude that besides increased population pressure, investment in education reinforced the newly developed migration pattern of the 1940s and 1950s.

RURAL LIVELIHOOD AND RURAL-URBAN LINKAGES

It may appear that developments in Buhera district match academic models of rural development in which societies dominated by migrant labour are seen as inherently transitional in capitalist development or economic modernisation. In these models of change, peasant agriculture is expected to disappear, and the semi-urbanised migrant workers to become (eventually) a permanently urbanised working class. These abstract models do not, however, capture the situation in Murambinda ward of Buhera district. Although labour migration from the area has been on the increase, even in the post-independence period (see appendix), migrant workers continue, like in the past, to maintain links with their rural homes, where in most cases their families reside and work the land. Only few migrants strive to become permanently settled in town with their (nuclear) families. But even this group of (relatively successful) migrants often retains a claim on land in the area, as the following portrait illustrates:

> Abinos was just a teenager when he went to Harare in 1976. He went to stay with his *mukoma*, a son of his father's elder brother, who had started his own tailor business after having worked in the textile industry. Abinos assisted his *mukoma*, who in turn helped to finance his study by correspondence. Things went well. By 1980 Abinos had passed two O-levels, and soon after he passed a third one. It got him a job at an insurance company. Abinos remained in town, married,

and progressed in his work. He was able to buy a car and obtained a stand in Mabvuku high density area, where he started to construct a house for his expanding family. On the small stand where he built his house, one can also observe a few maize plants. His wife planted them. Meanwhile, upon his marriage, he was also given his own land by the *sabhuku* (village headman) in his home village. Here he built his *musha* consisting of two round huts.

These days Abinos has plans to erect a big granary and a brick house on his *musha*, but only after he has completely finished his house in town. He does not, however, consider going home and staying at his *musha*, not even when he is pensioned. Abinos prefers life in town to a rural existence. The security of his urban income puts him in a position to stay in town permanently.

Nevertheless he develops his *musha* on which his younger brother, who is not yet married, stays together with his *muzukuru* (a son of his sister). His young *muzukuru*, who is from a big family, looks after Abinos' cattle. In return Abinos pays his *muzukuru*'s school fees, as his father has not enough money to send all his children to school. To his younger brother Abinos sends agricultural inputs (like maize seeds) and money to organise for people to work the land for him. When he visits his home area occasionally, he collects some of his thus produced maize to take with him to Harare.

One may question the economic viability of Abinos's decision to develop his *musha* and have his field ploughed by others whom he pays. However, this practice of hiring people to work the land (sometimes referred to as *kurimisa*) cannot be reduced to economic decision-making. Having land and a *musha* is, just as the practice of planting a little maize on one's stand in town, of significance from a socio-cultural perspective; it is part of one's identity.[23]

A largely urban existence thus goes together with a continued commitment to rural life. In Murambinda ward this continued commitment towards rural life is, as also the case of Abinos illustrates, exemplified by the large number of brick houses built by migrants or returned migrants.[24] But urban income generally also serves to pay for other household expenditures, as a survey among 86 migrant workers returning home for their 1995 Christmas holiday, demonstrates (Table 2). Apart from supporting the rural household, rural links are also of great importance for the migration process itself; rural and family ties continue to play an important role in the transfer of information and money homewards and in finding work and accommodation in town.

Rural-urban linkages and communal agriculture

Migrant workers' continued commitment towards a rural existence has resulted in urban income becoming a permanent, and often substantial, contribution to rural livelihood in the district. Many if not most, households have come to rely partly on urban remittances for their subsistence. The amount remitted is, however, hard to quantify for remittances sent home by migrants vary enormously, depending on factors such as the age of the migrant, his wage, number of dependants in town and at home, transport costs, etc.[25]

Table 2: Three major expenditures of migrant workers from Murambinda Ward (N=86)

Expenditure	Number of migrants that mention item	Percentage
School fees	50	29
Food	51	20
Clothes	51	20
Farming inputs (cattle, ploughs, seeds, fertiliser)	21	8
Constructing homestead *(musha)*	17	7
Support of family/relatives	13	5
Paying rent	9	4
Bridewealth *(roora)*	7	3
Transport	6	2
Other expenditures	27	10
Not answered	6	2
Total	**258**	**100%**

Yet, agricultural production remains important in the district and in years of good rainfall, reasonable surpluses are produced.[26] Urban linkages can play an important role in sustaining yields or new initiatives in communal farming. Not only are urban remittances invested in agricultural inputs, as Table 2 shows, these inputs are often brought from town.

Such continued interest of migrant workers in peasant agriculture may be based on various grounds. For some, as the case of Abinos illustrates, involvement in agricultural activities merely serves to make a social statement about oneself (Gatter, 1993, 181). For others, peasant agriculture is a form of 'social security'; a way of making a living when retired or retrenched from urban employment. Meanwhile they maintain their claim on the increasingly scarce land, by making sure it is cultivated during their absence. These migrants can employ workers, or have relatives work their land.

In most cases however, wives of migrant workers stay at home to farm during the agricultural season. Many migrant men are industrial workers whose incomes do not suffice to allow them to stay permanently in town, even if they would like this. They have to rely on agriculture to supplement their urban income. The story of Simukai illustrates this:

> Simukai went to Harare in 1976. He went to stay with his *babamunini*, a younger brother of his father, who lived in hostels in Mbare Township. His *babamunini* worked in the tobacco industry and helped Simukai to get a job as a general worker with the same company. As an employee of this company Simukai got his own place in the hostels. Ever since, Simukai is working in town for 8 months a year, from April to late December. This is the period of the year tobacco is processed. As a result he can not be home to help his wife during planting time in October and November; his contribution to agricultural activities is therefore limited.
>
> Since a few years, Simukai is busy constructing a brick house on his *musha*. It has taken him a long time to finally start with it, for he first had to spend much of his income on the schooling of his children. His eldest son has now finished and, like his father, got a job in the urban economy. This means more income for the household and Simukai's son can thus help financing his younger brothers' and sisters' education.
>
> Simukai earns about Z$900 a month, which does not allow for a fully urbanised existence for his (nuclear) family, if they would have wanted so. Hence, his wife stays at home to farm and he sends her some 200 dollars a month to meet cash needs. Simukai visits home almost every month (one return trip home costs almost Z$100 in transport costs), and it is expected of him that he brings things from town. He bought the hybrid maize planted by his wife from the farmers co-op in Harare. When returning to Harare Simukai often takes some maize his wife has produced. Simukai's wife is, like many other women in the district, responsible for agricultural production. Besides grains and groundnuts, she also grows vegetables.

A few migrants specifically aim at making a living from the soil. They invest large sums of their urban income in the development of their land, in order to return home to take up peasant farming. In order to generate enough cash they diversify their farm practices into vegetable production, orchards, or the production of marketable crops such as cotton and sunflower. They may get some of their ideas and new technologies to develop their farm enterprise from their social network in town.[27]

There is, of course, also a category of farmers in the district that is completely dependent on communal agriculture for their livelihood. This category that has no links to the urban economy faces a difficult existence, often having limited resources and being dependent on the unreliable climatic conditions in the district. Especially widowed or (de facto) divorced women can find themselves in a difficult position, as the portrait of Loveness may illustrate:

> Loveness is in her late thirties, living alone with her six daughters on a small *musha*. She was born in Buhera, not very far from the village of her husband, where she is now living. Since her parents had not much money, Loveness stopped school at Standard 3 level and married in 1975. She stayed together with her husband, on the land given to him by the *sabhuku* (village headman), until they had four children. In 1981 he went to look for work in Harare, in order to be able to pay *roora* (bride wealth) to Loveness' parents and to buy cattle to plough with. Loveness stayed on her husband's plot to plough, and occasionally her husband would send her some money which could be used to hire cattle to plough. It was, however, too little and thus Loveness started a vegetable garden to generate additional cash income. After several years he stopped remitting income and Loveness heard rumours — from her husband's *babamunini*, who helped him with accommodation and getting a job — that he had married another woman in town. On the very few occasions that he would visit home to "steal my grains", they would have many conflicts, whereby he would try to convince the *sabhuku* to chase her from the land. Loveness had to resist both her husband and the *sabhuku*.
>
> Since 1991, Loveness has not seen her husband. In order to generate cash she not only grows vegetables, but also moulds bricks, makes baskets, brews beer and works on other people's plots. Loveness' vegetable garden is, however, her prime source of income. It is very close to a river, where she gets the water. Cultivating so near the river is not allowed but when Loveness and other women were asked to move further away from the river, she did not go. The new *sabhuku* (the successor of the one who tried to chase her), with whom Loveness has a good relation, did not force her to quit her garden.

Loveness grows very little maize, not having the money to buy the
seeds. What maize she grows she got as drought relief from
government. She planted using a *badza* (hoe) as she has no cattle, and
she received the seeds only after the first rains, when it is very difficult
to borrow or hire someone else's cattle to plough. Unable to produce
much maize, Loveness tries to sell *nzungu* (groundnuts), *mhunga*
(bulrush millet) and *rukweza* (finger millet). But these 'traditional'
millets have no good market value and even selling them in the form
of beer is difficult; "people only buy your beer if you also come to
their beer parties", something for which Loveness has not got the
money.

The portrait of Loveness shows the difficulties of farming in the absence
of urban linkages. In the drought prone Buhera district, it is very difficult to
generate cash through crop production, causing poor people to revert to
traditional crops whose production does not require much capital, but at the
same time does not enable them to generate enough cash to engage in maize
production.

The cases presented show that communal agriculture is of varying
importance to rural households. Urban income is not simply a (substantial)
contribution to rural livelihood. Urban links can be found in communal
agriculture itself, and enable people to sustain particular cultivation practices,
such as for instance the production of hybrid maize.

Agricultural labour and crop choices

The (social) organisation of agricultural production is highly influenced by
labour migration practices. Most notably, the increased importance of labour
migration from the Buhera district has turned communal agriculture into a
female dominated activity. Women not only comprise the majority of
communal farmers in the district[28] but also have taken a bigger share in
agricultural tasks over the years. Ploughing for example, was traditionally
considered a male task, but nowadays often women also plough. Furthermore,
the increasing number of vegetable gardens in the district are generally worked
by women. For many, these gardens form a complementary source of income,
as produce is sold locally, notably at the market of Murambinda growth
point. For others, especially for widowed or *(de facto)* divorced women — the
case of Loveness — or women whose husband or children in town do not
or cannot remit much income, such gardens are of great importance for
survival.

Besides a process of feminisation, communal agriculture has also increasingly become an activity of elderly people, often retired male migrant workers who have returned home to stay with their wives. These elders generally have the largest lands, yet they are no longer able to work all of it. As a result of the land shortage in the area, some elders have already redistributed parts of their land to their sons while alive, or have their daughters-in-law assist on their own plots. However, in some villages, where land shortages are less acute, arable land may be left unused for years.

Like most people, the elders at home generally do not have to survive on farming alone. The strength of rural-urban (kinship) ties in Buhera district — after all the elders have often assisted contemporary migrants in finding employment and accommodation in town — often ensures assistance of relatives — like unmarried sons — in town upon retirement from urban employment. Given this specific composition of the agricultural labour force, it is understandable that some households in the district face labour shortages during specific periods of the agricultural season, such as planting time.[29]

The popularity of the drought prone maize crop in Buhera district can also be understood from the perspective of labour export from the area. Labour requirements of traditional food crops such as *mhunga* (bulrush millet), *rukweza* (finger millet) and *mapfunde* (sorghum) are often high, even after harvesting when many women leave the district to visit their husbands in town (during the cold dry season called *chando* and the warm dry season, *chirimo*). Though more drought resistant and better to store than maize, these traditional food crops have become of less significance as subsistence crops since the introduction of hybrid maize in the 1970s. Labour requirements of maize are far less and more evenly distributed in time.

To be sure, changing consumption patterns and increased cash needs are also of major importance to understand the current popularity of hybrid maize. Yet, maize was not a new crop in the area, nor is it the only crop that fetches good prices when sold. Groundnuts *(nzungu)*, a major cash crop up to the 1970s, still fetch reasonable prices, and the crop is less drought prone than maize. However, labour requirements of the crop are high and despite the fact that its cultivation does not require much land or capital, the crop has lost much of its significance as a cash crop. Hybrid maize, which cultivation is capital intensive (fertilisers, seeds) and risky from an agronomic point of view, is nevertheless preferred for its better market value. In good seasons it contributes substantially to rural livelihood, whereas in years of crop failure many, if not most, people have to rely for survival on their urban links supplemented by their small plots of traditional millets. Hence, labour migration has decreased risk-aversion in communal agriculture.

Yet another example can be found in soil conservation measures, often regarded as an important issue by agricultural interventionists. Nowadays, such conservation measures are often seen in relation to agricultural development, as a way of improving or at least sustaining yields. And indeed, many of these conservation measures will surely be beneficial to any peasant farmer. But again, for many people in Buhera agricultural production is not any more the mainstay of their livelihood. Limited labour availability and households' reliance on urban sources of income, make investments (of both labour and capital) into such measures — or into agriculture in general — not always as logical as it may seem to agricultural interventionists.[30]

CONCLUSION: THE INTERVENTION PERSPECTIVE AND COMMUNAL AGRICULTURE

This chapter described how circulatory labour migration to towns transformed the nature of rural livelihood in the Buhera district and particularly, the role of communal agriculture in it. Over the years, rural people have pursued livelihood strategies that consist of varying combinations of labour migration, subsistence and market-oriented agricultural production. Communal agriculture itself gradually changed towards an activity dominated by women and elders (often returned migrants). Younger men (and an increasing number of women) generally aspire a career in the urban economy, but maintain their rural links.

Nowadays, only a few households have to make a living from communal agriculture alone. This increased reliance on urban income has not reduced communal agriculture to merely food crop production for subsistence purposes. Increased cash needs have caused agricultural production to remain market oriented, while at the same time risk taking in communal agriculture — i.e. taking the risk of crop failure — has increased. Making a living from the soil alone is no longer a lifetime career, but is limited to specific social categories.

The observations of Buhera district presented in this chapter have important implications for agricultural intervention. Firstly, they may suggest the development of more differentiated strategies of intervention along lines of social positions of people involved in communal agriculture. Agricultural extension in the communal areas may thus become target group oriented, instead of area oriented as in Agritex's current block village based extension approach.

More important, however, is that the perspective of this chapter revealed the limitations of the interventionist perspective. Interventionists are committed to the improvement of agricultural production — i.e. increasing agricultural output in the communal areas to enhance food security (see Agritex, 1993). As a result, they cannot view agriculture as what it is to many people in the communal areas. People's involvement in agriculture is not solely geared towards the highest possible yields — i.e. full utilisation of agricultural potential. It is at the same time a way of maintaining rural links — that are of major importance in pursuing an urban career! — and sometimes merely 'an idiom of social expression' (Gatter, 1993, 181).

Assuming a (continued) preference for peasant farming, as intervention efforts inherently do, is to isolate communal agriculture from its social and economic context. Agricultural decisions are informed by a number of considerations that are not related to farming activities. Since labour migration is nowadays a major source of livelihood in communal areas such as Save Communal Land, it should be no surprise that farming related decisions have become largely a function of the way in which this labour migration is socially organised.

NOTES

1. The research for this chapter was funded by the Netherlands Foundation for the Advancement of Tropical Research (WOTRO).
2. The use of 'communal agriculture' and 'communal areas' in this chapter is to indicate geographical areas. It does not necessarily refer to forms of agriculture in which the land is communally worked.
3. Rather, Africans from other areas were pushed into Buhera district by the colonial administration.
4. The Land Apportionment Act divided land along racial lines and institutionalised the Native Reserves. Initially, Sabi Native Reserve was part of the Charter district. Since 1946 this area (of approx. 5,360 km^2) constitutes a separate district (wrongly) named Buhera, after the ruling inhabitants of the area, the Vahera of the Shava Museyamwa clan. Nowadays, Buhera is the only district in the country that consists entirely of communal land, and it forms one of the largest communal areas in Zimbabwe.

5. The term peasant is used here in the general sense of rural producer without any theoretical or historical connotations as regards peasant modes of production or the origin of peasant production in Zimbabwe (for a discussion see: Wolf, 1966; Bernstein, 1977; Friedmann, 1980; and Palmer and Parsons, 1977; Ranger, 1978).

6. One example is the Food Production Drive policy implemented in the early 1950s, in which government invested in fertilisers, introduced new crop varieties and propagated cultivation practices such as winter ploughing, and manure application. The policy concentrated on specific areas, including some 20,000 acres in the Buhera district (file: S160 AGR 3/1E/52, National Archives of Zimbabwe/NAZ).

7. Native Commissioner (NC) Buhera, annual report 1952. File: S2403/2681, NAZ.

8. As from the 1920s onwards the colonial state developed policies that aimed to reorganise African land use. Before that, state policy was largely pre-occupied with the mobilisation of African labour for the expanding white economy.

9. Such general descriptions do not, however, account for differences within the district. Conditions for agriculture vary from Natural Region III in the northern and western parts of the district to Natural Region V in the south-east, towards Birchenough Bridge. Such physical differences have to some degree determined the kind of agriculture practised. Cattle and goats are relatively more important in the drier south-eastern part than in northern and western Buhera.

10. File: N/3/11/5, NAZ.

11. File: S160 DB 104/1, NAZ.

12. File: S2588/2004, NAZ.

13. This is not to say that migration from the area started during the colonial era. Already in the pre-colonial history of the area, its main inhabitants, the Vahera peoples of the Shava clan, moved to settle in other areas (see for instance: Beach, 1994; Aquina, 1965, 6).

14. Population data of this ward were obtained from unpublished documents of the Central Statistical Office, Harare.

15. For the economically active population (age category 20-65) the figures are respectively: 58.4 men per 100 women in Murambinda ward and 59.5 in Buhera district, against 91.5 in Zimbabwe as a whole. In the highest age category (65+) the difference between Murambinda ward, Buhera district and national figures was smaller (82.4 for Murambinda ward, 74.3 for Buhera and 88.9 for Zimbabwe as a whole), which may be an indication that still many men return home upon retirement. (Data sources: CSO, 1994a; 1994b.)

16. Files: N9/4/23 to N9/4/28, NAZ. These monthly reports (1910-1915) of the Native Commissioner (NC) of Charter district (of which Buhera was a sub-

district), indicate that most passes to work were issued from March to June. This is not to say that the dominant pattern of labour migration from the area was seasonal migration. People could stay away for several years. Yet, these figures indicate that migrants timed their out-going with the agricultural season.

17. The Assistant Native Commissioner (ANC) Buhera also noticed this preference to stay home instead of going out to work, even in years of food scarcity: 'The Natives themselves are in many instances to blame as they will not go out to work, thus necessitating the feeding of a far larger number than would otherwise be necessary. There is however no cause for anxiety. The natives have cattle which they can trade for food; there is also work to be obtained if they overcome their antipathy to such.' File: N9/4/43, September 1922, NAZ.

18. Files: N9/4/25, Monthly reports Charter Nov.-Dec.1912, NAZ; N9/4/43, Monthly reports ANC Buhera, July-Nov. 1922, NAZ. In periods of scarcity, the Native Commissioner Charter and the Assistant Native Commissioner Buhera often made notice of inhabitants of the Buhera sub-district exchanging cattle for grain with (European) traders or the government.

19. Harare faced a growing housing shortage from the 1940s onwards: in the period 1941-1951, 'The number of Blacks in employment in Harare (including the peri-urban areas) increased by 136 percent (..) from 32,000 to 75,000. At the 1962 census the total number of Blacks in Harare had increased to 215,810.' (Zinyama, 1993: 21).

20. File: S1542/D7, NAZ: ANC Buhera to NC Charter, dd. 5 Nov. 1937, requesting for agricultural demonstrators for the Sabi Reserve.

21. Conflicts over land generally involve village leaders (sabhuku) contesting village boundaries (e.g. boundary conflicts are at the same time leadership struggles). In the books of Headman Murambinda's court (dare), no mention is made of such conflicts over land in the 1960s and 1970s, whereas in the 1990s, these conflicts have become a common phenomenon in the area. Hence, it is argued here that growing land shortage in the district was not an uniform process. In some areas it occurred earlier than in others. Such differences may even vary from village (i.e. bhuku) to village. In Murambinda ward this is clearly the case; villages vary considerably in size and number of inhabitants, whereby the leaders of villages of non-Vahera clans generally date the problem of land scarcity earlier than their Vahera colleagues do. It needs to be noticed however, that differences in perception may also relate to the type of agricultural practices people refer to when making statements on land scarcity.

22. In 1962 the first national population census including all population categories was held. In the period 1962-1992 the population in the district more than doubled from 94,670 to 203,739. This increase implies an average annual growth rate of 2.57% (data sources: CSO, 1964; 1989; 1994a).

23. Furthermore, having land and a homestead is of major importance in case of death. Like in most areas in Zimbabwe, people have a strong preference to be buried in their area of origin (*kumusha*), even if they have stayed most of their life away from their 'home', in town.

24. A survey in four villages *(bhuku)* in Murambinda ward found that of 105 homesteads 44 had at least one brick house with iron or asbestos roofing. 37 (84%) of these were owned by (ex-) migrant workers.

25. For instance, one migrant in Harare claimed that he only remitted $200 a month in cash (for him almost a week's pay); but he also went home with a bag full of groceries and left food for his son who was looking for a job in town.

26. In good seasons during the late 1980s maize intakes at the Grain Marketing Board (GMB) depot Buhera (to which particularly producers in the north-western part of the district deliver) could amount to over 15,000 tonnes per year.

27. Like for instance hand-driven pumps to install on wells, ideas for collecting and storing water, fruit trees, etc.

28. This dominance of female farmers in the district is, for instance, also reflected in Agritex figures of graduated Master Farmers in Buhera district. For the period 1987-1992, Master Farmer graduates consisted of 69% of women.

29. As a way of solving such labour shortages, many children assist on the land before going to school, or do not go to school at all for some days during planting time. Another 'solution' to cope with labour shortages is, of course, leaving land uncultivated or staggered planting. The latter practice may, however, also result from the fact that the cattle or donkeys used for ploughing are generally weak at the end of the dry season when grazing is poor.

30. I disregard for the moment the highly politicised historical context in which such conservation policies were developed and implemented during the colonial era (for a discussion on this issue see: Beinart, 1984; Phimister, 1986; see also: Ranger, 1985; and Van der Zaag, this volume).

APPENDIX

Figure 1: Population of Buhera district, 1982 and 1992

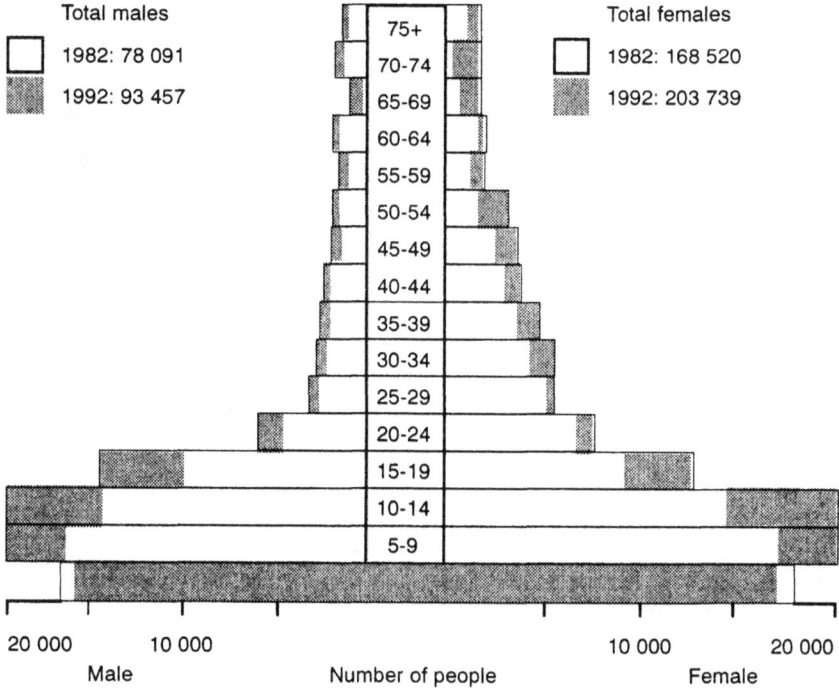

Total males		Total females
1982: 78 091		1982: 168 520
1992: 93 457		1992: 203 739

Age groups (top to bottom): 75+, 70-74, 65-69, 60-64, 55-59, 50-54, 45-49, 40-44, 35-39, 30-34, 25-29, 20-24, 15-19, 10-14, 5-9

Scale: 20 000 — 10 000 (Male) — Number of people — 10 000 — 20 000 (Female)

Comparison of the 1982 and 1992 population pyramids shows that the number of economically active men (age group 20-64) residing in the district has hardly increased as compared to the increase in the number of women. This suggests an increase in the rate of male labour migration from the area. The exceptions to this general trend are found in the age groups 20-24 and 25-29.

In 1992 many more men between 20 and 24 seem to remain in the district compared to 1982. This is probably due to the fact that secondary education has become widely available in the district during the 1980s. The number of women in the age group 25-29 has hardly increased over the same period. It seems an increased number of them has moved out of the district. This may indicate that nowadays more women seek an urban career, or that they join their husbands who work in town.

Part II
Technology Transfer

Chapter 5

Of trees and grassroots interactions
Forestry extension in Chinyika Resettlement Scheme, Zimbabwe

B.T. HANYANI-MLAMBO[1]

INTRODUCTION

Since time immemorial, trees were an integral part of, and enjoyed a central role in Zimbabwean agriculture. Trees and forests shade and shelter crops from the wind and the destructive impact of the rain, facilitate the recycling of nutrients in the soil, assist the process of soil formation itself, improve infiltration of rain water and reduce soil erosion. Forests protect watersheds and tend to stabilise micro-climates, thereby enhancing the chances of adequate rainfall for agricultural activities. In many rural areas, forests and farm trees play an important role in household food security. Here, tree or forestry products supplement crop production and household diets, as well as spread harvests across the seasons. Trees are also important to livestock through the provision of browse and through their influence on grass quality and quantity. Trees further provide the household with firewood, building materials, and supplementary income. Trees may further give households additional security of tenure where land is held under 'communal ownership'.

An issue that is increasingly becoming a matter of concern in Zimbabwe and throughout the tropics is the disappearance of trees. Deforestation becomes a problem when the rate of harvesting exceeds the sustainable yield rate. In Zimbabwe the concern raised by forest and woodland destruction started about 60 years ago and is even more acute now that the rates have reached an alarming 1.5% per annum (Gore *et al.*, 1992), compared to 0.9% for the Southern African region.[2]

Much literature on conservation forestry[3] is still abound with material that characteristically holds population pressure, with related ramifications, as the primary cause of forest destruction.[4] However, deforestation also occurs in areas with reduced population densities (Plumwood and Routley, 1982;

157

Kajembe, 1994; Mackenzie, 1992). This is what I call the apparent deforestation paradox. Other authors have looked at deforestation in terms of a lack of distinct land and tree property rights (Casey and Muir, 1986; Nhira and Fortmann, 1991, 1993), while Lue-Mbizvo and Mohamed (1993), in a study of Makoni district, critically analyzed the institutional and legal frameworks in natural resource management. According to many policy makers, ignorance is the major factor behind deforestation. The need for awareness provision makes forestry extension very important in conservation forestry efforts.

This chapter is based on the problem of continued deforestation in the context of a resettlement area despite reduced population pressure on the land and intensified extension efforts. The remainder of this introduction will briefly sketch the context of the research on which this chapter is based. The next section will look into the 'boardroom' version of the forestry extension programme in the resettlement area studied. After an account of the forestry situation in the resettlement scheme, follow sections that provide a grassroots perspective on the forestry programme. A final section analyses the institutional weaknesses apparent from the bottom-up perspective, and attempts to discern the main lessons to be learned.

Forestry conservation and resettlement: the Zimbabwean context

Early forestry conservation policies in pre-independence Zimbabwe served the needs of the mining industry. With the progression of time 'protection forestry' policies were developed, and large areas of forest land were set aside as conserved forests. Protection forestry also came to be allied to 'production forestry', where communal areas were planted with exotics to produce timber and other wood products to satisfy rural demand. During the pre-independence period, peasants were often faced with a complex, constantly changing and sometimes contradictory body of rules. Conservation programmes were usually rigidly enforced in communal areas. Farmers therefore tended to equate conservation with coercion.

Immediately after independence in 1980, government launched a resettlement programme aimed at relieving population pressure in the communal areas, declared state ownership of all communal and resettlement land, expropriated more land for the creation of protected forests and forest reserves, strengthened existing protection forestry legislation and introduced an extensive Rural Reforestation Programme. The main objective of this programme is the growing of trees, mainly gum trees (*Eucalyptus spp.*), to provide communities with sources of fuelwood and poles for construction, the idea being that if people's attention shifts to gum tree use, the few remaining indigenous forests would be saved. To achieve this feat, the programme has been coupled with a strong forestry extension component. The emphasis on forestry extension in present day conservation forestry and the existence of an apparent deforestation paradox acted as an impetus for a detailed study of forestry intervention processes at the grassroots level.

Research context

The study was conducted in the period July–September, 1994 in Chinyika Resettlement Scheme. The scheme, established in 1983 as part of the government's land re-distribution exercise, is located in Makoni North District in the north-eastern part of Zimbabwe. To date, the pioneer scheme extends over 120,000 hectares of land and stands as one of the largest resettlement schemes in Zimbabwe. The first indigenous settlers arrived in the Chinyika area in the summer of 1979, but it was only in 1983 that the scheme was officially demarcated and resettled. Some of the early settlers were included in the scheme. Others, however, were left out and are now referred to as 'squatters'.

Rainfall figures for the scheme average 800 mm per annum and half the scheme experiences uneven rainfall and suffers from mid-season dry spells. The vegetation is the mixed bush savanna type with *Julbernardia globiflora (Munhondo), Combretum hereroense (Mutechani), Brachystegia spp. (Mupfuti, Musasa)* and *Acacia spp. (Minunga)* as the dominant tree species. This ever-dwindling resource substantially contributes, directly or indirectly, to the household economies of the indigenous Maungwe people, who still comprise the majority of resettled farmers. The four dominant crops in Chinyika are maize, virginia tobacco, groundnuts and sunflower. Other minor crops include rapoko, sorghum, roundnuts, beans, rice and cotton.

The study was carried out against a background of continued deforestation in the resettlement area, despite settlers' access to land which is three to four times larger than the communal average and intensified extension services. The average arable area in the resettlement area is six hectares per household, compared to an average figure of two hectares in the communal areas. In Chinyika, households are furthermore allocated 0.5 ha for a homestead (split into 0.25 ha for a residential site and 0.25 ha for a garden or orchard), and are also entitled to a share of 0.8 ha in a communal woodlot. The extension worker-to-farmer ratio in the resettlement area is 1:200 while the respective ratio for communal areas is 1:800.

Approach and methodology

The objective of the study was to establish the factors behind the paradox of continued deforestation despite reduced population pressure and intensified forestry extension efforts, by analysing and documenting the *de jure* (official) and *de facto* (on the ground) forestry extension policy and programme objectives, instruments and strategies. The actor oriented perspective (Long, 1992) formed the theoretical underpinning of the study. Basically, this approach emphasizes the central role played by human action and consciousness and takes full cognisance of all aspects of social systems, particularly social actors and the dynamic interactions between them. These social actors are considered knowledgeable and capable. By identifying and categorising social actors, and analysing interfaces (conflicts, negotiations and accommodations) between social actors, the approach allows for a critical analysis of intervention programmes, giving insight into their strengths and weaknesses.

Field data were gathered using formal and quantitative methods such as questionnaire surveys, and informal and qualitative methods which included formal and informal discussions with informants, and anthropological techniques of recording oral histories and participant observation in case study situations. The surveys were based on the responses of 10 per cent selected households in each of 11 villages (10 per cent of the scheme's 112 villages) while the case studies focused on the activities and interactions of six representative key informants. These included a Forestry Extension Officer, an Agricultural Extension Worker, a successful tobacco farmer, an average grain producer, a poor squatter and a neighbouring communal farmer. Case studies were used as the major analytical tool while surveys provided preliminary data for sharpening of the research focus. The review of published and 'grey' literature served as an important source of secondary data and research guidance.

FORESTRY EXTENSION: THE 'BOARDROOM' VERSION

The Zimbabwean government is well aware of the problem of rural deforestation and has put in place various policies to conserve the few remaining forests and initiated programmes to counter and reverse the deforestation problem. Government's interest in conservation forestry is largely (though not exclusively) directed through the Forestry Commission, a parastatal agency within the Ministry of Environment and Tourism. The agency is responsible for the development of general forest policy in Zimbabwe in consultation with its parent ministry. The present policy of the Forestry Commission is divided into three main areas: (a) regulation, (b) giving advice on forestry matters, and (c) commercial forestry activities (Katerere *et al.*, 1993). This entails the Commission to undertake a variety of non-commercial activities (in addition to its commercial operations) such as forest conservation, forestry research and forestry education. The Forestry Commission also implements the state-initiated Rural Reforestation Programme.

Forestry research

The Rural Reforestation Programme initially concentrated on the planting of three *Eucalyptus* species — *E. grandis*, *E. camaldulensis* and *E. tereticornis*. Trial species were to include other *Eucalyptus* species, various *Acacia spp.* and others, with emphasis on multipurpose species with coppicing ability and agroforestry potential, and indigenous species where possible, as well as fruit and fodder trees. Also, given likely problems of acceptability of *Eucalyptus* fuelwood and potential problems with its integration into agroforestry systems, further research was to concentrate on the suitability of other species. Due to the high initial degree of dependence on the *Eucalyptus* species, a final research area was on its specific pests and diseases. It was proposed that through project monitoring, and less via formal research, the social acceptability of the *Eucalyptus* species could be determined. Specifically, research was to be centred on village wood users' responses to the types and quality of wood provided by the fast growing *Eucalyptus* species and to the way in which it is planted and managed.

Forestry extension

As outlined in the Rural Reforestation Project document, the overall success of the project in achieving its objectives and attaining its production goals was going to depend on the efficacy of the extension services provided (Forestry Commission, 1982). The main aims of the extension effort were:

i) to promote tree planting and management as an integrated part of agriculture;
ii) to develop and maintain farmers' awareness of the needs to increase rural wood supplies and to conserve indigenous wood resources; and
iii) to develop and maintain an awareness of the need to use wood efficiently.

Specific extension work was designed to include: advising farmers on the planting and tending of *Eucalyptus* woodlots with focus on the principles of sustained yield, coppice management, etc., and the establishment of this and other species in other contexts, such as windbreaks and in agroforestry. The task of making farmers acknowledge woodlots as a long-term crop with an added financial advantage was incorporated as part of the extension programme. It was also expected that much of the extension thrust was to be directed towards rural women, who are traditionally the collectors of firewood and poles from the forests. However, Master Farmers, because of their perceived comparative affluence and better knowledge and understanding of agriculture and the need for conservation, were to be the prime targets for extension.

According to the Rural Reforestation Project main report, demonstration woodlots were to be set up near Forestry Commission-established nurseries as part of the extension service to demonstrate to resettlement and communal area farmers the correct methods of plantation establishment, maintenance and exploitation (Forestry Commission, 1982). In addition, such woodlots were intended to incorporate research trial plots, as part of the Forestry Commission's research into the selection of species and management techniques suitable for rural woodlots.[5]

A great deal of extension work was to be targeted towards schools, both to students and their teaching staff. Students were to be taught how to grow seedlings from seed, how to care for them, how, when and where to plant the trees, and to subsequently manage them. They were to be encouraged to grow trees around the school, with the hope that they continue this practice outside the school environment and in later life.

Extension services were to be provided by the Forestry Commission's Forestry Extension Officer, Provincial and District Woodlot and Nursery Managers, while Nursery Supervisors, with the help of agricultural extension workers from Agritex, were to chip in by using Forestry Commission nurseries as focal points for forestry extension to rural dwellers. It was hoped that through these combined efforts, peasants would 'learn of the value of trees' and 'the need to integrate tree growing with agricultural practices, where rural farmers begin to perceive trees as a crop to be managed as a part of their farming routines' (Forestry Commission, 1982).

The crucial process of monitoring and evaluating the effectiveness of extension was to be carried out jointly by forestry extensionists and agricultural extension workers, with the use of reports from the Woodlot and Nursery Managers and extension agents, through normal channels of authority. Proposals were that such monitoring and evaluation could be carried out through the extension organisation by employing the Training and Visit (T and V) method of extension, where regular training, feedback and reporting was to be undertaken as a matter of routine. Liaison between the top level officials of involved divisions was to enable the implementation of necessary changes.

Throughout the life of the reforestation programme, the demand for seedlings, the areas of woodlots planted and the care given to woodlots were to indicate how effective the extension programme was. In addition, the effectiveness of extension was to be evaluated by monitoring physical performance, goal achievement and the extent to which forestry and agriculture were being integrated.

DEFORESTATION AND AFFORESTATION IN CHINYIKA

The deforestation reality

Compared to surrounding communal areas, Chinyika Resettlement Area is still moderately forested but evidence from 1975, 1978, 1986 and 1987 aerial photographs, combined with physical inspections during 1994, and stories from the locals tell a different story. The woodland has been considerably thinned since 1975. As would be expected, the least altered vegetation cover is on the mountains. On small and large hills, the plant population has in many cases been reduced from forests to woodlands or mere shrublands. The evergreen vegetation which used to grace river areas and important in stabilising the river beds, has also been victim to this thinning. However, the hardest hit areas are the ones under settlement and the plains where the vegetation cover has been almost entirely cleared for cultivation. Survey data collected on settlement and cultivation trends over the past 10 years show a positive correlation between demographic increases, increases in the area under cultivation and the rate of deforestation. This large scale clearing has especially been so with the 'clean field' craze, an idea instilled into the people since the late 1920s when the ex-missionary Emory D. Alvord introduced his 10 rules for permanent agriculture. One of these rules clearly stipulated that there should be thorough stumping and clearing of fields to ensure easy tillage (Alvord, 1958). Furthermore, though farmers do not openly say it, a lot of them have extended their fields over the allocated six hectares meant as arables into the reserved grazing areas and areas demarcated as waterways, a situation aggravated by the subdivision of arables to cater for children and squatter relatives.

Apart from land clearance for cultivation purposes, a host of village level activities contribute to deforestation. To begin with, for most rural folk, wood remains the most accessible form of raw material for their industries. The Provincial Development Officer for Manicaland once commented, 'The problem is that of wood based technology, where you have to destroy a tree for you to proceed' (Nkomo, pers. comm.), referring to hut construction, brick making, homestead and garden fencing, crop and food processing and preparation. However, it is the last two activities, fuelwood collection for food preparation and crop processing in the form of tobacco curing, which are causing a stir in the resettlement area.

Also disturbing is the illegal harvesting of fuelwood by people from neighbouring communal areas and the commoditisation of the woodlands. The 'stealing', as local residents put it, of fuelwood by villagers from neighbouring communal areas is especially rife in resettlement villages that are bordered by these communal areas while the selling of fuelwood is a more widespread occurrence. What started as a survival strategy during the 1982/83 drought period has become a booming trade in wood today.

Reforestation realities

The first phase of the Rural Afforestation Programme actually took off the ground in Chinyika in June 1983. For woodlot establishment activities, the recommended three *Eucalyptus* species have been and are still being planted. About 80 per cent of the resettlement villages have established communal woodlots averaging 1 ha in size. Within these villages, 70 per cent of residents are engaged in communal woodlot establishment and management activities, while only 30 per cent of Chinyika residents own individual woodlots (where a woodlot comprises anything from a few trees to above and beyond). But the great majority of the residents (at least 90%) have at one time or another attempted to establish an orchard and/or woodlot though most said they gave up after the trees were destroyed by livestock or termites.

Looking at current figures, the average area under woodlots and orchards is 0.124 ha per household. This figure falls far short of the officially recommended planting rate of 0.201 ha per household annually during seven years. The situation is even worse when one considers that the field statistic of 0.124 ha is a current combination of areas under orchards and woodlots whilst the recommended figure was meant for woodlots only, and recommended for each year (in a seven-year rotation). In addition to a failure to meet the planned and projected reforestation rate, the resettlement villagers' rate of tree planting is much lower than the rate at which indigenous trees are being destroyed.

The establishment of traditionally planted fruit trees and exotic orchard trees is, however, progressing at an encouraging rate. Unlike woodlots, where establishment is usually a result of encouragement and pushing by Forestry Commission, the Natural Resources Board (NRB) and Agritex agents, orchard tree planting was mainly a product of individual initiative, often done immediately after resettlement. Fruit trees established in orchards include mango, guava, paw paw, *muzhanje (wechingezi,* Mexican apple), oranges, lemons, grapes, bananas, tamarind, mulberries, granadilla, peaches and avocado pears.

Low reforestation rates can be attributed to a number of weaknesses in the reforestation programme. According to the Forestry Extension Officer, the major hiccup met by the programme is that of failure to generate enough interest in farmers. Farmers are, however, aware of the need to plant trees and are interested in the reforestation programme. The central problem, though, stems from not involving the local people in the planning stages of the programme and lack of a strong feedback mechanism in the extension system. One offshoot of this is the assertion that the gum tree is the best bet. However, peasants' value for trees goes beyond just that for fuelwood or poles which explains why they prefer indigenous and general orchard trees instead of gum woodlots. Some farmers believe in maintaining the landscape with indigenous trees, and loath the idea of changing the natural tree vegetation for pines and gums. According to Kunaka (1990) and discussions the researcher had with farmers, gums have a number of negative characteristics:

i)   Gums are water guzzlers and in some areas they lower water tables appreciably.

ii)   The gum requires top class soil, non-sodic and at least one metre in
      depth and a rainfall of not less than 500 mm per annum. Incidentally
      these are the soils for maximum crop production and not many farmers
      would sacrifice their croplands for gums.
iii) Underneath gums, cereals, groundnuts, *brassicas*, etc. do not grow well.
      With limited crop lands, a tree which does not have a suppressive action
      on crops is clearly the better choice. Gums may also be directly toxic to
      some crops.
iv)   On combustibility, the eucalyptus fare badly as they burn faster,
      consuming more wood, compared to indigenous trees.
v)   Gums do not provide strong poles as the *Terminalia pericopsis*, *Dichrostachys*,
      Mopane, etc. Gum poles also show no resistance to weevils. In carpentry,
      gums do not form good heads, for example, for axe handles.

Reforestation agencies have not done enough to support individual farmers
with the establishment of orchard and indigenous trees. Planting material
for some orchard and indigenous trees is also not readily available to farmers.

FORESTRY EXTENSION AT THE GRASSROOTS: THE FARMERS

Now that we have a general picture of the deforestation problem in Chinyika,
I proceed with a more detailed assessment of forestry extension. This requires
an understanding of the role played by the various actors involved. The two
main intervening actors in the forestry programme in Chinyika are agents
from the Forestry Commission and from Agritex. The group of 'beneficiaries'
consists of the resettlement farmers. This is, however, not a homogeneous
group. According to the farmers' own categorisation, there are basically three
types of farmers, namely: tobacco growers, grain producers and squatters.
According to the interventionists' perspective, however, farmers fall into
two broad groups: the innovative or receptive farmers, whom they tend to
concentrate on, and the uncooperating or 'lazy' farmers, often neglected by
extension agents. To escape from the interventionists' simplistic and
dichotomic labelling and also from the simplified farmers' categorisations,
here I wish to embrace both types of classifications. In this section I will
introduce four farmers (a tobacco grower, a grain producer, a squatter, and
a neighbouring communal farmer). In the subsequent section I turn to the
institutional environment in which a forestry extension officer and an
agricultural extension worker have to implement the forestry programme.

(a) The tobacco farmer

According to the survey I carried out, 19% of Chinyika farmers grow tobacco. These tend to be largely the younger farmers (in their 30s and 40s) who are into virginia tobacco growing mainly because of the attractive producer prices and larger gross margins compared to other crops. High earnings from this crop mould this group of residents into a class of their own. However, because most tobacco farmers still rely on fuelwood for the curing of their crop (as against coal), they are largely blamed for the deforestation going on in the resettlement area by both civil servants and fellow settlers. Ideally, all tobacco farmers are supposed to establish at least one hectare of gum trees for their future wood requirements. This group is also encouraged to take up special loans from the Agricultural Finance Corporation (AFC) to be used in the purchase of coal which they are advised to use as stand-in fuel while the gums grow. Despite all these persuasions, only a handful of tobacco farmers have established individual woodlots; even then, they did so more out of community pressure than out of individual desire. Considering the coal alternative, up to 90% of tobacco farmers bought coal but less than 10% of them are using it. For the majority, the loads of coal strategically placed next to their tobacco curing barns are just for window-dressing, i.e. to maintain good relations with the extension workers, other farmers, and protect themselves from the Natural Resources Board (NRB). To understand better this paradox, I carried out a case study of a tobacco grower.

Mr A. was born on a commercial farm in Rusape in 1955. He never had the chance to go to school and spent most of his early years working on the farm. 'This got me interested in farming, but not in trees,' he remarked. In 1981 he and his wife moved to the Chinyika area as squatters, getting a stand and field in 1983. Mr A. remembers that they started like most other settlers, with no cattle. Agricultural income helped them to buy cattle, establish a grinding mill, a store and a beer outlet within 10 years after arriving in the area. Mr A. furthermore managed to get himself three more wives in the process. Today he stands as an example of a very successful tobacco grower.

In 1984, through extension worker encouragements, Mr A. established a 0.6 ha gum woodlot in part of his field. The objectives then were to create a future source of woodfuel, poles, and cash through wood and pole sales. Ten years down the road, he is only using his gum produce in his tobacco barns as cross poles on which tobacco is hung, while wood from indigenous trees still fuel his barns. In questioning why he and fellow farmers were shunning away from gum and coal use, the following was revealed to me. First, farmers are well aware that gum wood quickly burns out in the 'oven' and leave no embers unlike wood from indigenous trees whose embers can provide heat long afterwards. In trying to free themselves from a general agricultural price squeeze, farmers are also putting less and less financial resources into their production, an example of which is the use of indigenous wood with no financial costs since resources from the forests are conceptualised as 'free'. On the other hand, coal use requires large investments in buying the coal, getting ash collecting rails and vents.

On woodlot establishment, Mr A. and most other tobacco growers said, the permits they had meant that the land was not theirs which made them feel insecure. Another snag is the fact that woodlot establishment is not something born out of the farmers' own initiative but something that was introduced by the government. Farmers thus perceive them as the government's, which is why there are cases of woodlots labelled as the extension worker's, Forestry Commission's, or Agritex's. Actually, the general impression given by a number of tobacco growers who did not have individual woodlots, those who have attempted and failed and those who have them now, is that raising gum woodlots is doing the government a favour. They do not perceive it as something for their own good or benefit. On the other hand, the state, represented by agents of the Forestry Commission and Agritex, conceptualise gum woodlot establishment as being in the long term interests of the people. This in itself is the beginning of an 'interface' between differential perceptions. This finding may also shed light on why many established woodlots are grossly neglected.

(b) The grain producer

Grain producers constitute the majority of settlers in Chinyika Resettlement Scheme. Many of them are elderly men and women who cannot cope with the technical and labour requirements of tobacco growing. All in all, this group looks down upon tobacco farmers whom they blame for the destruction of local woodlands. It comes as no surprise therefore that the general opinion among grain producers is that there should be sterner measures against wood cutters, as a solution to the resettlement's deforestation problem. However, this group of grain farmers, like all other farmers, are also involved in forest destruction. Having migrated from communal areas where the lands are almost devoid of vegetation, many perceive the vegetation as still plentiful. In other words, settlers have a difficulty in visualising a situation where the resettlement area will be deforested like the surrounding communal areas.

Few individual tree planting activities by members of this group were observed in the field. Most farmers argued that this would imply diverting crop land for woodlot development with a negative impact on the family's food security. Some farmers have gone around this problem by establishing woodlots at their homesteads, a feat enabled by 'illegally' extending their homestead. In the survey that I conducted, crop activities proved to be highly prioritised by all farmers with livestock farming and tree growing (and conservation in general) categorised as 'necessary but not basic'. This aspect of priorities also explains why most grain producers (and other farmers) complained of woodlot destruction by livestock, stressing the need for fencing material while doing nothing about it. Excuses have always been that of lack of money. Though some cases are entirely sincere and genuine, it is rather questionable for some successful grain producers (*hurudza*) who sell about 30 tonnes of maize and make a gross income of up to Z$28,000 per season (an income greater than the annual average for formally employed people — 1994 figures). Because trees, unlike crops, take a long time to mature, tree growing activities also have a major economic disadvantage in that they do not bear immediate benefits.

However, some grain producers may be considered exceptional, in that these farmers set up their own woodlots and orchards even before the intervention by the Forestry Commission and Agritex. Constantly strategising to maintain and upgrade their orchards, they nurture their relationship with the extension workers, taking advantage of the assistance that is available. Their demand for services dove-tails with extension workers' inclination to favour such innovative or receptive farmers.

One such a farmer is Mr B. He has a 0.5 ha orchard, up to one hectare of gum trees and raises his own seedlings from scratch. Born in a rural village near Rusape in 1920, Mr B. did his primary education and spent most of his early life in that village. In 1945 he went for further education at a mission school in Old Mutare. There he became interested in exotic trees during part-time jobs to raise his school fees. During the holidays he 'cultivated' his newly found interest by planting numerous exotic fruit trees in the family orchard.

At the end of the war, Mr B., who now had a family of his own, migrated and settled in Chinyika. In 1983 they were allocated a field which they use for grain production and a stand where they built the homestead. Mr B. says he feels quite tenurially secure. Apart from the gums on his two woodlots (one at the homestead and another at the edge of his field) Mr B. raises mango, guava, orange, lemon, tamarind, peach, paw-paw, banana, mulberry and avocado trees in his orchard. Trees at his homestead also include the jacaranda, *Mupfuti*, *Mutohwe* and *Mukute*. These trees provide the family with shade, fruits and wood. What makes Mr B. exceptional is the fact that all trees (both exotic and indigenous) were planted as seedlings from his own nursery. His experimentation with indigenous trees impressed the Forestry Commission. Mr B. admits that the 'achievement' path was not all rosy, since he had to continuously adapt to changing conditions and further strategise when new aspirations arose.

Individual orchard and woodlot establishment is hampered by severe competition from crops (limited land), water shortages, attacks by insects and livestock (especially during the gum's early stages of growth), as well as the time and labour constraints. Innovative farmers go around these problems by establishing individual woodlots at their homesteads made possible by 'illegally' extending their homesteads (into the grazing area), collectively raising capital through savings clubs, digging and establishing their own wells as well as collecting waste water for trees, and working in locally initiated mutual help groups which give them time for forestry activities.

Mr B. developed his own way of combatting termites in his trees. Technology development through trial-and-error experimentation and integration of new knowledge thus has always been a necessary part of his farming life. Such locally generated knowledge is disseminated through farmer-to-farmer interactions. Mr B. is part of a small network of four farmers who share and try out each other's discoveries. At the same time, experimenting farmers such as Mr B. maintain good links with forestry and agricultural extension agents. They pragmatically believe that new and old ideas and values should be incorporated into forestry and other agricultural programmes. A problem I experienced with farmers such as Mr B. is that they emphasise their own ignorance and that they have to learn a lot from extension workers, to the point of losing some of their own initiatives. Possible explanations of such a showing is the humble nature of the resettlement farmer and the top-down approaches that have been used for so long by extension agents.

(c) The 'squatter'

Squatters are people living on lands which were not allocated to them. They are known locally as '*vakaberekwa*', which literally translates to 'those on other people's land'. These 'illegal' inhabitants are often found on land demarcated for grazing, usually far from the villages or usable roads. The numbers involved are much larger than the official estimates suggest. Most of the squatters tend to be poor, a condition aggravated by their limited access to land and the fact that what they are allocated by their 'accommodative' relatives or what they manipulate for themselves is not the best of lands. Because of limited land, many of them tend to be grain producers who cannot dream of the luxury of tobacco growing and virtually no squatter possesses an individual woodlot or partakes in communal woodlot establishment-cum-management activities. Just like tobacco growers, squatters are heavily blamed for the deforestation that is going on in the resettlement scheme, only this time the blame is mainly from civil servants who feel that there is little control over these inhabitants.

Mrs C. is a 'squatter' living on a portion of her uncle's stand. Born in 1952 in Chiduku Communal Area near Rusape, Mrs C. and her husband came to the resettlement scheme in 1981 during the time when a lot of current residents also settled in. They failed to secure residence when the scheme was officially demarcated in 1983. The family moved and squatted in a village in the south-eastern part of the scheme. Their home, together with those of other squatters, were destroyed twice by government authorities. It was after the second 'attack' that Mrs C. decided to seek refuge at her uncle's place. Frantic efforts to get officially resettled in Chinyika or an alternative resettlement scheme have proved fruitless.

Mrs C. said she feels bad about the seclusion of the so-called 'squatter', the use of that epithet and the unfair blame they get for causing deforestation. Squatters are excluded in locally organised groups since these are only reserved for farmers with resettlement entitlements. She also had never had any contact with the local extension worker whom she blamed for shunning squatters. As concerns their association with deforestation, she said it was simply a matter of trying to survive. In her perception, squatters have in general simply taken over 'vacant' and unused lands.

Mrs C. acknowledged that tree planting might give them more secure tenurial niches but she doubted this because of the constant insecurity all squatters now have. She however said she would plant and manage trees if

the land belonged to her or if she had individual control over it. To her, the
whole problem boils down to the issue of uncertainty.

(d) The Communal Area farmer

Farmers from neighbouring communal areas are not residents in the
resettlement scheme but nevertheless actors in conservation forestry since
they are very much part of the conservation problem. Much of this
involvement is in the form of illegal[6] harvesting of wood (both poles and
firewood) from resettlement woodlands.

As noted earlier, the selling of wood has become a booming trade. This is
especially facilitated by the presence of a ready market in the form of civil
servants housed at Chinyika's eight Rural Service Centres, policemen and
soldiers at the former Inyati Mine, policemen from Mayo Resettlement
Scheme (where fuelwood shortages are already acute) and people from the
Social Welfare Department who buy wood to provide fuel to refugee camps.
Trees sought after are mainly the *Julbernadia globiflora* (*Munhondo*), *Brachystegia
spiciformis* (*Musasa*) and *Brachystegia boehmii* (*Mupfuti*). Unlike cases with normal
fuelwood collection where women do most of the job, men are exclusively
involved in this wood trade. This activity is especially rife during the winter
months.

According to the discussions I had with Mr D., a Chendambuya villager,
who saw me as a potential fuelwood customer, most of the collected wood
is for household fuelwood use with occasional 'special' trips for fuelwood
he sells to teachers at schools in the area. Part of his harvest finds its way to
larger fuelwood markets of Rusape, Marondera, Mutare and even Harare.
Some fuelwood traders from these urban areas also induce communal (and
resettlement) farmers to cut and sell them firewood. Compliance in most
cases is a matter of sheer poverty. Resettlement residents feel that communal
farmers are just being stubborn and pushed by a misinterpretation of 'new
found freedom' (being independent). On the other hand, communal dwellers
argue that there are virtually no trees left in communal areas and perceive
themselves as having an equal right to trees in resettlement areas. The latter
argument is quite strong since they have always had (some) access to these
tree resources, even during the colonial period when these areas were under
private farms (though the legality of this access is another matter). Usually
overlooked is also the resentment Maungwe communal area farmers have
towards especially 'newcomer' resettlement people since they expected to be

given first priority in the scheme. Fuelwood harvesting from the resettlement scheme for them, the acclaimed owners of the land, is then a question of exploiting what traditionally belongs to them.

FORESTRY EXTENSION AT THE GRASSROOTS: THE EXTENSIONISTS

Having provided sketches of four farmers and how they deal with forestation, I now look into how two extensionists from the Forestry Commission and Agritex attempt to implement the Rural Reforestation Programme.

(e) The Forestry Commission extension officer

Work in Chinyika Resettlement Scheme was in three stages. The first stage involved 'awareness' campaigns through short courses and one-day workshops, and the training of Forestry Aiders who then became helpers to the Forestry Extension Officer, Mr E. This was followed by the establishment of satellite nurseries and the consequent setting up of community and individual woodlots. The third and final stage of the programme involved general management of the established woodlots before these were handed over to village communities and schools for management, with the Forestry Commission continuing to give marketing and technical advice. Mr E. is the only Forestry Commission worker for the entire Makoni North District. The district comprises three resettlement areas — Chinyika, Gwindingwi and Mayo, and three communal areas — Tanda, Chendambuya and Chikore. This makes Mr E.'s task a cumbersome one.

Mr E. told stories of the first (unofficially also the last) general awareness training courses he held for groups of farmers. These farmers were identified by agricultural extension workers from villages around Chinyudze and Bingaguru. Mr E. also recounted nursery establishment activities, the low-key interest shown by farmers during times of free and heavily subsidised seedlings and the daunting task of village woodlot establishment when they proved to be not so popular with villagers. This gave him the impression that farmers had no interest in rural reforestation. Mr E. is well aware of the tendency in most farmers to resist any message that comes from the government. He calls this an 'in-born nature' which he attributes to the enforcement of conservation during the colonial era. This, transport problems and the fact that there is only one forestry extension officer for the entire district, encouraged him to work through agricultural extension workers and the forestry aiders he established in some villages.

Although the Rural Reforestation Programme blueprint stipulated Woodlot and Nursery Managers and Supervisors as providers of extension services at ward and village levels, these never existed at the grassroots level. Thus, the establishment and training of forestry aiders is both a local initiative and a survival strategy of the incumbent forestry extension officer. According to Mr E., these forestry aiders seem to be effective in executing conservatory extension work and rekindling the reforestation drive since the settlers respect each other more than an 'outsider'. Rewards for the forestry aiders is by making them feel important which comes in the form of introductions and letting them make speeches on his behalf at big events such as ministerial visits. This way, Mr E. says, he is assured of their cooperation. His other major instrument is by spreading the 'message' through schools.

Despite the commission's stated commitment to interactive extension work, real participation and incorporation of indigenous knowledge is still at 'preaching' stage at the ground level. The use of demonstration woodlots and nurseries as focal points of extension is also not evident on the ground. Moreover, the Forestry Commission has to enforce provisions of restrictive legislation such as the Forest Act and encourage farmers to plant trees, at the same time. This gives the forestry extension officer conflicting regulatory and extension roles, with detrimental effects on the whole forestry conservation programme.

(f) The Agritex extension worker

As earlier highlighted, the Forestry Commission is very thin on the ground. Thus most of the groundwork falls into the hands of Agritex, the extension arm of the Ministry of Agriculture. This department is therefore the *de facto* implementer of the Rural Reforestation Programme. In Chinyika, Agritex is staffed by three Agricultural Extension Officers, three Agricultural Extension Supervisors and 21 Extension Workers. The mandate of Agritex in the resettlement area is to provide general extension services and train farmers in the use of new technology, all aimed at attaining and maintaining high levels of production as well as ensuring that resources are conserved. Taking a closer look at conservation, extension workers, who are the frontline staff in Agritex, emerge as the key social actors within the department.

Born in Nyamaropa (Nyanga) in 1959, Mr F. developed an interest in agriculture at a very early stage. As a youngster he was involved in grain, burley tobacco and cotton production which was carried out in the family fields. Driven by this early interest, he enroled for agricultural training in Esigodini. Mr F. joined Agritex soon after training and was posted to Chinyika as an extension worker, where he got involved in the planning, surveying and land demarcation exercises that were going on in early 1983. His current extension work covers eight villages, which gives a ratio of about one extension worker to 130 families (households). This is better than the original planned ratio of 1:200.

Within the villages Mr F. is preaching the gospel against deforestation which he blames on the farmers' irresponsible ways. He helps with the establishment of individual and community woodlots by urging and identifying village groups interested in woodlot establishment, encourages the growing of fruit and ornamental trees and is engaged in what one can call 'reactivation' of agroforestry, especially the use of the *Leucaena spp.* Conservation work is also marshalled through Agritex's Master Farmer Training Programme which has 'orchard establishment' as one of the conditions to be met by trainees. As an individual, Mr F. also sends contributions to agricultural education radio programmes in which he highlights success stories from his area. He furthermore organises field days where the 'exemplary' farmer is given a platform to share experiences, as a way of encouraging other villagers or settlers.

At times, contradictions between what the extension agency stipulates as objectives and what it recommends hampers progress. An illustration here is Agritex's desire to help the villagers put vegetation back on the land and the promotion of clearfelling and destumping of fields to facilitate the tillage by tractors.

Mr F. feels that the farmers' interest in conservation is very low, hence a need to constantly encourage them. In cases where farmers show interest, like the planting of *Leucaena* trees for fodder provision, programme implementation is very slow. He admitted being constantly pre-occupied with time schedule worries, since extension workers are required to follow self-drafted annual workplans. Work progress is reviewed at the end of each month, the result of which is put in writing. However, the quantity of reports is usually more important than their quality, in a system where report writing has become more a routine than a basis for evaluating progress.

According to Mr F., his task as an extension worker is to 'educate or teach farmers crop and animal production, how to establish and manage viable projects and training Master Farmers trainees'. The extension worker thus perceives his own role primarily as an educational one. However, the extension workers have many other roles, including reporting conservation-regulation offenders to the NRB, assistance in drought relief and drought recovery programmes, enrolment in population census and voters' registration, and helping farmers in acquiring inputs.

In the field, faced with farmer 'resistance', multiple tasks, limited mileage and general transport problems in addition to low salaries and subsistence allowances, and poor housing, Mr F. feels that he has preached the gospel well enough and now has 'resigned' — i.e. he concentrates only on the 'good receptive' farmers. In trying to meet organisational expectations in the face of field limitations, Mr F. has devised his own approach to local level extension. In a nutshell he reviews past methods and explains his current strategy thus:

> The contact farmer method is not working but trying to cover every farmer is practically impossible; so my strategy has always been working with village-based groups, where all residents in a target village constitute a group.

What started as a secret field strategy of an individual extension worker has now been adopted and recommended as an experimental provincial extension strategy. The approach has been dubbed the 'Blocked Village Based Extension Approach' (BVBEA). Here, an extension worker is encouraged to concentrate his/her efforts (80% of time and resources) on one village for a period of time — one or two years; after which the extension worker moves to another village.

Trying to know the people, the area's constraints and potentials before an intervention and involving the farmers from the word go, are the major strengths of the approach. However, in practice farmers are still treated as passive recipients and their own wealth of ideas and knowledge are rarely incorporated. Furthermore, statements like, 'It is part of my job to identify farmers' problems and find solutions to these' show and instil such an attitude in extension workers.

Despite an overall negative picture about extension efforts, Agritex has achieved a lot in the resettlement scheme. Of significance to conservation forestry is its success in the promotion of private orchards, and individual and communal woodlots. Remarkable work has also been achieved in assisting farmers to arrest (reduce) soil erosion, and increase the resettlement's agricultural output. The relative success of Agritex compared with other intervening agencies, can be attributed to decentralised decision-making, individual innovativeness and the presence of a local-level network of agents.

THE PARADOX OF DEFORESTATION AND FORESTRY EXTENSION: A REVIEW

Unlike the much anticipated simple intervener-recipient, or extension system-target group scenario, the foregoing reveals a complex multi-actor situation in Chinyika. In reality, the situation is even more complex: also institutions such as the Natural Resources Board (NRB),[7] the Department of Rural Development (DERUDE),[8] local government organisations,[9] schools,[10] and a number of non-governmental organisations take part in afforestation activities, but cannot be dealt with in any detail here because of lack of space.

In this concluding section I will first attempt to deconstruct the deforestation paradox in resettlement schemes from a farmers' perspective. The assumption of reduced population pressure gets a different dimension by considering (tree) resource use on the ground and focusing on prevalent levels of insecurity of tenure. Failure to address these issues severely limits the effectiveness of intensified forestry extension efforts. I will further show that underlying assumptions of extension agencies tend to overlook and ignore farmer diversity and poor livelihoods, again resulting in low levels of effectiveness. I will then briefly review the extension approach followed in Chinyika regarding afforestation, and specifically look into its limitations. I will conclude this section with some recommendations for a change in approach.

Deconstructing the deforestation paradox

First of all, it should be noted that many farmers share a sense of insecurity with respect to their tenure in the settlement scheme. This is most obvious in the case of squatters, but also at the other extreme of the spectrum, with the entrepreneurial tobacco growers. The latter are reluctant to invest in long-term, capital-intensive projects. A related issue concerns access to forestry resources in the resettlement area: most actors involved perceive access to tree resources as entirely 'open'. This means that in effect no farmer feels he/she owns it, hence nobody (except some state institutions) assumes responsibility for it. This gives rise to the free-rider scenario with respect to indigenous woodlands in Chinyika.

If the above finding is correct, parachuting as many extension messages as possible may not solve anything, as the basic unfavourable boundary conditions (ill-defined rights to land and forest resources) remain unchanged. Making suggestions to resolve the above tenurial issues falls outside the scope of this chapter; but is nevertheless urgently needed. This also demonstrates the limitations of forestry extension, and should be borne in mind in the remainder of this chapter.

The thrust of extension efforts have shown a bias against squatter groups who cause extensive environmental damage by illegally settling and cultivating on grazing lands. They are ignored by the officials and extension agents for political and 'official' reasons. Official, in the sense that officially squatters do not exist in the resettlement scheme. This tendency continues to reinforce underestimations with regard to population pressure on (tree) resources in Chinyika resettlement scheme. Thus one end of the deforestation paradox, that of assumed reduced population pressure, can be declared void.

Women farmers, whom the extension component of the Rural Reforestation Programme greatly intended to target, are also shunned by extension services. This continues despite the fact that women are the major agricultural producers and traditionally the collectors of firewood and poles from the forests, and are among the first to experience the consequences of deforestation. Obviously there is a paradox here: Although tree planting may be in women's immediate interest, in most of the households interviewed homestead trees are perceived as men's property and concern. This is seen to be in line with the fact that, in most resettlement villages, men are the *de jure* heads of households, they are registered on the permits and thus control most of the family's farm resources. The resettlement scheme's non-accommodative stance for secondary rights force women to perceive resettlement as relatively insecure, compared to for instance their communal area counterparts. Here I return to the tenure issue with which I started this section. Extension efforts alone are unlikely to be sufficient to resolve this problem.

Turning to the intensified forestry extension efforts, the other end of the deforestation paradox, I observe that the efforts can be characterised by a number of unfounded assumptions.

First, we have seen that, unlike the ignorance assumption held by many interveners, most farmers are concerned about their environmentally unfriendly activities and the dangers of land degradation. Evidence of the effects of removal of natural tree cover in the resettlement area, as observed by farmers, include the reduction in overall rainfall amount and reliability, lowering of water tables, reduction in river flows especially in the dry season, and an increase over time of surface run-off and soil erosion. The existence of words such as '*gwenga*' in the vernacular, which means desert, also portrays farmers' awareness of the hazards of deforestation. Despite the diversity of farmers and their strategies, there is a general consensus that deforestation persists because of a lack of alternatives.

Afforestation is actively pursued by some. However, most farmers find that this should not be expected of them since their land base is limited; they do not own the land, they are short of labour and lack financial resources. It is important to note that woodlot establishment always requires a minimum financial outlay for the purchase of pignet wire for fencing. Meanwhile, farmers continue to strategise to keep going, be it in terms of securing their tenure in the scheme, maintaining a good image to the rest of the community or simply in meeting their felt needs.

Second, farmers perceive outside institutions pushing conservation and forestation programmes in a particular way. This perception is largely informed by how the colonial state coerced farmers to follow conservation regulations, but is reinforced by the tendency of state agencies to follow similar heavy-handed and top-down intervention strategies in the present day. In this context it becomes understandable that farmers establishing a woodlot may think that they do the state a favour rather than themselves.

Third, although farmers share a general perception on conservation issues, these differ in detail, depending on the minute differences in individual farmers, their resource base and their farming strategies. The introduction or imposition of standardised solutions that do not take into consideration this diversity nor the complex nature of the local context results in differential impact and problems of poor adoption.

As we have seen, the affluent and receptive farmers receive relatively more attention from extension services. Poorer farmers, though, may share many of the aspirations of their wealthier colleagues, but they simply cannot achieve these aspirations out of sheer necessity to first survive before being able to embark on more innovative farming strategies. As a result, many of the poorer farmers indeed do not establish woodlots, which gives rise to extensionists labelling them as lazy farmers or *nyopes*. This labelling process presents a typical case of 'double hermeneutics', and has a tendency to become a self-fulfilling prophecy: outsiders attribute meaning to a local setting, the interpretation of which is fed back into the community through various ways (e.g. labelling), is re-interpreted by the people concerned and consequently affect their behaviour.

Limitations of the forestry extension approach

Forestry extension by the Forestry Commission and Agritex, who perceive their role as educators and awareness providers, is based on the Transfer of Technology (TOT) model. In the model, participation at 'stages' of the technology development process, such as development, dissemination and utilisation, is restricted to a specific set of actors — researchers, extensionists and farmers respectively. Technical recommendations are still dispatched from one group of actors (researchers) by way of another (extensionists) to the utilisers (farmers). Here, an almost linear (one-way) relay of information occurs, in a system which assumes the target group, the so-called utilisers, as passive recipients of recommendations.

Past research efforts (for example, Drinkwater, 1989), and failures with the World Bank funded Training and Visit extension system, established weaknesses of this classical diffusionist model. The lack of a feedback mechanism in the system and the straight-jacket kind of research has resulted in non-incorporation of indigenous knowledge in research and extension, inappropriate recommendations and low adoption levels by individual farmers.

Notwithstanding the TOT model having been changed at grassroots level and made more effective, farmers are still treated as passive recipients. The fact that extension staff see themselves as solving problems on the farmers' behalf reflects a deeply ingrained top-down orientation in the organisational cultures of agencies such as the Forestry Commission and Agritex. Furthermore, agricultural extension workers feel that the farmers' interest in conservation is very low, hence the need to constantly push rather than to encourage them. In cases where farmers show interest, like the planting of *leucaena* trees for fodder provision, programme implementation has been very slow. In-house monitoring and programme reviews are constantly in progress but no major programme evaluation has been carried out.

The multitude of actors in Chinyika Resettlement Scheme, with different backgrounds, experiences, etc. and the resultant differential lifeworlds, perceptions, real objectives, practices and strategies has produced interfaces (struggles, negotiations and accommodations) at the various points of interaction. These interfaces occur between intervening and local actors. The result is not the largely assumed single intervener-farmer interface but a much more complex situation. Such emergent interfaces yield programme implementation different from its planned version and outcomes which are not as expected. Phrased otherwise, the emerging interfaces partially explain the limited success of the forestry extension programme and overall forestry conservation project. One example is reflected in the fact that the programme is still pushing for the establishment of gum woodlots at the expense of indigenous trees, despite many problems with *eucalyptus* such as the little interest by peasants who prefer indigenous trees found in the forests. Most established *eucalyptus* woodlots, which give the programme its partial success status, are currently grossly neglected. Another example is the 'compartmentalisation' of (the once holistic traditional) agriculture into forestry, cropping systems, animal husbandry and water resource management by the intervening agencies. The long lines of communication in the participating agencies compound the problems.

Recommendations for extension

Considering the limited success of current forestry extension and conservation programmes, it becomes imperative to adopt changes in extension approaches. First, unlike current top-down extension approaches, the extension agents in the new approach should cease to be educators or awareness providers and assume a new role of facilitators in farmers' own forest resource management and tree growing activities. Here the extension system would facilitate the exchange of knowledge between researchers, innovative and utiliser farmers, thereby establishing the currently missing feedback system in forestry technology development processes. By the same token, extensionists should strengthen farmers' informal knowledge networks by assisting farmer-to-farmer knowledge exchange processes through platforms such as field trips, field days and innovative farmer workshops. Finally, given the positive complementarity between the formal and informal knowledge networks in the conservation forestry sector, it is apparent that the success of the overall forestry conservation programme will be enhanced if the concerned authorities exploit this complementarity by integrating the two networks.

NOTES

1. This chapter is based on Hanyani-Mlambo (1995).

2. The estimated annual rate of tropical deforestation is 15.4 million hectares per year (which represents 0.8% of the total area in the tropical world). This rate is 4.1 million ha (0.7%) per year for tropical Africa and 1.3 million ha (0.9%) per year for tropical Southern Africa (FAO, 1990b).

3. Conservation forestry is used here to encompass the combined conservation of existing forests and reforestation activities meant to restore destroyed or degraded forests and woodlands.

4. Some studies emphasise increased fuelwood harvesting that results from increasing population pressure on the resources (Banks, 1980; Katerere, 1985; Makoni, 1990). Other studies attribute deforestation to a broader view of wood extraction and agricultural encroachment (Whitlow, 1979a; 1979b; 1980; Gore et al., 1992; Grundy et al., 1993).

5. It was furthermore anticipated that some extension effort was to be directed towards District Councils. As current and potential controllers of land allocation and thus management in communal and resettlement areas respectively, these councils needed to be made aware of rural reforestation and the consequent need for nursery establishment, woodlots, etc., so that they would readily assist in making land available for these purposes.

6. It is important to note here that according to the Forest Act (1982), communal farmers may not collect fuelwood outside their communal area; similarly, resettlement farmers are restricted to fuelwood collection within the resettlement area.

7. The Natural Resources Board in the Ministry of Environment and Tourism is the public trustee of all natural resources (trees included) and is engaged in the direct supervision of usage of natural resources, including enforcement of conservation measures, and various laws: the Natural Resources Act (1942); the Forestry Produce Act (1981) limiting the amount of wood which can be possessed or transported at any one time; the Parks and Wildlife Act protecting indigenous and other protected trees; and the Forestry Act of 1948 (amended 1982) controlling the cutting of timber, regulating its trade as well as prohibiting the burning of vegetation. In all acts, penalties for non-compliance are included. This criminalisation policy has its roots in the country's colonial history. In the present day effective policing through legislation has presented problems since the enforcement element in it has tended to carry over the ill feelings towards conservation coercion farmers had during the colonial days. To address this problem, the NRB policy has started to shift from policing to a more integrated approach that also includes educational and extension work as a means of

managing forest and woodland resources. However, NRB is very thin on the ground with offices and personnel only at district level. Thus its educational work in Chinyika Resettlement Scheme is mainly through Agritex; the NRB also has to rely on the Zimbabwe Republic Police to make follow-ups for fines for any environmental wrong-doing, though the police do not always do so.

8. At the time of fieldwork in 1994, the Department of Rural Development (DERUDE), of the Ministry of Local Government, Rural and Urban Development, acted as the administrator of the resettlement area through its network of Resettlement Officers (ROs). The ROs, who were initially involved in resettlement planning and responsible for settler selection and allocation of residential and arable holdings, are now, among other things, expected to carry out surveys and censuses necessary for planning purposes, to advice settlers on government policy on resettlement and the conditions of their stay on the schemes, and to foster the planning and implementation of second stage development projects such as woodlot establishment by groups and to facilitate community participation in project implementation.

9. Local institutions such as Village Development Committees (VIDCOs) are used as "vehicles" for interventionist programmes such as forestry extension, where they act as intermediaries between interventionists and local actors. In Chinyika, VIDCOs are acting as Village Conservation Committees. Ideally all villages should have separate conservation committees but these were non-operational in 1994. VIDCO members described their forestry conservation functions as discouraging residents from cutting trees during periodical village meetings and mobilising villagers in communal woodlot establishment.

10. Chinyika Resettlement Scheme's formal education system, consisting of 17 primary and six secondary schools, help with the management of woodlots which were once community woodlots and now in their care or simply those which they established on their own. Encouragement for most schools, apart from use in construction activities and the income they get from periodic harvests and sales, is through competitions sponsored by the Forestry Commission and the NRB.

Chapter 6

The bench terrace between invention and intervention
Physical and political aspects of a conservation technology

PIETER VAN DER ZAAG[1]

INTRODUCTION

Biriwiri Valley in Chimanimani district is characterised by bench terraces scattered all over its hills. Unlike other communal areas, here no contour ridges are found. The stone-walled terraces in Biriwiri were constructed since the early 1950s. These terraces are both a cause and a result of drastic agricultural changes. At the beginning of this century, people in Biriwiri practised shifting cultivation, mainly concentrating cultivation on the lighter soils found in the hills. This system was known as *nhomu* (new land) as it required the regular clearing of new fields. The flatter soils along Biriwiri river were only partly used as gardens, and other parts were considered too heavy to till with the hoe. Then, as early as the 1920s, some farmers acquired the plough, and this technological innovation, together with the use of kraal manure, spread quickly. With the plough, the flatter alluvial soils, very suitable for maize cultivation, could now be easily cultivated. In the mean time, the demonstrator (agricultural extension worker) started to discourage people to cultivate the steeper slopes and to stop practising shifting cultivation, which culminated in the 'centralisation' of these lands during the 1940s.

In the late 1940s the demonstrators attempted to introduce contour ridges as in other parts of Zimbabwe. However, in Biriwiri this was difficult to implement because of the many stones in the fields. As a compromise, farmers were encouraged to lay the stones along properly pegged contours, forcing farmers to plough along the contours. Gradually, local farmers found that terraces were very appropriate to cultivate, again, the much steeper hill sides, but now as a form of permanent agriculture. This made many more lands potentially available for arable use. This was very welcome, because people kept on arriving in the Biriwiri area, thrown out of nearby Melsetter Intensive Conservation Area but refusing to be resettled in the lowveld.

The politics of conservation

The bench terrace is an atypical conservation technology in the Zimbabwean context. Analysing its technical and political content, and confronting it with the mainstay technology, the contour ridge provides an opportunity to contribute to on-going discussions on conservationism, colonial intervention and ideas about development (Beinart, 1984; Phimister, 1986; 1993).

Central to the argument presented here is that technical interventions are 'not themselves socially neutral' (Beinart, 1984, 83). Beinart argues that ideas about soil conservation in Southern Africa were 'invoked, elaborated and applied . . . first in relation to settler, then also peasant, agriculture.' (p. 82). Phimister argues, in contrast, that the 'ideas and prescriptions associated with conservationist thinking were first put in practice in relation to peasant agriculture, not settler farming as Beinart claims' (Phimister, 1986, 263). The main thrust behind the effort of the Rhodesian government to develop and enforce conservation techniques was, according to Phimister, to be able to squeeze more people into the now 'centralised' communal lands. Another motivating factor was to boost maize production, which had become scarce during the Second World War, and continued to do so when White farmers shifted attention to the booming tobacco crop. Government, then, was not so much motivated by a conservation concern. Phimister concludes that for the study of soil erosion and conservation it is crucial that the political and economic contexts be taken into account (p. 274-75). I will argue that it is also important to look at the techniques themselves that were promoted; under what conditions they were developed, introduced and imposed, and with what consequences.

Without having the pretension to solve the Beinart-Phimister discussion, it would be erroneous to suppose that 'the colonial state' in Southern Rhodesia had only one face. As Beinart points out, during the 1940s a technocratic approach to soil conservation emerged which led to an emphasis on the role of the expert, on unilateral interventions and centralised planning (Beinart, 1984; McGregor, 1995; Wilson, 1995). This new trend coincided with a large influx of well-trained conservation officers from South Africa.[2] This may have contributed to Land Development Officers understanding their tasks

primarily as development oriented, rather than instrumental to segregationist
and racialist policies. It may have blinded many of them to see that non-
adoption of soil conservation measures by rural people formed part of 'their
political struggles . . . geared to retain control over their land, cattle and
settlements' (Beinart, 1984, 83).

A biography of a technology

The above implies that the technical expertise used in the communal areas
had both a technological and a political content. This chapter attempts to
demonstrate how political and technical aspects intertwined in the case of
one particular technology: the bench terrace. I will follow this technology[3]
in three phases of its 'existence':
1.   its *development*;
2.   the *appropriation* of the technology by the users;
3.   its *use*.[4]

In other words, I try to write a 'biography' of the bench terrace. This way
of analysing an agricultural technology will help to explain adoption or non-
adoption by farmers of this and other technological innovations, and will
reveal the ways in which physical and political aspects interlock.

The chapter sets out by describing how soil conservation techniques were
developed during the 1930s through the 1950s by the formal agricultural
research institutions in Rhodesia. Then the way in which soil conservation
techniques were introduced in Biriwiri is discussed, as well as how farmers
'appropriated' the bench terrace technique. The section that follows describes
how the terraces are currently being used. Finally, the merits of studying the
biography of an artefact are briefly discussed. Such a biography may help us,
researchers, to critically reflect on the political and social consequences of
our professional practice.

Methodology

Seventeen farmers were interviewed (of whom 11 were male) in Biriwiri and
Mutsangani wards, Muusha Communal Area, during the period 3 October–
18 November, 1994, with the help of an interpreter. Some of them were
later revisited. Eight retired officers and workers of the agricultural extension
service were also interviewed, who, during the 1940s through the 1960s had
worked in the study area. In order to get an idea of the formal research
carried out on the issue of soil conservation, and of the official guidelines

with which these extension workers were sent out to farmers, a literature review was carried out in the libraries of the University of Zimbabwe and the Ministry of Agriculture, and the National Archives were consulted. *The Rhodesia Agricultural Journal* (*RAJ*) proved to be a valuable source. Aerial photographs of 1950, 1963, 1971, 1981 and 1986 were studied in order to assess the rate of expansion of the terraced fields in the Biriwiri area.

DEVELOPMENT

Biriwiri farmers cultivate the steep slopes of the hills on an extensive scale. They have constructed the *marozhi*, or, what technicians would call 'bench terraces'. A bench terrace simply consists of a stone wall laid out along the contour of a field, the area behind which is filled with soil. The slope of the soil contained by the wall is much reduced, or even completely flat. A field in Biriwiri normally consists of many parallel bench terraces. Together they form giant 'steps', each step being roughly between 0.50 and 1.75 metres high. The width of a terrace varies greatly, and basically depends on the original slope of the hill before clearance.[5] This coincides with the fact that on steeper slopes the soils are generally shallower. A large proportion of farmers in Biriwiri now cultivate on terraced lands which originally must have had slopes of between 20% and 40% (11°-22°) (Figure 1).

Figure 1: The bench terrace (schematic)

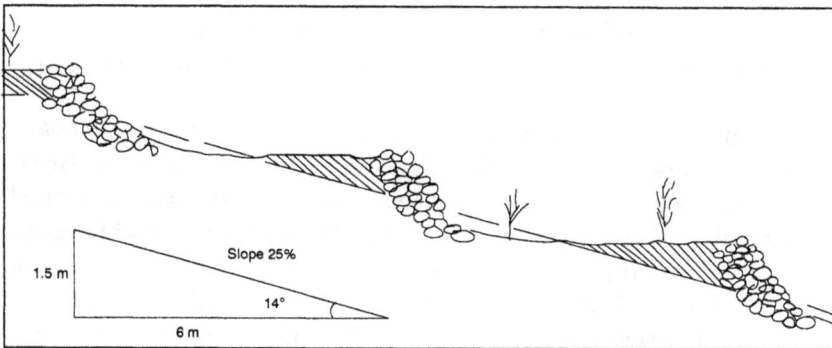

Apart from reducing the slope of the fields, the bench terraces solve another urgent problem which farmers in Biriwiri face: their fields are literally covered with stones. These stones are removed from the soil and made into walls. In this way the soil is much better suited to arable cultivation. It may however be difficult or even impossible, to cultivate these lands with the ox-drawn plough, during the first year or ever. For farmers, removing stones seems to be the primary reason to build the stone walls.

All interviewed farmers in Biriwiri acknowledge that it was the agricultural extension workers during the 1950s, the *madhomeni*, who introduced the technology of the bench terrace to them. All farmers maintain that it was a completely new idea to them. In this section I try to find out where the *mudhomeni* got the technology from. This is not an easy task, as the bench terrace was not the standard measure used by government officers to protect the fields against erosion.

The formal research

Let me start by briefly discussing the formal research efforts undertaken during the period preceding the emergence of bench terraces in Biriwiri, up to the date when the technique seemed to have been assimilated by the farmers, somewhere towards the end of the 1950s.[6] The basis of this discussion is formed by articles published between 1927 and 1957 in the *Rhodesia Agricultural Journal* referring to soil erosion and soil and water conservation. I assume that these articles reflect the activities of the formal agricultural research institutions during the period.[7] What is clear is that one technical solution was dominant: the contour ridge. Before I go into the contour ridge phenomenon, I first want to briefly mention two important features of the 'official' research effort of the time: the contour ridge technique was developed for relatively flat fields, in relatively high rainfall areas.

Soil conservation techniques as proposed in various articles in the *Rhodesia Agricultural Journal* were originally designed for soils found on White farms. The three experiment stations where soil conservation research was carried out were all sited in the midst of the large-scale commercial areas.[8] Research was not specifically geared towards the possibilities and needs in the communal areas. One clear example is that the experimental plots at Henderson had slopes of 3%, 4.5%, 5%, 6.5% and 8% (Nyamapfene, 1987, 14). Generally speaking, slopes steeper than 1:15 (7% or 4°) were hardly cultivated by White settlers (Haviland, 1934, 438). However, in many communal areas, farmers often had no choice but to cultivate on much steeper slopes, in spite of this being officially discouraged (Aylen and Hamilton

Roberts, 1937, 33).[9] As a result, the standard design procedures for contour ridges did not consider these steeper conditions. As to rainfall conditions, the contour ridges were designed to drain off excess water from the fields. Recent insights in water conservation techniques conclude that under the semi-arid conditions prevailing in most communal areas, water retention techniques should have been developed (Hagmann and Murwira, 1995, 7).[10]

The contour ridge

What then is the contour ridge? Aylen (1939b, 464) defined contour ridges as 'low substantial banks of earth, either with or without a trough or water channel on the upperside'. He stated further that:

> the primary object achieved by contour ridges is obtained directly by the ridges themselves and indirectly due to the fact that all farming operations on a contour ridged land must be undertaken on contours; thus every little furrow or plant row forms a check to the movement of water and so reduces erosion (p. 481).

Throughout the volumes of the *Rhodesia Agricultural Journal* from 1927 to 1951, this technique of physical soil protection emerges as the panacea of soil conservation.[11] An article by P.H. Haviland, Irrigation Engineer (Matabeleland) spelt out standard designs, design criteria and dimensions, construction methods and cost estimates, including estimates of how many feet can be 'excavated per native per diem [day]' (Haviland, 1934, 436).

The first comprehensive treatise of soil conservation was probably that by Aylen and Hamilton Roberts (1937), which was 'issued by authority of the Minister of Agriculture and Lands'.[12] This was a detailed publication of 80 pages, illustrated with some 38 figures, plates and design drawings. A few years later, Aylen published in the *Rhodesia Agricultural Journal* a sequel of articles which pretends to update his previous work with Hamilton Roberts, from which he quotes extensively (Aylen, 1939a; 1939b; 1939c; 1940c). The basics of the contour ridge remained the same, and no design recommendations were given for lands steeper than 1:15. The last major contribution on physical soil conservation works in the sample of the *Rhodesia Agricultural Journal* volumes consulted is by Cormack (1951). The design procedure is now refined to the use of complex graphs which incorporate long-term data on rainfall intensities in Rhodesia. Yet again, slopes steeper than 10% (6°) are not considered.[13]

From 1953 onwards it would appear that research was increasingly directed towards detailed quantification of soil loss under various conditions (see Hudson, 1957). Gradually the emphasis shifted from mechanical measures against erosion control to 'the importance and place of good farming and proper land use in conservation' (Rhodesia, 1952, 465). The establishment in 1950 of CONEX (the Department of Conservation and Extension in the Ministry of Agriculture) also appeared to have been linked to this new trend (Weinmann, 1975, 147).

The bench terrace

The question remaining is: what if the slopes of fields were much steeper than the 10% limit, as was the case in many parts in Biriwiri? The first suggestion in the colonial era to build terraces from stones on hillsides appears to have been made in 1930 by a White settler in Shamva.[14] The only technical description of a conservation technique suitable for steeper slopes is given by Aylen in an article on contour planting and terracing of orchards, which according to him is suitable for slopes up to 1:5 (11°) (Aylen, 1940b, 199). He maintains that 'as the method is costly it is obvious that it is not economical for large areas and is really only suitable for small orchards in cases where the slope is steep but fairly regular over the whole area'. Aylen gives four different methods of constructing these terraces, the most satisfactory method being 'the evolution of specially designed contour ridges to form bench terraces' (p. 198).

A year later, Aylen wrote a brief article entitled 'Who built the first contour ridges' in which he expressed his admiration for the abandoned terraces in Nyanga. However, the article does not contain any technical data, and does not in any way suggest that these old terraces by then were being studied for the development of a contemporary form of terracing.[15] Again a year later, a soil conservation officer suggested for Rusitu valley:

> smaller ridges spaced at closer vertical intervals, which could eventually become bench terraces, could be adopted to permit the safe cultivation of much steeper slopes than usual but per acre costs would be high.[16]

A study probably available by the early 1950s in some of the local libraries was that by Tempany (1949), who compared the various soil conservation measures in 'the British Empire'. What he wrote on bench terraces is reminiscent of the terraces of Biriwiri:

> On very rocky land use can be made of available rocks and stones to
> build contour stone bunds or terraces. They are constructed in the
> form of low, strong, contoured stone walls. [. . .] In time contour
> stone terraces may become converted into bench terraces as a result
> of gradual filling up (Tempany, 1949, 15).

By 1950, then, only a few scanty references were available on soil
conservation techniques suitable for steep conditions. Obviously, the
mainstream technical solution was the contour ridge, which by the 1940s
had got a kind of paradigmatic status: most research was geared to sophisticate
its design, and the agricultural departments, both for the Black and White
farming sectors, were obsessed by how many miles of contours had been
constructed each year. There seems to have been little space for research
into situation-specific soil conservation techniques, let alone for the study
of soil conservation practices by Black farmers. Soil conservation techniques
were apparently considered neutral tools which could be applied anywhere
(cf. Beinart, 1984, 83).

The contour ridge technology, however, was not a 'neutral' technical
solution, as it was developed for specific topographic, pedologic and climatic
conditions. As to topography, the ridge was a technical solution limited to
slopes less than 10%. As to the soils, many experiments were conducted on
deep red soils (see for instance Robertson and Husband, 1936, 162). Trials
were also conducted on white granite sands (sandveld soils) which were more
likely to reflect communal conditions: but these trials mainly focused on the
tobacco crop needing relatively steep gradients of the ridges (Hudson, 1957).
The tobacco crop, then, had very different requirements compared to maize.
As to climate, the ridge was designed to drain off rainstorm water, hence the
importance of hydraulic calculations of discharge capacities for establishing
the dimensions of the ridges required. In many communal areas, however,
rainwater was too precious to simply 'evacuate' it from the land.

Conservation technology was also developed under a particular political
constellation, and the physical works were promoted in different ways in the
commercial and communal sectors: whereas White farmers could obtain
cheap loans from government, Black farmers could not; and where White
farmers in principle were free to choose, Black farmers were, in practice,
forced.[17] Even the problem the contour ridge was supposed to combat, i.e.
soil erosion, was, at least in part, created by the politics of the day (see Cliffe,
1988). The contour ridge, then, had not only become a tool in the fight

against soil erosion, it had also become an artefact which justified extensive government interference at grassroots level all over the country. Or, in Beinart's words, 'the welfare of the soil often emerges as the cutting edge of justification for intervention in peasant agriculture' (1984, 53). And this in turn sowed the seeds for the contour ridge, during the 1960s, getting yet another meaning: that of a symbol of colonial oppression (cf. Phimister, 1993, 227–28; Beinart, 1984, 80–81).

APPROPRIATION

In this section I describe how the bench terrace technology emerged in Biriwiri, and how farmers took 'possession' of it. First, the side of the interveners will be highlighted. In the second part of this section, the focus is on the farmers, and on the process of constructing the terraces.

Introduction of conservation measures: The model

An important article by Aylen describes the way soil conservation measures were introduced and implemented in the then Native Reserves (Aylen, 1942). As already said, the basis of conservation works in most of these areas was the contour ridge, being introduced on the arable land immediately after the centralisation exercise.[18] Aylen described the process of implementation as follows:

> Soil conservation is not forced on the natives, but is only adopted where the majority vote for it, when it becomes compulsory for all. Tactful and careful explanation of the benefits of soil conservation by the Native Commissioners and Chiefs does much to educate the natives, but when demonstrations are added as a proof, it is believed that the positive conversion will be more readily attained, thus facilitating the supervision of maintenance (p. 158).
>
> The works will be set out for the natives, and under the powers conferred by the National Resources Act they will be ordered to construct them. They will be lent one hundred "Evans-type" land levellers fashioned from old railway steel sleepers, and a supply of discarded shovels. [. . .] Next year no native will be allowed to cultivate unprotected land within this area.
>
> In order to encourage adequate construction and ensure completion of the work a small bonus (based on yardage) will be paid to the natives of each village or community when all the lands

belonging to that group are fully protected by completed works built up to the full size (p. 160).

This 'model' implementation process of contour ridges as painted in 1942 by a White government officer gives only part of the story: how these conservation works were actually implemented, and how farmers perceived these interventions, are of course quite different things. What is striking in the above quotation is the way the government officer hovers between force and tact, stick and carrot, when describing the nature of the intervention. Farmers in Biriwiri, when asked about the way the conservation works were implemented, also have problems in giving a straightforward formulation, reflecting a similar ambiguity:

> People were pressured but not forced. But we were not allowed to plough without contours. So we had to [Interview with Mr Mangwende, 4/10/94].

What may lie behind this ambivalence is the fact that conservation works were, practically speaking, impossible to implement without the active cooperation of farmers. The labour requirements were enormous, and no government would have been able to undertake such works by itself. It is perhaps because of this fact that the Land Development Officers came up with this mixed strategy of coercion and cooperation; but in the last instance it was coercion that shaped the process (cf. McGregor, 1995, 265).[19]

Introduction of the bench terraces in Biriwiri

All Biriwiri farmers interviewed concur that the technology of the terraces was introduced to them by the *mudhomeni* (demonstrators). But, as one demonstrator of nearby Mutambara, Mr Dunga, recalled:

> We were not trained in stone ridges. When we came to places where you could not dig the contour ridges we introduced stone work. But we were never trained to do that, because the schools were not in mountainous areas [Interview with Mr Dunga, 12/11/94].

What these demonstrators knew very well was how to peg and construct contour ridges. It seems that the local demonstrator first started to introduce the idea of contour ridges around 1950 with 'his' 16 'plotholders'.[20] These were farmers from Biriwiri, Mhakwe and Matano. The selected farmers appear to have been those considered 'progressive' who generally had plots on flatter lands, and probably owned a span of oxen and used the plough. It is likely that the contour ridge concept was rejected from the start as the ridges were hard or even impossible to be dug on these soils. But the *mudhomeni* had to find something that would make his plotholders plough along the contour. It is likely that the idea of collecting the stones and laying them along contour lines emerged as the logical compromise, because there are some scanty references that farmers already practised such a system (see below). For the progressive farmers it solved part of the difficulties they experienced with ploughing the stony fields, while the *mudhomeni* complied with the ploughing-along-the-contour paradigm which he had to implement.

However, these emerging stone walls were not all that urgent for the 16 farmers. Only later would these walls be embraced by the poorer peasants who opened up the steep hills for agriculture. For the moment, however, many farmers resisted the conservation measures, as is vividly remembered by a local farmer who headed the peggers' gang:

> There was a lot of noise because of these contours. Many people were fined because they were resistant, and wanted to plough where the lines had been pegged. It was just forced, a directive from the government that everyone's piece of land should be pegged. Neither did the chiefs and the headmen like the idea, but because it was a directive from the government they had to accept. They eventually accepted the idea and taught the importance of it to these people [Interview with Mr Mahosa, 2/11/94].

The demonstrators, then, had to act strategically in order to get people accept the stone walls. In the Mutambara and Biriwiri areas, under the same Land Development Officer, the demonstrators worked out the following strategy:

> It was a matter of persuasion, to show them. And to group them, so that they could help each other. There was so much work, and people would find it very hard, and in the group they would tell each other that it was good. So, I first organised a lecture to explain, and then divided them into small groups of 15. You know, in the same way they did threshing of *mhunga* and rapoko. They would just brew sweet beer and the people would gather and help. We call it *nhimbe*, group work. The chiefs were the hardest people teaching them, persuading them. I wanted the chief to give me his son to work with me. The son of the chief would be ahead and I behind. I made sure that the

son of the chief was involved. In this way I managed to convince the
people [Interview with Mr Dunga, 12/11/94].

The second demonstrator in Biriwiri, who succeeded his predecessor in
1955, recalled:

> We started this system of going to the chief, we convinced Chief
> Muusha, and we asked for his son, Mr Masungani, and we gave him
> the title 'supernumery demonstrator'. I supported his name to [the
> Land Development Officer], and he to the District Commissioner.
> We managed also in difficult areas to introduce the terrace, because
> some areas were difficult, where the farmers did not want to adopt
> this new system [Interview with Mr Manase, 23/06/95].

The terrace: Invention or intervention?

The above seems to point to the important role played by the demonstrators
and the peggers' gang. Farmers help to maintain this impression, by stressing
to outsiders that their terraces have been properly laid out by the *mudhomeni*,
thereby implying that they are officially approved. Farmers know that the
cultivation of steep lands is officially prohibited.[21] The role of government
staff, however, should not be exaggerated. Mr Manase, the Biriwiri
demonstrator in 1955, said:

> Before 1955, there were already terraces in Biriwiri, so-called terraces
> I mean. The piling of stones was not on a pegged level. The farmers
> had the system of removing stones in order to plough. Some knew,
> others did not [Interview with Mr Manase, 23/6/95].

And Mr Sisimayi, a well informed teacher, agriculturist and conservationist
in Chimanimani:

> When the *mudhomeni* were working with Africans, they had seen already
> that Africans were cultivating their plots on the mountains and making
> their piles of stones all over. The *mudhomeni* did not want to give
> illustrations on steep stony hills. They gave illustrations on plots in
> flat lands. People think that Biriwiri farmers were instructed by
> government demonstrators. But I say no. There came creation in the
> minds of people putting stones in piles. It was the need of more
> land by the same primitive cultivators to shift from piles to lines
> [Interview with Mr Sisimayi, 7/8/95].

The above may explain the speed with which people started to adopt the
'improved' bench terrace concept in new areas hitherto uncultivated. Take

for instance Mr and Mrs Rupiya. The Rupiya family originated from Biriwiri
and settled in nearby Mutsangani in 1957, when they cleared their current
fields and started making terraces on a hillside with slopes of 20% or more.
The farmer said he had copied the technique from other farmers located
near Biriwiri mission, and that the peggers' gang never came to measure his
fields and peg the contour lines. Using his common sense and craftsmanship,
he determined himself where to put the stones. As they were the first to
cultivate the steeper slopes in that part of the Biriwiri valley, subsequent
farmers copied in turn their techniques. Now, in 1994, nearly all the hill
slopes in Mutsangani are terraced and fully cultivated.

Labour requirements

The bench terraces found in Biriwiri are not constructed in one go. When
somebody decides to take up a new piece of land, or extend his or her
existing fields, first trees are stumped, small bush is cut and finally burnt.
Then a beginning is made with the construction of walls of stones along the
contour, by removing stones from the top soil and dumping these at 'lines'
at regular intervals along the contour. This activity could typically be
undertaken during the months of August, September and October. When
the farmer judges that there is enough soil cleared from stones, he or she
may decide to plant, with the hoe, the first crop of maize intercropped with
beans. During this first year, the slope of the fields so cultivated is still steep,
and very much similar to the original slope. However, after the rains, some
of the soil may have accumulated against the stone wall, and the slope would
be reduced slightly. After the first harvest, new stones will have surfaced and
removing them and incorporating them in the stone walls takes another three
or more months. In so doing the field gradually begins to take the first
appearance of a terraced structure. It is only after a number of years of
cultivation and clearing of stones that the final shape of 'proper' bench
terraces is reached. The very big stones are simply left untouched by the
farmers, as these are considered 'stubborn' (*matombo asikanzwi*), and
incorporated in the terrace structure.

   The above process of terrace construction is a laborious affair. Farmers
stress time and again that it requires a lot of their energy or power (*simba*),
and that one needs to be determined (*chido*), otherwise the whole project of
clearing a steep piece of land may fail. The example of the 1 hectare field of
Mrs Sauti is illustrative. It should be noted that this field is situated on one
of the steepest sites encountered, the original slope being some 40%.

In 1975, Mrs Sauti first cleared the land, which took her some 50–100 labour days. Then, in August 1975, the *mudhomeni* came and the peggers laid out the contours. From August to October she worked very hard to remove the most obvious stones, laying them along the lines. She was helped by her cousin sister on a full-time basis, and during the weekends, also her husband gave a hand. By that time she had no children to take care of and she emphasised that she was working full-time. This work took not less than 100–150 labour days, but perhaps more. Subsequently, she and her cousin sister had to plant, which also implied removing of stones. Then, in three consecutive years, she worked again during the August, September and October months to further clear the fields of stones, and maintain and improve the stone walls. This may have been less intensive than the first year, but surely took at least another 50–100 labour days per year, which adds up to an additional 150–300 labour days. After three years, she said, the terraces had taken their current, articulate, shape [Interview with Mrs Sauti, 18/11/94].

In sum, it may well take between 300 to 550 labour days or more before a one hectare field is a bit in shape (cf. Tempany, 1949, 15-16; Wilken, 1987, 116–117). This figure does not include the maintenance work done on a routine basis every year during planting, when stones which appear on the surface have to be removed. This implies that hoeing takes much more time than it would normally take on fields without stones. This extra effort, however, is highly productive, as is explained in the next section.[22]

The large amount of labour required has some implications. It will strengthen the relationship which the farmer establishes with his or her plot. In the process of construction, the technique is assimilated as something self-evident. By then it has become 'our' technology.

Once the terrace technology was available to farmers, new horizons were opened up. While the first terraces were constructed on the flatter lands which were already cultivated, farmers such as Mr and Mrs Rupiya soon found out that the terraces could also make the steeper hills productive. So, the first fields to be terraced were the ones nearest to the river, but later on, much steeper plots up the mountain were also established. This pattern perhaps is linked to farmers becoming, with time, better versed with the terrace technique, under increasingly more critical conditions of land shortage. During the 1960s and 1970s many hillsides were cleared and terraced, by sons (and more so perhaps by unmarried daughters) of Biriwiri who wanted to have their own plots of land, but also by people who were pushed off the White farms of nearby Melsetter (now Chimanimani), and by people coming from Ngorima Tribal Lands, pushed out of the overcrowded but fertile Rusitu Valley.

USE

The final part of the biography of the bench terrace should be concerned with its use. In this section I want to discuss only one aspect: its sustained use made possible by soil characteristics and cultivation practices.

Farmers cultivating the bench terraces for 20 years or more insist that yields are stable, provided, of course, that annual fluctuations in rainfall patterns are taken into account. This has to do with the fact that the terrace soils are relatively young, and that these soils, tentatively classified as belonging to the fersiallitic-lithosol group (Bromley *et al.*, 1968, 37), are capable of providing sufficient nutrients for a reasonable maize crop during a decade or more without being depleted.

If one looks at the external nutrients brought into the soil, the following picture emerges: Some farmers will apply manure regularly on their plots, and chemical fertilizers, often alternating on a particular plot manure in one year and fertilizer in the other. Other farmers, such as those higher up the mountains who tend to own less cattle, lack sufficient manure. Many of them will use some fertilizer (especially top dressing). Quite another external source of nutrients can be provided by *leguminosae*. A number of farmers intercrop maize with beans planted in the same hole at the same time, and they say that they may not even harvest the beans. This may indicate that they skilfully use the beans for nitrogen fixation.

If however, one asks farmers the reason of stable yields, they do not firstly mention the importance of the external nutrients applied to the soil, but rather stress the internal source of nutrients available in the soil: while planting the seed with the hoe, they explain, you remove the stones you encounter, and the soil below is 'unexploited' or 'virgin', and therefore you get higher yields. This observation may point to the importance of the process of weathering of parent material as a major source, and a sustainable one, of nutrients.[23]

This may explain why so many farmers maintain that the yields on the bench terraces are very similar, if not higher, to those obtained from the few flat plots found in the valley bottom in Biriwiri. The latter consist of thick layers of relatively rich alluvial soils, where the weathering process will provide a negligible amount of nutrients to the crop. The potential importance of weathering to sustainable crop yields on bench terraces have yet to be established by soil scientists. For the moment I take it to be one characteristic which may help to explain the fast expansion of the terraces from the 1950s up to the present day in the Biriwiri area.

CONCLUSION

I structured this chapter around three stages of the 'life' of the bench terrace: its development, its appropriation, and its use. It now appears that this was highly artificial, for the bench terrace was not developed in the laboratory or on research stations, but on the fields of farmers. In other words, the stages of development and appropriation were intertwined. The bench terrace was an emergent phenomenon, a result of the interlocking of the different agendas of land development officers, demonstrators and farmers.

High on the agenda of the land development officers was soil conservation, whereas farmers pushed out of fertile lands were in search of alternative plots. The demonstrators found themselves somewhere in-between. For farmers the bench terrace was not first of all a conservation technique, but a technique with which they could open up new, marginal, lands. Bench terraces thus served the colonial powers well, as it enabled the settlement of more people in Biriwiri Valley without the government having to relinquish any land (cf. Phimister, 1993, 1986). Clearly, then, technologies may mean quite different things to different people. The discussion whether conservation policies and techniques either originated from a concern with settler farming practices or from a political economy perspective of marginalising peasant agriculture, the lines along which Phimister (1986, 263) wishes to organise the debate, appears to miss the point. Implementation was a multi-faceted and to some extent contingent process, involving different actors. In particular contexts, specific in time and space, different actors struggled with each other's agendas and with the specific physical conditions, which gave rise to particular outcomes. In Biriwiri, the social, cultural and political consequences are complex and fall beyond the scope of this chapter. The material outcome is evident: the bench terrace.

The data presented in this chapter on how the terraces were developed in Biriwiri are of a preliminary nature.[24] The Land Development Officer of the area most probably adopted the idea of putting stones along the contour from one or two local farmers and started to promote it. The demonstrators were trained in pegging out contour lines, so they pegged some of the land and explained to some farmers what to do. The farmers, for their part, insist that the demonstrators taught them the technique. I, however, doubt whether this is entirely true, and rather understand this as a strategic pose. By emphasising the role of demonstrators in developing the benches and pegging the contours, Biriwiri farmers imply that the cultivation of the steep hills was officially endorsed by the state. Given the large labour requirements, and the time it takes before a contoured land is properly terraced, I concluded that eventually it was the farmers who fully developed the technology, and appropriated it as if it were a by-product of the construction process. This finding could be quite relevant for the current land tenure debate in Zimbabwe, for Biriwiri farmers of Muusha Communal Area are willing to make substantial investments in their fields, even if they perceive access to them as not secure enough, and in the face of restrictive state rules prohibiting the cultivation of steep hillsides.

Whereas the development and appropriation stages of the bench terrace overlapped, this was not the case with the contour ridge. The contour ridge was developed on research stations and on White farms and surely *not* in the communal areas. Its forced implementation in the communal areas eventually led, during the 1960s, to it becoming a symbol of colonial oppression. Farmers started to actively resist the conservation measures (see e.g. Beinart, 1984, 80–81; Whitlow, 1985, 319; Phimister, 1993, 227–228), and resorted to 'Freedom Farming' which consisted of ploughing up and down the slopes, and ignoring buffer strips (Bulman, 1970, 16). In contrast, in Biriwiri no destruction of bench terraces occurred. Rather the opposite was the case: the greatest expansion of terraces in Biriwiri valley took place during the 1960s and 1970s, as evidenced by aerial photographs. Even during the years 1975 through 1979 the area under terraces expanded, when all inhabitants of the valley were forcefully moved to the *keep* ('protected village') at Mhakwe. For farmers this meant that they had to 'commute' to their plots, at one hour walking distance or much more![25]

The bench terraces were not destroyed in Biriwiri, as farmers considered them useful.[26] This is related to the fact that farmers played an important role during the development of this technology. The relevance of studying the biography of a technology, then, is not only to describe under which conditions a new technology is developed, how it is introduced to users, how the latter reject it or adopt and adapt it through a process of appropriation; it may reveal the linkages and discontinuities between these stages (Long and Van der Ploeg, 1989). And this, in turn, may explain its eventual success. This conclusion is relevant to the research practice: research findings are often not universally valid, but heavily biased by the material, social and political contingencies under which such research is carried out. This realisation may help us, researchers, more critically assess our research agendas.

Apart from the locality specific nature of (emerging) technologies, the Biriwiri case shows that 'Knowledge emerges as a product of the interaction and dialogue between specific actors' (Long and Villarreal, 1993, 147). This finding departs from the prevailing 'transportational paradigm' in the understanding of knowledge processes, exemplified for instance by Rogers' diffusion metaphor and Havelock's linkages (Dissanayake, 1986, cited in Long and Villarreal, 1993, 145). It calls for the re-organisation of the interaction between farmers, agricultural extension officers and researchers.

NOTES

1. An earlier version of this paper was presented at a University of Zimbabwe/ZIMWESI workshop entitled 'Extension Intervention and Local Strategies in Resource Management: New Perspectives on Agricultural Innovation in Zimbabwe', held at Mandel Training Centre, Harare, from 10 to 12 January 1995. I thank the following persons for their valuable assistance and guidance: Mr Chikukwa and Mr Chaparapata (Agritex Chimanimani); Mr Pamenus Tuso (Biriwiri); Mr Ben Madondo, Mr Joe Chivizhe and Ms Simbabure (Agritex Manicaland). I benefitted much from the discussion during the workshop, and from comments and suggestions from Dr Henry Elwell, Dr Dayo Ogunmokun, Eng Jeremy Ascough, and Ms Katherine Verbeek. Furthermore, I am grateful to Mr Jens Andersson, Mr Alex Bolding, Mrs Mavis Chidzonga, Mr Emmanuel Manzungu and other members of the ZIMWESI team for their critical contributions.

2. Many of these White officers were of British extraction. They were young war veterans who had gone to war before entering university, and immediately after the war were given the opportunity to follow special intensive three or four year courses at the University of Witwatersrand. After the election victory of the National Party in 1948, many decided to come to Rhodesia, fearing that the new government would only offer jobs to Afrikaners (Interview Mr Daryl Plowes, 31/7/95; cf. Weinmann, 1975, 147).

3. The bench terrace is here considered an artefact, which is part of, and a manifestation of, a particular technology. In the present context, I take 'technology' to include the human skills, and the physical tools, implements and instruments, with which agricultural production is made possible and/or improved (adapted from Edquist and Edqvist, 1978, 9).

4. I wish to acknowledge the influence of the work of Peter Mollinga and others at the Department of Irrigation and Soil & Water Conservation, Wageningen Agricultural University, on my understanding of technology. The term 'appropriation' is inspired by the work of Miller (1987) and earlier used by me in an irrigation context (van der Zaag, 1992), and is used in a similar way by Mackay and Gillespie (1992).

5. Suppose, for instance, that the height of the stone wall is 1.5 m, and the terraces are meant to become completely level, then the width of a terrace on a slope of 50% (27') would be 3 m, of 25% (14'): 6 m, 20% (11'): 7.5 m, 10% (6'): 15 m, etc.

6. I do not pretend to write a history of soil conservation in Zimbabwe. Much on this history has already been done: Rowland (1974), Stocking (1978) (both cited in Stocking, 1985), Phimister (1986), Nyamapfene (1987), Mpofu (1987), and Whitlow (1988) provide a general picture; Sutton (1984) and Beach (1995) concentrate on the Nyanga terraces, while Beinart (1984) gives an incisive historical overview for Southern Africa, and Stocking (1985) for the entire African continent, both with various references to Zimbabwe; see also the critical contribution by Cliffe (1988) on the conservation issue in Zimbabwe.

7. Consultation of the NADA (Native Affairs Department Annual) journal for the period under review revealed that hardly any article was published on conservation issues.

8. These stations were: Glenara (1934–1936), Henderson (1953–1963), and Hatcliffe (from 1974 onwards). (Robertson and Husband, 1936, 16; Nyamapfene, 1987, 13–14; Weinmann (1975, 145) state that Glenara research station was operational from 1933 to 1938).

9. For instance, in Zimunya Communal Area, in 1949 22% of all lands cultivated was on lands steeper than 8° (14%), and by 1981 this percentage had increased to 33% (Whitlow and Zinyama, 1988, 37).

10. Already in the 1930s researchers were aware of the importance of water retention at the field in the drier regions of the country. They therefore recommended to put the contour ridge at a flatter gradient of between 1:800 to 1:2,000, but this again meant that the ridges to be dug became disproportionately bigger. See for instance Robertson and Husband (1936, 166), Aylen and Hamilton Roberts (1937, 45), Aylen (1939b, 466–7), and Cormack (1951, 160).

11. The area protected by means of contour ridges increased rapidly between 1927 and 1944. In 1930, Hamilton Roberts (1930, 841) reports that several hundreds of miles of contour ridges had been constructed, some 3% of the cultivated lands on commercial farms having been fully or partially protected. By 1944 more than 220,000 acres or some 43% of the lands under summer crops had been protected (Rhodesia, 1945, 324, 326). An additional 300,000 acres had been protected by the same year in the Communal Areas (Haviland, 1947, 345).

12. The Minister, F.E. Harris, wrote in his foreword, 'I trust this Bulletin will be studied by all owners of land, and that it will be translated and explained to all natives in our Reserves.' To my knowledge, the translation of this bulletin never materialised.

13. With the refinement of the design of the contour ridges came more specialised equipment to construct them (e.g. Aylen, 1940a). Haviland (1934, 441) noted that the use of implements drawn by oxen or a tractor was more economical than hand labour if ridges were constructed on a larger scale. Most of these implements were developed by White farmers. In addition, there was the Native Department Ditcher: 'a cheap implement recently developed by the Native Department . . . capable of good work in light sandy soils' (Aylen and Hamilton Roberts, 1937, 65, 55).

14. Letter from Captain J.M. Moubray, Chipoli, Shamva, to CNC, dated 28 October 1930, National Archives of Zimbabwe (NAZ) file S138/72. I am indebted to Alex Bolding for making me aware of this reference. The letter reads, in part:

> As you know the natives at the present day collect loose stones from their land and throw them in heaps. I have got some of my natives to make short terraces of these stones along the contour instead of the useless heaps. In the ordinary hillside it is astonishing to find what a number of such small terraces can be made from the ordinary rough stones that 'grow up' out of the lands. It is no more trouble for them to make short terraces, once they have got the idea, than useless round heaps — and you sir would be astounded if you saw the soil, that would otherwise be washed away, that such short terraces impound and conserve.

15. I found in the RAJ one other reference to proper bench terraces in an article by Close (1943), in which he gives some experiences from Asia. The author concludes that 'the real solution of the soil erosion problem lies in universal terracing' (p. 196). In a footnote the editor of RAJ gives the Senior Conservation Officer the opportunity to qualify this statement: 'Terracing is undoubtedly the ideal soil conservation measure, but unfortunately makes the use of modern implements most difficult . . . A satisfactory compromise is contour-ridging . . . In Rhodesia complete terracing is restricted to orchards, market gardens and irrigated lands where the benefits are many times the additional costs'. Apparently, the RAJ editor wanted to make clear to his readers that in Rhodesia there was only one feasible solution.

16. Letter from Mr Mackenzie, Soil Conservation Officer to Director of Irrigation dated 7 October 1942, NAZ file S1729/486/1942.

17. In the communal areas, farmers bore the brunt of the conservation works in the form of their own physical labour, but also in the form of a 'maize bounty', a tax levied on marketed maize. In contrast, the White farmers received, from as long back as 1929, subsidies and free technical assistance for physical soil conservation measures (cf. Aylen and Hamilton Roberts, 1937, 6). Still in 1957, mention is made of subsidies on approved conservation works of 33% in the higher rainfall areas and 50% in the drier areas (Cormack and Whitelaw, 1957, 52).

18. The Rhodesian centralisation scheme, pioneered in the late 1920s, was the first general attempt at 'villagisation' (Beinart, 1984, 76) and consisted of the demarcation of 'arable and grazing areas, with kraals or villages between the two' (Rhodesia, 1952, 109).

19. McGregor (1995, 264–265) has argued that local leaders may have colluded with the Land Development Officers.

20. Mr West, the Land Development Officer based in Biriwiri, reported for Muwushu that from July to September 1949, 6,111 yards contour ridges had been dug by hand, protecting 51 acres, at a total cost of £39-13-2, of which some 75% was for wages. During the same period, 18-inch high stone walls had been built across gullies at a cost of £210. NAZ file S160/M1, Melsetter-Concession, 1944–1949.

21. This rule, dating back to the 1930s, states that slopes of over 8% steepness are marginal to cultivation, and mainly suitable for perennial crops, and slopes of over 12% should not be cultivated at all (Whitlow and Zinyama, 1988, 32).

22. Hoeing is also expensive, as various farmers told me that they have to buy a hoe every year, as the stones wear the hoe rapidly.

23. This weathering could be in the order of 150–400 kg/ha/yr (Owens, 1979, cited in Nyamapfene, 1982, 287) for dolerite parent material, or even as much as 1 ton/ha/yr (Elwell, n.d., cited by Nyamapfene, 1982). See also Close (1943), who, when discussing experiences from Asia with terraces, notes that 'In the Philippines mountain slopes have been successfully terraced in heavy rainfall areas. The principle is that if rain falls on a dead-level surface it does not flow off but soaks in. The result of soaking in is to decompose the sub-soil rock and to create new soil' (p. 195).

24. Further study in three areas could be worthwhile. The first is technical analyses of the soils contained in the terraces, with a view to determining their fertility, and the role weathering plays. A second important area of further study is the way in which the terrace technology was assimilated and appropriated by the

Biriwiri farmers, including the role played by the local authorities. It has already been noted that farmers now master the art, but it is worthwhile to understand the details of this process, as this may shed light on the more general theme of 'technology development and transfer'. A third area of inquiry is to review a time series of hydrological data from Biriwiri river on discharge and soil content, and a quantification of soil loss directly from the field. Such a study will assess the effectiveness of the terrace technology in containing soil erosion on steep fields.

25. Stones, however, did play a (small) role in the freedom struggle in Biriwiri. In June 1964, stones were put across the main road between Wengezi and Skyline junction, near Biriwiri, as one of the first acts of resistance; one White farmer was killed (Sinclair, 1971, 175).

26. A similar experience is found in Machakos, Kenya (Tiffen *et al.*, 1994, chapter 11).

Part III
Local Knowledge and Networks

Chapter 7

Women farmers make the markets

Gender, value and performance of a smallholder irrigation scheme

CARIN VIJFHUIZEN[1]

In this chapter I explore how agricultural produce in Tawona smallholder irrigation scheme obtains its value, by focusing on groundnuts and tomatoes. Detailed quantitative data on use and exchange value presented here are limited to groundnuts produced in two households. This material allows me to show that agricultural produce not only obtains exchange value in terms of cash through sales, but also through labour, barter, and gifts. Besides that, the use value of the produce is obtained through consumption and seeds for reproduction. How produce obtains its use and exchange value is shaped by actors through their various transactions in different contexts (Long, 1994, 7). Some of these actors operate in the locality, others may be considered external actors, establishing links between the locality and more distant markets. It is, however, not only the external actors who determine the value of agricultural produce. In negotiating deals, local farmers actively contribute to determining the value of products. The value agreed upon is often influenced by the social relationships between the actors involved. Value cannot, therefore, be entirely explained by economic considerations alone. Value also is a social phenomenon.

The material presented in this chapter demonstrates that women farmers are not only important players in transactions concerning food crops such as groundnuts, but also in transactions concerning cash crops such as tomatoes. They actively explore and open up opportunities for transactions, whereby they negotiate about the exchange value. These transactions are not limited to cash deals. This finding has some relevance to planners, economists, irrigation engineers and extension officers. The actual output of smallholder irrigation schemes is much higher than cash transactions reveal. Smallholder schemes, therefore, perform much better than conventional statistics want us to believe.

THE FARMER AND THE CULTIVATION PROCESS OF GROUNDNUTS

Janet, the woman farmer in this chapter, was married to Abraham. They were a nuclear family, which means that they lived together with their unmarried children in one home (*musha*). They had four daughters and four sons, who were between one and 20 years old. At the time of doing the research for this case study (1994/95), most children went to school, except for the eldest daughter and two youngest sons. Abraham, the husband, was a builder and sometimes had piecework in the village. Sceptical and worried Janet said: How much piecework? (*Mapiecework mangani?*) She explained that her husband only found work a few times per year. This did not help the family much financially because he also spent the money on beer. Cultivating crops on the irrigation plot was their livelihood and all the work on the plot was organised by Janet. She took the decisions regarding agriculture. She said: I talk, he listens (*Ndinotaura, anoteerera*). His piecework was perceived as complementary to her work in agriculture. Janet organised agriculture and managed the home.

In the Tawona scheme, groundnuts were cultivated on irrigated plots from October/November up to the beginning of April. The farmers also cultivated maize in that period and therefore dedicated only a few *madhunduru*[2] to groundnuts. Janet started growing groundnuts in 1994 because: 'If you have your own groundnuts you do not admire (*kuyemura*) from others.' She explained the desirability of the crop as follows: 'Your child can go and steal it.' Farmers had a strong desire to grow groundnuts, because they roasted and ate it or prepared peanut butter (*dovi*) to be mixed with vegetables or to put it on bread. The general perception was that groundnuts are good for all people because it contains oil (*mafuta*).

Janet had not grown groundnuts before at her husband's place, so she had no seeds. She received the seeds from *maiguru* (wife of an elder brother of her husband) when she helped her with the harvest of groundnuts for one week in 1994. She was given one bucket of unshelled groundnuts as payment for her labour. Then she shelled the groundnuts and used part of it to sow two *madhunduru* (0.08 ha).

Groundnuts were sown behind the plough and one by one the seeds were dropped in lines. The women who sowed were all related to Janet: a daughter, the mother of her husband, and four *maininis* (wives of younger brothers of her husband). They sowed two *madhunduru* in three hours. Weeding was only done once and at the same moment the roots were covered with soil. It was done by two sons, three daughters and one *mainini* of Janet. After that, Janet applied fertilizer. The irrigation of groundnuts was mainly done by Janet herself, her daughter and son. Before the final harvesting, farmers uprooted several groundnut plants to eat them fresh, like they do with maize and millet. In Janet's case a very small amount was consumed fresh. The final harvesting of groundnuts was a labour intensive process. First the plot needed to be irrigated to make the uprooting of plants easier. After uprooting, the picking of groundnuts from the roots started. Janet was assisted by one daughter and five adult women who harvested the groundnuts in six days.

USE AND EXCHANGE VALUE OF GROUNDNUTS

Harvesting was done from 29 March to 3 April 1995. The payment of labour, consumption, gifts and barter all took place in the same month of April. Sales (exchange for cash) occurred at the beginning of May and the selection of seeds in October 1995. The total net amount of groundnuts harvested amounted to five bags (approx. 200 kg).[3] Because of the multiplicity of transactions, and the sometimes casual way in which these are performed, I only managed to get information of how 125 kg (60% of the harvest) were used and exchanged. I will come back to this point later. The following account of the various destinations of Janet's produce only refers to the recorded transactions. I will first detail those deals involving labour, consumption, barter and gifts. I then look into sales, and to those groundnuts set aside as seeds for the next season. The section ends with a brief discussion of the yield obtained.

Labour

The groundnut pickers were paid in kind. Janet explained:

> Harvesting is a lot of work. I paid with groundnuts (*Ndabhadara nenzungu*). I did not tell them before the work what they would earn. I decided on my own, after observing how they worked. The women just came to help. I had no specific reason to work with them. They

just came and helped me, because they like to have groundnuts. They do not want money.

Three women were paid one bucket of groundnuts each for six days harvesting. One woman was Ndaabari, who was not related to Janet and her family. Janet said: She is my worker (*Mushandi wangu*). Last year Ndaabari approached Janet in the scheme and said: 'Let me help you to do your work so that you give me something.' Janet agreed, because Ndaabari did not have an irrigated plot. Since last year they work together and they intended to continue doing so. Sipiwa, the mother of the husband of Janet also received a bucket. The third woman who received a bucket was Madinda, who was not related to Janet. She was a neighbour and a friend. She approached Janet and asked if she could help harvesting groundnuts. Two *maininis* (both wives of brothers of the husband) received half a bucket each, because 'they arrived when we were about to finish' said Janet.

In total four buckets of groundnuts, which was 22% of the monitored harvest, were paid for labour.

Consumption

On the first day of harvesting two cups were consumed on the irrigation plot by the groundnut pickers and one cup at home by the family. On March 30, two cups were consumed on the plot and two and a quarter cups at home. On the 31st of March the workers on the plot consumed three cups. The 2nd of April, three cups were consumed by Janet, her husband and eight children. On April 4, 7 and 8, the family consumed three cups. On April 9, two cups and on April 20 one cup were used for making peanut butter for the family. On May 7, Mrs Chikwanda, a friend of Janet, was given half a cup, but the groundnuts were roasted and consumed together in the home of Janet. On June 18, two cups were used to make peanut butter for the family.

In total two and three quarters of cups were consumed on 13 occasions. When she prepared groundnuts for seeds in October 1995, another eight cups were consumed. Thus 22% of the monitored harvested groundnuts were consumed in the household.

Gifts

On the 3rd of April Janet gave two cups of groundnuts to a woman not related to her. The woman had given Janet a bottle of cooking oil earlier in the year. Janet explained:

> The woman received cooking oil for free from the *povo* (drought relief)
> which she gave me. Later I had to give that woman two cups of
> groundnuts. It was to show her that I was happy what she had done
> to me. *Akandibatsira nedambudziko rangu* (she helped me with my
> problem). I do not measure that in money. It was *kubatsirana* (helping
> each other).

Also on the 3rd of April she gave one and a quarter cups to a wife of a
younger brother of her husband. Another wife of another brother of the
husband was given one cup a day later. Janet explained:

> They are given because they are *maininis* and they do not have
> groundnuts. I was also given groundnuts by *maiguru* (wife of an elder
> brother of the husband) when I did not have groundnuts.

On the 4th of April Janet gave Mr Sithole one and a quarter cups, because,
as she said, he was a friend. Later he gave her dried meat and a basket. The
sister of the husband (*vatete*) received three cups of groundnuts. Janet
explained:

> I gave three cups because she is *vatete*. Furthermore, *vatete* provided
> three scotch carts of manure at the beginning of this year. The three
> cups was just a gift. She gave me manure which is expensive to buy
> from other people. *Vatete* said 'My brother has no money to buy
> fertilizer.' We are just helping each other (*kubatsirana*).

On the 7th of April Janet gave me two cups of groundnuts. I did not
remember that I gave her something earlier this year, but her daughter was
one of my research assistants. Maybe the mother was happy about her
daughter's employment. On the 5th of May, two cups were given to Mrs
Chikwanda. She was a good friend of Janet and Janet said: 'I gave her because
she also gives me a lot of things.'

When she prepared groundnut seeds in October 1995 she gave away another
four cups. Thus 13% of the monitored groundnut harvest was used for
gifts. It should be noted that more than half of these gifts remained within
the *dzinza* (patrikin) and was given to wives of the husband's brothers. All
gifts could be traced to be reciprocations of earlier gifts received by Janet.

Barter

On the 3rd of April Janet bartered 3 kg of groundnuts with one plastic cup
of vegetable seeds. She bartered with Isaac, a son of Ndaabari's husband

who was her workmate in the field. Janet and Isaac knew each other very well. Janet said:

> Isaac came to me because we are used to each other (*kuwirirana*). *Tinotamba tose, nekunamata tose* (we play and pray together).

Janet said that they are both members of the Zion church. Janet explained the barter as follows:

> Isaac had a small cup of vegetable seeds which had a value of ten dollars. I gave him a *poto* (size 6 pot; almost half a bucket) of groundnuts, because it is more or less ten dollars.

That was the only barter Janet did. Janet said:

> Isaac needed groundnuts and I needed vegetables. I cannot finish a *hozi* (granary) with *kuchinjanisa* (barter) only, because I also want to bath.

Janet meant to say that she bartered a little (2% of the monitored groundnuts), because she also wanted to sell groundnuts, to obtain cash for buying soap to bath.

Sales

Isaac also bought one and a half buckets of groundnuts from Janet on March 30, for Z$45.[4] He sold them in small portions at a school where a football match was organised. Janet explained that he needed groundnuts whereas she needed money to buy salt and soap.

On May 1 and 6, two buckets and two and three quarter buckets were sold for Z$45 and Z$48 respectively, to a buyer who came to the village. The buyer only wanted big groundnuts and selected them with the farmers in their homes. The buyer came from Birchenough Bridge, 12 kms from the place where Janet lived. The sister of the buyer's mother appeared to be the real trader.[5] The buyer explained that she used small groundnuts for peanut butter. The bigger ones were cooked with two cups of salt in a *hari* (claypot) for an hour. She used a claypot because groundnuts would discolour in a tin. After boiling, the groundnuts were dried in the sun and later measured into small plastic bags. The plastic bags with groundnuts were sold for Z$1 each. With a friend she then sold the groundnuts at the market place in Birchenough Bridge. They put the small plastic bags in a dish and carried it above their heads to sell them to passengers in buses. One bucket with groundnuts was equivalent to 54 small plastic bags of one dollar. The trader bought the buckets with groundnuts from farmers for Z$24 per bucket. Hence the trader gained Z$30 per bucket.[6]

In total Janet converted 42 kg (34% of the monitored groundnuts) into Z$163 in cash through selling (equivalent to nearly Z$4 per kg). She spent the money mainly on consumption goods for the family (sugar, soap, etc).

Seeds

By October 1995 three buckets of groundnuts were left. In the third week of October Janet shelled the groundnuts for seeds. She selected the best seeds, whilst the others were consumed (an estimated eight cups) and given away (an estimated four cups). Thus, 7% of the total monitored harvest was set aside for the next season as seeds.

Table 1: Use and exchange value of groundnuts; case 1 (0.08 ha)

	Weight (kg)	Percentage	Frequency of transactions
Sales	42	34	3
Labour	27	22	1
Consumption	28	22	13
Gifts	16	13	7
Barter	3	2	1
Seeds	9	7	1
Total accounted for	125	100	26
Unaccounted for	75		

Yield

The quantitative data of these transactions are summarised in Table 1. Janet harvested a gross six bags and one bucket from two *madhunduru* (equivalent to over 3 t/ha). After drying and selecting the groundnuts, five bags remained. Janet perceived those five bags as her harvest, because it was these groundnuts that subsequently were exchanged and used in various ways. This amounts to a net yield of 2.5 t/ha.

As indicated earlier, it was not possible to observe exactly how all the five bags obtained their value, because the daughter, my research assistant, did not manage to follow exactly the groundnut transactions of her mother. She may have missed some gifts and barter deals. I observed that often women coming from the fields with their vegetables passed through Janet's home to give her some of their produce. Janet then would respond by giving them groundnuts. Janet may have forgotten to inform her daughter of some of these quick and regular groundnut deals. This does not necessarily mean, however, that these taken-for-granted exchanges are not valued by her.

The average yield of groundnuts for Chipinge district is 0.4 t/ha for dryland communal areas (rainfed agriculture) and 0.6 t/ha for dryland small scale commercial areas (1993/1994, Agritex files). For irrigation schemes, the extension service indicates that for cultivating groundnuts a learning process needs to be considered and therefore they use a projected yield for groundnuts that increases during the first three years of cultivation and obtains a maximum at year 4 of 2.14 t/ha (Table 2).

Table 2: Projected groundnut yields for smallholder schemes

Year	Yield (t/ha)
1	1.50
2	1.71
3	1.93
4	2.14

(Agritex, 1994, 575)

The average groundnut yield for the irrigation scheme in which Janet cultivated was not recorded in the past years. An extension worker explained that the extension service did not record the groundnut yields, because the crop is grown on a small scale. In the summer season of 1994/95, a significant 9% of the scheme was planted to groundnuts.[7] The farmers in general said that they harvested two bags per *dhunduru* (0.04 ha), some 2 t/ha.

ANOTHER GROUNDNUT EXAMPLE

For comparative reasons, I wish to briefly present another case of groundnut production and use. Two wives of one husband, a polygamist, also recorded how they distributed groundnuts, which they harvested from one *dhunduru*, 0.04 hectare (Table 3). The husband had a permanent job and the two wives also cultivated a garden along the Save river, besides the two acre (0.8 ha) irrigated plot. The table shows that they did not sell any groundnuts.[8] The groundnuts were mainly consumed within the family (two wives, many small children and the husband) and given away to wives of brothers of the husband and to *madzitete* (sisters of the husband). The fresh groundnuts consumed before harvesting were included in the consumption. Barter in this home was more often recorded than in Janet's home. On three occasions barter took place where one cup of groundnuts was bartered with okra plants, water melons and millet. Half a cup was bartered for vegetables and two cups for one kg tomato seeds. Just like Janet, the two women farmers did not have any problem in mobilising labour to harvest groundnuts. Sisters of husbands and wives of husband's brothers were assisting, but the two women farmers explained that 'Those who helped were not given a lot, otherwise we would be left with nothing.' However, those relatives who helped with the harvest also later received groundnut gifts.

Table 3: Use and exchange value of groundnuts, case 2 (0.04 ha)

	Weight (kg)	Percentage	Frequency of transaction
Sales	0	0	0
Labour	4	9	1
Consumption	17	39	9
Gifts	14	32	11
Barter	5	11	5
Seeds	4	9	1
Total	44	100	27

SOURCES OF KNOWLEDGE CONCERNING GROUNDNUT CULTIVATION

Janet referred to *maiguru*, the wife of the elder brother of her husband, as the first source of knowledge concerning groundnut cultivation. She gave Janet seeds in 1994 and explained how to sow and grow the groundnuts in the irrigation scheme. The second source of knowledge was her own experience which she derived from her own parents' home (25 km from the present home), where they also cultivated groundnuts. Lastly, Janet said:

> The extension officers also talk (*Vadhomeni vanotaurawo*). The extension worker has knowledge; everything needs a teacher (*Mudhomeni ane ruzivo; zvinhu zvose zvinoda mudzidzisi*).

Janet explained that the extension workers taught farmers about sowing in rows, how to weed and when to apply fertilizer. While Janet was explaining, she demonstrated the extension messages in the sand, by putting small sticks in lines and heaping the sand around it. Janet said that the messages were spread through meetings, which were mainly attended by women farmers.

The local extension worker explained that farmers receive knowledge from both sides, i.e through their own experience and from the extension service. He explained that it is mainly women who attend extension meetings. He was happy to work with women, because:

> Women can easily cope with new ideas. Women take up extension, because women are in the fields and men migrate. If men are around they only drink beer.

GROUNDNUTS AND TOMATOES

To put the cultivation and the way groundnuts obtain value into perspective, I want to briefly show that tomatoes obtain their value in a different way. Men were more interested in tomatoes, mainly due to the cash sales involved with Cairns Tomato Company. The polygynous husband of the second case obtained tomato seeds by bartering groundnuts. He himself prepared a seedbed for the tomatoes. After that his wives transplanted the small tomato plants into the irrigation plot, weeded, irrigated, harvested and sold the tomatoes to private buyers and picked and heaped the tomatoes for collection by the tomato company. One of the wives said:

> Men are interested in tomatoes because they are always after money. We are ruled by men. If you want to plough one *dhunduru* for groundnuts he will say: it is for you women, but when 5 *madhunduru* or more are sown with groundnuts it is his.

After the tomato sales the husband obtained the money and discussed together with his wives how to spend it. Janet also cultivated and sold tomatoes and she explained:

> Groundnuts are for women (*Nzungu ndedzemadzimai*), because they are cultivated on a few *madhunduru*. When we grow groundnuts it is the season of maize and therefore we have to request the husband if we can grow a few *madhunduru* of groundnuts. For tomatoes, in the winter season, I do not request *madhunduru* because the whole field of tomatoes is for the husband.

Farmers in general did not perceive groundnuts as a women's crop and tomatoes as a men's crop. Both crops were mainly cultivated by women farmers who were also actively involved in the transactions concerning both agricultural products.

NETWORKS, VALUE, PERFORMANCE AND EXTENSION

Kopytoff (1986, 67) argued that a biography of a thing is as useful as making a biography of a person. Biographies of things can make salient what might otherwise remain obscure. I have taken that perspective to look at the groundnut, where it started and where it ended. In this section I take a closer look into this 'career'. It will emerge that women farmers actively decide what to exchange for what, when and where and what to dedicate to gifts.[9] In the final part of this section I compare groundnuts with tomatoes, and show that the scale of agricultural production influences what types of exchange prevail and who controls the cash proceeds.

Labour

In groundnut cultivation, hired labour was paid in kind. The women who assisted Janet in harvesting were given groundnuts. Janet explained that the quantity given was according to their performance. Groundnuts are much appreciated because of their nutritious value. Especially those not able to cultivate groundnuts themselves prefer to work in exchange for groundnuts rather than cash. Janet did not need to actively mobilise her labour force. They all came themselves and offered to help with the harvest. However, two different types of labour networks could be distinguished in Janet's case. One originated from Janet's own arrangements and friendship, namely a workmate and a friend. Another network originated from marriage, namely the patrilineal extended family (*dzinza*) represented by the husband's mother and wives of the husband's brothers. This type of labour organisation was called *kushandira*, which means that the workers were given something in return for their labour, in this case groundnuts.[10]

Gifts

In the first case 13%, and in the second case 32%, of the groundnuts was given as gifts to other people. Gifts are not a one way and free transaction as economists often assume (e.g Sithole, 1995, 55, 56 and 122). Gifts are reciprocal. Based on the perceptions of the farmers, it emerged that reciprocal gifts do not have the same value ('do not balance') and are not valued in terms of cash; and that they are exchanged at different moments in time, without negotiations about it. The counter-gift is typically delayed.[11] The fact that no overt negotiations take place about the exchange value, implies that gifts do not take place in a market-like context. Gifts, therefore, cannot be perceived as commodities (see Kopytoff, 1986; Long, 1992, 149). Farmers said that a gift is based on helping each other (*kubatsirana*) and the love you feel for each other (the value of affection).[12] Thus gifts occur between people who already know each other. In both of the above cases gifts occurred mainly in the *dzinza* of the husband.[13] In Janet's case the sister of the husband (*vatete*), the wives of the husband's brothers and mother in law received groundnuts. In addition, Janet gave also groundnuts to her friends.

Cheal (1996, 98–99) observes that gifts are mainly practised by women. He points out that:

> part of women's work consists of 'the work of kinship'. Kinship work is an important part of family life. It is necessary for the maintenance of extrahousehold ties to significant others.

Thus, the social meaning of gifts is to establish or maintain relationships, including relationships of mutual dependency. Komter (1996a, 3–12) reviewed different perspectives about gift exchange. Important perspectives are that gift-giving creates the expectation or obligation to give in return, but is also perceived as a means to achieve a common culture and social cohesion. Gratitude is, just like faithfulness, a powerful means to establish social cohesion. Gratitude is, in the sociologist Simmel's words 'the moral memory of mankind', and as such, essential for establishing and maintaining social relations. A gift is an instrument to assign symbolic meaning to fundamental dimensions of personal relationships, such as love, power, or dependency. Komter (1996b, 120) concludes that women's autonomous role in gift exchange and their concomitant strategic power position have remained unrecognised by most anthropologists. Below I will argue that due to women's active negotiations about use and exchange value, they wield power in the agricultural production sphere. In addition, the gift may be an important strategy of women to ensure the household's food security.

Barter

When women farmers bartered, they negotiated at that very moment about the relative value of the products, expressing these values in cash terms. They looked for a proper balance between the cash value of the different products. Groundnuts were often exchanged for another crop, or for seeds. For example, Janet said that a small cup of vegetable seeds had the value of ten dollars. Therefore she gave Isaac a size 6 pot of groundnuts which was also valued at ten dollars. Farmers perceived barter as independent from love and assistance and therefore barter could take place with any person. Barter occurred when both exchanging parties were in need of something.

Farmers explained that barter (*kuchinjanisa*) and gift (*chipo*) both fall in the family of *kupa* (to give), but remain different. One farmer said: *zvakatifananei*, meaning that barter and gift are alike, but not identical (*zvakafanana*). In conclusion, barter is based on cash value of goods, which are exchanged instantaneously, without restrictions with whom to exchange. In contrast, gifts are not explicitly valued in cash terms and reciprocal gifts are not perceived by people to have the same value. A counter-gift comes later, without overt negotiations. A gift is seen to have a strong normative element, as it is interpreted as an act of helping out or an expression of affection.

Sales

Selling groundnuts is in principle done to any person, provided that the farmer who sells has a need for cash to buy essential consumption goods such as sugar, salt and soap. Janet perceived the barter of vegetable seeds with groundnuts as a better transaction than the sales of groundnuts for money. However, she had to sell some because she needed cash for the purchase of some essential consumption goods. She had no alternative source for obtaining cash. One third of the harvest she thus sold (Table 1).

In the second groundnut case the women farmers did not sell any groundnuts. Apart from the groundnuts, the women also cultivated a vegetable garden. They sold most of these vegetables. They thus had a regular source of cash, which was supplemented by remittances from the husband. Apparently, the women had enough cash at hand and preferred to consume the groundnuts themselves and distribute groundnuts through gifts (Table 3).

Exchange and use value

The above demonstrates the exchange value of groundnuts, and that this valuation differs across the various transactions taking place. Labour mobilisation is based on the high nutritious value attributed to groundnuts. Barter occurs when the exchanging parties are interested in each other's products, the deal being negotiated in terms of the cash value of these products. Gift is based on the value of affection and helping each other.

Sales are based on cash value, which implies that cash is needed by the one who sells. However, the 'career' of the groundnut was not only determined by its exchange value. Its use value is obtained through consumption (22% for case 1 and 39% for case 2) and seeds for reproduction (7% and 9% respectively, see Tables 1 and 3).

Groundnuts and tomatoes

The scale of cultivating tomato in the winter seasons (mainly viewed as a cash crop) and groundnut in the summer season (considered mainly a food crop) is different. On average, groundnuts are cultivated on 1 or 2 *madhunduru* (0.04 or 0.08 ha) and tomatoes on 18 or 20 *madhunduru* (0.72 or 0.8 ha). Tomatoes are mainly sold to Cairns tomato company. Women farmers said that if there was a company which would buy the groundnuts, then most probably the groundnuts would also be grown on a larger scale.

Most of the men were not interested in the cultivation of groundnuts, which they left in the hands of their wives. They did of course benefit in some way from this arrangement.[14] The men were more interested in tomato production. Whereas at most a third of the groundnuts harvested were exchanged for cash, nearly all tomatoes were sold (some 80%). The balance was used to pay labourers who assisted with harvesting, and for gifts. It is important to note that the bulk of the sold tomatoes were marketed through the tomato company (70%) and only a small part through local traders (30%).

The chairperson of the Irrigation Management Committee (IMC) signed the contract with Cairns tomato company. Included in the contract were 230 farmers in 1995. The contract implied that farmers received hybrid tomato seeds of the company and were obliged to deliver the tomatoes to the company. The company was then obliged to collect the tomatoes from those farmers. The names of the farmers in the contract were those of the registered plotholders. That meant that mainly men were noted in the contract, whereas women are mainly doing the agricultural work of transplanting, weeding, and irrigating. It is also mainly women who harvested the tomatoes and then decided which boxes they put aside for collection by the Cairns trucks and which boxes or buckets they would sell or barter with private buyers. Members of the IMC recorded the number of boxes each farmer delivered to Cairns. The farmer was paid by Cairns for the number of delivered boxes, on the basis of the IMC list. The price of the boxes was established after the truck with the tomatoes was weighed on a weigh bridge. The price was Z$600 per tonne (1,000 kg). In general a box weighs 20 kg, hence the farmer received Z$12 per box. Members of the IMC visited Cairns once per month in the tomato season (June up to September) and collected the total cash for one month and the lists with the calculated earnings per farmer. The IMC members paid the farmers in the village. On those pay days, more men than women were present, whereas on the irrigated plots the sex ratio is the other way round. The reason why mainly men collect the cash is that mainly men are the plotholders and therefore registered in the contract.

In general, women were not happy about the control of company cash by their husbands, since they do most of the work. There was, however, little they could do: in the majority of cases the irrigation plot was registered in the husband's name, his name was therefore used in the crop contract. The husband would thus not only claim to be the 'owner of the plot', but also the controller of the cash proceeds of the tomato crop (Vijfhuizen, 1996). This generalised picture especially applies to most polygynous homes in the village of the research (34% out of 400 homes). In nuclear homes (18% of the homes), it was often the husband who decided about the cash after a discussion with his wife, like in Janet's case. Unmarried and widowed women (26% of the homes) often decided about the cash themselves.

Thus women were involved in most transactions concerning both groundnuts and tomatoes. A remarkable difference was the control over cash. Women operated on all markets, but controlled the cash from small scale crops or 'small sales' in case of tomatoes. Hence, women did not only work in the fields but also negotiated and decided about exchange value. They decide to sell to private buyers or to deliver to Cairns. When the tomato production is at its peak (July), farmers mainly deliver to Cairns because private buyers are not present, offer low prices or only want to barter with goods. Women's involvement in both crops is maybe a reason why farmers did not perceive groundnuts as a women's crop and tomatoes as a men's crop. Thus a cash-crop is not necessarily a men's production domain as Boserup stated (1970). It can be a matter of control over cash, and then again not in all homes. If formal markets are related to cash crop and informal markets to food crops, we can conclude that women operate on both markets and not, as is often assumed, on informal markets only.

Irrigation technology and extension

Boserup's (1970) influential book was the first one to analyse the position of women in socio-economic development. Boserup argued that when

agriculture is modernised, e.g with ploughs, cash crops and irrigation, the
agricultural production system will change from predominantly female to
predominantly male (see Boserup, 1970, 15–36). She argued that as a result
of the attitudes of the extension services, who only approach men and teach
them the modern methods of cash crops, the gap between the labour
productivity of men and women continues to widen and as a further result,
women will want either to abandon cultivation and retire to domestic life, or
to leave to town (Boserup, 1970, 53–56). In accordance with Boserup's thesis,
Mvududu (1993, 22) and Dikito (1993) found for the Zimbabwean context
that training and extension services were usually offered to men.

However, modernisation through irrigation in the scheme under study did
not result in women abandoning agricultural production and retreating to
the domestic sphere, nor did men take over agriculture. The reality is that
mainly men are registered as plotholders and therefore claim to be the cash
controllers. Nevertheless the agricultural production remains predominantly
female. It is therefore not surprising that the extension workers explain that
they mainly work with and teach women farmers. Women thus operate in
public spheres. Women are important in irrigated agricultural production as
negotiators in the transactions about exchange and use value of all those
small-scale and few larger scale crops. That also means that women wield
power in that sphere. I have argued elsewhere that women and men are both
leaders (*vatungamiri*) in a home, each of them having their own spheres of
power. Women wield power in the agricultural production, consumption
and exchange spheres and men in the homes and for example in the courts
(Vijfhuizen, 1995). Thus the importance of Shona women does certainly
not lie in the spheres of domestic activities and children only, as Holleman
(1974, 114–115) suggested:

> if you want to find out about the importance of women, you will
> have to explore those aspects of African life in which women figure
> naturally and prominently. And these lie, obviously, in the sphere of
> domestic activities and the rearing of children.

Madondo (1992) and Dikito (1993) argued that the introduction of
irrigation contributes to heavier tasks and marginalises women. However,
women farmers in Tawona smallholder irrigation scheme perceive irrigated
agriculture as a relief and not as a burden (see also Chimedza, 1989, 35;
Mvududu, 1993, 17). This is mainly because they live in Natural Region V,
where it is difficult to safeguard the household's food security without an

irrigated plot. Schrijvers (1985, 13) points out that marginalisation is perceived as a process in which one category of people, in this case women, are pushed out of those sectors in the production spheres which wield power, towards the periphery of the subsistence sector, the informal sector and the margins of the economy. Such a process did not occur in Tawona irrigation scheme. Women are not marginalised by an irrigation intervention, but wield power, particularly as decision-makers about use and exchange value.

Economic performance of irrigation schemes

Janet cultivated her groundnuts for the first year and her yield was much higher than the extension service assumes for a learner. This may be attributed to Janet's exceptional farming skills. It may also indicate that the figures used by the extension service are based on inaccurate data.

The yields of small-scale crops are not recorded by the extension service. Not only the production of groundnuts is ignored in the statistics, which was 9% of the land in the 1994/95 summer season, but also other small-scale crops such as wheat (winter), onions (winter), vegetables (summer/winter) and, since 1995, tomatoes (summer). Hence small-scale crops are not included in the calculation of the scheme's economic performance, despite the fact that, taken together, they represent a considerable value.

Only the yields of large-scale crops are recorded in Tawona irrigation scheme, that is tomatoes in winter and maize in the summer season. For the large-scale tomato cash crop, the sales of farmers are recorded. All the other transactions in which the large-scale tomato cash crop also obtains use and exchange value are excluded. The maize crop is the only food crop in Tawona irrigation scheme for which production data are collected by the extension service, because it is grown on a large scale. The extension service records the yields by asking the farmers how many bags they harvested. Farmers and Agritex both know very well that farmers in this smallholder irrigation scheme in Natural Region V always report a lower number of bags than they really harvested. One reason is that irrigation farmers do not want to be excluded from the drought relief programme by the government. Thus they strategically do not reveal what they really harvested, and why should they?

In sum, in a smallholder scheme such as Tawona, much more is produced than is officially recorded. Hence its overall performance is much better than the figures compiled by the extension service, and the economic reports based on these data, would make us believe.

CONCLUSION

This chapter has shown that women farmers are actively involved in the cultivation of agricultural produce and also in the transactions concerning its use and exchange value. The extension personnel are appreciative of the role of women farmers in production, and training and advice is mainly directed to women. The moment women farmers search for and open up opportunities for exchange, and negotiate about the exchange value of agricultural produce, they, as it were, help to make the markets. The difference between groundnuts and tomatoes is the control over cash.

When preparing crop budgets and calculating yields, the extension service tends to concentrate on larger scale crops and the produce through sales. The smaller scale crops are often not recorded. Extension personnel collects data which often does not include non-cash transactions. As a consequence, the performance of irrigation schemes may be significantly under-estimated, as well as the returns for individual farmers. In discarding the small-scale crops, and ignoring the complex transactions made, the official picture dismisses those dimensions that are so much valued by farmers themselves: the crops and transactions that are crucial ingredients for food security in the household and for maintaining networks of relationships.

NOTES

1. I would like to thank the following persons: Locadia Makora for her valuable explanations and support; Servie Mwoyounotsva for observing the groundnut transactions; the farmers who provided useful information and perspectives; Agritex for providing valuable information; Dr Olivia Muchena, Dr Jan den Ouden and the editors of this book for their useful perspectives.

2. One *madhunduru* measures approximately 0.04 ha; 10 *madhunduru* make up 1 acre (0.4 ha).

3. In the local volumetric exchange system, one bag contains six buckets, and one bucket equals seven enamel cups. One enamel cup with dried unshelled groundnuts weighs approximately 950 grammes. So, according to this system, a bucket should weigh 6.7 kg; and a bag contains 40 kg of dried unshelled groundnuts.

4. In March 1995, one US dollar was equivalent to 8.5 Zimbabwe dollars (Z$).

5. In the area of the research, traders in groundnuts are mainly women.

6. The trader's net profit was much lower, as she incurred costs: transport to the village (Z$6 per person and Z$6 per bag), firewood (Z$30 per week), claypot

(Z$10 per week); 2 cups of salt for every cooking (Z$2.40), plastic bags, and an unknown reward to the trader's assistants.

7. Tawona smallholder irrigation scheme (Chipinge district, Manicaland province) irrigates 169 ha and has 260 irrigators. The average plot size per farmer is 2 acres (0.8 ha = 20 *madhunduru*). It is therefore called a 'comma-hectare' scheme. Many farmers cultivated 1 or 2 *madhunduru* (0.04–0.08 ha) groundnuts. The total number of *madhunduru* cultivated with groundnuts was approximately 380 (c. 15 ha) in the 1994/1995 cropping season, or 9% of all the irrigated land of the scheme.

8. The women recorded the groundnut transactions themselves. They also missed many taken-for-granted exchanges, mainly those concerning consumption and gifts (hence the low recorded yield). However, they did not record sales, because they did not sell groundnuts.

9. Verschoor (1992, 178) argued that 'commoditization processes (how goods obtain exchange value), then, should not be seen as the sole outcome of external forces impinging upon individuals, households or enterprises. Rather, I view these processes as social constructions subject to constant negotiation by social actors in specific time-space situations'.

10. Another type of labour organisation was *mushandirapamwe*, which means that farmers assisted each other by working in turn on each other's plots. Hence with *mushandirapamwe* labour is exchanged. This type prevails with the harvesting of maize.

11. Compare with Bourdieu (1990, 105–6), who observes that 'the counter-gift must be deferred and different, because the immediate return of an exactly identical object clearly amounts to refusal. The interval between gift and counter-gift is what allows a relation of exchange'.

12. A Shona proverb says: *Kandiro kanoenda kunobva kamwe* (A plate goes where another plate comes from).

13. Marriage is virilocal. That means if a woman marries she will go and live with the family (*dzinza*, patrikin) of the husband.

14. The men shared in the consumption. They also used the consumptive items bought by their wives with the groundnuts proceeds. The men also benefit from the good relationships within the *dzinza*, which were cemented in part by the gift-giving of groundnuts by their wives.

Chapter 8

Knowledge is like live coals, it can be borrowed[1]

How indigenous knowledge and Shona proverbs can be used in agricultural extension

OLIVIA N. MUCHENA[2]

INTRODUCTION

Indigenous knowledge is unique in that it is generated in response to the natural and human conditions of a particular environment and context. It is dynamic and creative due to continuous experimentation and evaluation stimulated by adaptation requirements and external influences. Rural people's knowledge covers the whole range of human experiences, and straddles different academic disciplines. It would include, for example, the physical sciences and related technologies (including among others agriculture, ethnobotany, ecology, medicine, climatology), the social sciences (including economics, social and cultural organisation, political science, religion), and the arts and humanities (for instance communication). This observation facilitates the understanding for those used to the Western academic tradition, but it should be noted that indigenous knowledge is not compartmentalised. An indigenous knowledge system can thus be defined as a learned way of knowing and looking at the world unique to a geographic or ethnic context.

Warren (1989) shows how 19th century social sciences contributed to negative values and attitudes towards indigenous knowledge systems through ethnic stereotypes and prejudices. In Africa, colonisation and the alienating effects of Western education systems contributed much to this process. Indigenous knowledge was side-lined because of the exclusive nature of Western science. For Western science 'reason constitutes the only legitimate

---

[1] Translated from the Shona proverb '*Njere moto unogokwa*' (see text for an explanation).
[2] This chapter is based on Muchena (1990).

way of knowing and is the only arbitrator of truth' (Sardar, 1988). Western science has been criticised for its mechanical, reductionist world view. The supposed neutrality and universality of positivism has been questioned. In contrast, indigenous knowledge is seen as holistic and inclusive in how it approaches reality. Agriculture, for example, in many indigenous knowledge systems is understood in both technical and social terms, as it encompasses cognitive and affective dimensions. Land management would include technical knowledge such as terracing as well as taboos, often phrased in religious terms, and sanctions regulating land use.

It has been argued (for instance by Njoroge and Bennars, 1986) that there was nothing like knowledge for its own sake in pre-colonial African societies. However, anthropologists such as Levi-Strauss have refuted this. Richards (1985) found that Sierra Leonean farmers, when asked why they planted a certain type of rice, responded, 'For experiment'. The interest in experimenting goes beyond the utilitarian value for, as Richards indicates, 'Playing with rice is the national sport of rural Sierra Leone' (p.145). Thus, knowledge is actively created, and should therefore be understood in terms of a process: cognitive mapping, acquiring, coding and decoding information. Understanding this process of how indigenous knowledge is produced and refined can provide a basis for genuine dialogue between farmers on the one hand and researchers, extension workers and agricultural educators on the other (Freire, 1973; Howes, 1985).

The aim of this chapter is not to reconstruct and maintain a romantic static past, but to examine ways that the contribution of indigenous knowledge in sustainable agriculture can be effected. An emic-etic approach to agricultural education and extension is a possible route. This is explained in the section that follows. Subsequently I will focus on Shona proverbs as a source of indigenous knowledge. I will explore how these proverbs may play a role in agricultural extension programmes.

THE EMIC-ETIC PERSPECTIVE: TOWARDS A NEW CURRICULUM FOR AGRICULTURAL EXTENSION

Synthesising the cognitive models underpinning both indigenous knowledge systems and Western agricultural education would result in an emic-etic approach to agricultural extension education. According to Goodenough (1981, 16),

> . . . when we describe any socially meaningful behavioral system, the
> description is an emic one to the extent that it is based on elements
> that are already components of that system, and the description is an
> etic one to the extent that it is based on conceptual elements that are
> not components of that system.

The Shona perception of and values on land and its management is an
example of an emic view because it is based on the internal, functional and
structural elements of a particular cultural system. On the other hand, the
Western view of freehold (private property) as a basis or motivation for land
development is an etic perspective when applied to the Shona cultural
situation.

Ruttan (1988) indicates how in development literature cultural endowments
are often viewed as obstacles to technical or institutional change. One reason
for this view is that cultural endowments are analysed from an etic perspective
using predetermined general concepts. A large component of the emic
perspective comprises religion, taboos, myths and related ethnic ideologies
and values. These elements of indigenous cultural systems may appear
irrational from an outsider (etic) perspective.

An emic-etic framework for analysing the potential contribution of
indigenous knowledge systems would reveal what is genuinely useless, what
has comparative advantage over Western knowledge and what has intrinsic
value for unique knowledge generation. The challenge is to identify those
cognitive, affective and practical elements in a knowledge system that provide
the basis for genuine participation and collaboration in agricultural extension
education and natural resource management, and would signify an important
contribution towards sustainable development.

The holistic nature of indigenous knowledge systems appear to have
elements of several curriculum orientations: its humanistic orientation is
apparent in the affective and cognitive dimensions; its behaviourist and
technological orientations are apparent in the physical, biological and social
dimensions. A mere integration of few elements of indigenous knowledge
systems in existing agricultural extension curriculum is a band aid approach.
An emic-etic perspective (Figure 1) includes the recognition of the integrated
nature of indigenous knowledge systems. The emic perspective provides
the internal conceptions and perceptions of agriculture while the etic
perceptions provide the framework for determining the effects and/or
significance of beliefs on human ecological behaviour (Lovelace, 1984).

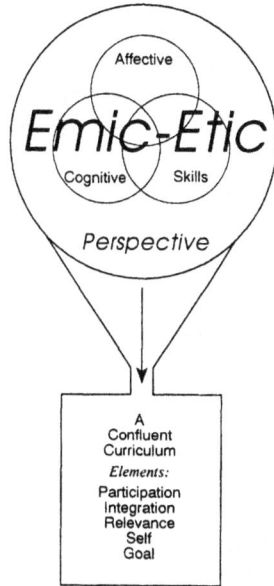

Figure 1: An emic-etic perspective for a confluent curriculum approach to indigenous knowledge systems

The above requires a systematic, well-structured process of synthesising elements of current agricultural education and extension with the holistic, integrated nature of indigenous knowledge systems. Current literature on indigenous knowledge validates its technical merits. The barriers to its recognition and appropriation by professionals, institutions and alienated farmers themselves are rooted in the affective domain: negative attitudes and value judgements (Thrupp, 1988; Marsden, 1989). An initial emphasis on the affective domain of an emic-etic based curriculum would be aimed at countering these professional and personal negative attitudes and institutional barriers to indigenous knowledge systems.

A curriculum that combines the cognitive psychological aspects of indigenous knowledge with emic-etic perspectives is likely to produce extension personnel who can effectively manage knowledge and information systems from the farmers and research stations. Such a curriculum could be called a 'confluent curriculum'. According to McNeil (1985, 11),

> The essence of confluent education is the integration of an affective domain (emotions, attitudes, values) with the cognitive domain (intellectual knowledge and abilities).

It is important to note that ignoring the affective domain variables in learning tends to lead to failure in cognition and psychomotor domains as the learner may fail to conceptualise or appropriate a value set. Whether agriculture is viewed as business or as a way of life, there are values and attitudes that go with each respective view. One possible reason for the neglect of the affective domain in agricultural education and extension is that affective variables are seen as difficult to identify, promote and measure. An emic-etic framework of analysis and a confluent curriculum could provide some solutions to these problems. The elements of a confluent curriculum essential in the process of designing an emic-etic training approach are highlighted below.

Participation
Marsden (1989) argues that people's participation should not be limited to decision making roles or to some consultation. Farmers should be the co-generators of knowledge. Freire (1970), Chambers (1983), and Richards (1985) have noted how researchers and extension professionals assumed ignorance of the farmer, which then becomes a major obstacle to dialogue and exchange of information. A curriculum approach that promotes participatory values through teaching and learning methodology, and content (teaching students how to identify and record indigenous knowledge systems, for example) is likely to have lasting effect.

Integration
This principle involves interaction, interpenetration and integration of feeling, thinking, and action (McNeil 1985, 12). Western trained extension workers have learned to compartmentalise the domains of their knowledge. This is contrary to the African world view, for example, where knowledge is not just a matter of cognition but of feelings, values, morals, and ethics as well.

Relevance
Relevance of what is learned to the learner is important especially in adult extension programmes. Universal prescriptive solutions (agroforestry, for example) typical of the dominant knowledge system may fail because of factors in the context that were not planned for. Knowledge that is situation and locality specific is a characteristic of indigenous knowledge systems that need to be appreciated for its relevance.

Self

This principle of confluent theory espouses self as a legitimate object of learning. The alienating effects of Western education need to be counterbalanced by what people can contribute to the process of experimenting, generating, synthesising and hybridising knowledge in agriculture and natural resource management. Participation in knowledge generation leads to appropriation and ownership of what is learned.

Goal

The social goal of a confluent curriculum is to develop the whole person within a given society. There is need to make African extension programmes holistic rather than just looking at the farmers as a recipient of technology to increase productivity for national and international markets.

In addition, the elements of a confluent curriculum and the nature of indigenous knowledge as defined above lend themselves to the principles of sequencing in curriculum development. Thus, starting with the farmers' indigenous knowledge, extension educators can move from the simple to the complex, from the familiar to the unfamiliar, from the concrete to the abstract. Such an approach will ensure a genuine two-way information exchange system.

SHONA PROVERBS AS A SOURCE OF INDIGENOUS KNOWLEDGE ON NATURAL RESOURCES

Proverbs belong to the genre of oral literature. Proverbs as a tool for communication are used for emphasising ideas, warning, correction or summarising an argument. Though their metaphorical meaning may be important, this chapter concentrates on the literal text of proverbs as a source of indigenous knowledge on natural resource management. Trees and forests are high risk natural resources in Zimbabwe, and therefore proverbs were selected with reference to trees and forests. Most of the proverbs mentioned in this section were selected from the Hamutyinei and Plangger (1987) collection.

Roth (1987) identifies the biophysical, sociocultural and environmental management as the three constellations that environmental management should be concerned with. According to Roth,

> Environmental management education is concerned with an individual's self-understanding, and understanding of the co-

inhabitants of the earth and inter-relationships within and among
each of these constellation of concerns (p.30).

The policy document 'National Conservation Strategy, the Zimbabwe Road
to Survival' (Zimbabwe, 1987), focuses mainly on the biophysical and
environmental management aspects. The approach appears to be technocratic
with little attention to the socio-cultural elements. But the objective of the
strategy to make conservation 'an integral part of the consciousness of every
member of society' cannot afford to neglect the cultural element.

Natural resource management, and environmental conservation, then,
require popular support and cooperation from the people of Zimbabwe,
something which up to independence in 1980 had been lacking in Zimbabwe
(see e.g. Beinart, 1984; Cliffe, 1988). The task of making environmental
education interesting, meaningful and integral to the life of the young
Zimbabweans is a challenge. Humour is a pervasive aspect of Shona proverbs.
This can be used to capture the attention and interest required. By their
nature, proverbs have an important pedagogical potential. They combine
the metaphorical, and abstract reasoning, with a strong imagery and a typical
structure. These aspects lend themselves to the teaching and learning of
concepts like analogy, inferences, causality and generalisations (Penfield and
Duru, 1988). The learning principle of starting with the known to the
unknown applies here, based on the nature of the concrete imagery of Shona
proverbs.

In traditional Shona society, parents, relatives and community elders were
responsible for educating the young, through various means. Through
apprenticeship, particular age groups were given responsibility for specific
tasks such as hunting, herding or collecting forest products. Initiation
ceremonies and work parties (*nhimbe*) provided further opportunities for
learning. Proverbs will have played an important part in oral presentations.
In Shona a proverbial expression is usually prefixed by '. . . *ndosaka vakuru
vakati* + proverb' (that is why the elders said + proverb). Hamutyinei and
Plangger (1987) point out that '*vakuru vakati*' (the elders said) is indicative of
both the power of sanction by elders and approval by the generalised other.
Makina (1981, 35) rightly concludes that,

> Compared with the present day efforts in adult conservation
> education, there was more done in traditional society. I contend that
> today's adult is less informed, less knowledgeable about wildlife
> conservation than the same adult a hundred years ago.

The traditional heritage incorporated in proverbs may provide a base for contemporary education programmes.

Knowledge as contained in selected proverbs

Observation through the senses

The process of knowledge acquisition in indigenous knowledge systems involves various modes and levels of observation as is the case in any scientific knowledge generation process. Careful observations, through the senses, yielded knowledge about the presence or absence of certain species and their qualities:

> *Mazhanje ari musango/asi handi masango ese ane mazhanje*
> The mazhanje fruits (wild loquats) are in the forest, but not all forests
> have mazhanje fruit trees

> *Kuona onde kutsvuka/imo mukati mune honye*
> A fig can be red (brown and ripe) outside/yet there are maggots
> inside

The landscape, with its plains and forests, is an important element in many proverbs. Attributes of the vlei (plain), with its sparse vegetation, makes it an unlikely place to hide in case of trouble compared to the forest.

> *Zvandiwanira pabani/dai manga muri musango ndaihwanda mumuti*
> It has found me in the vlei/if it was in the forest, I would have hidden
> in the tree

However, under different circumstances, even the forest is no place to hide because it is common property.

> *Dondo harina mbikira*
> A forest provides no place for hiding (something)

The danger of big fires whose origins are small and unnoticed is recognised in the following:

> *Kamoto kamberevere/kakano pisa sango mberi*
> A small and sneaking fire burns forests far ahead (can cause a big
> fire)

Cognition by inference

Knowledge about plants conveyed in the proverbs text include plant physiology:

Kuyevedza kwemaruwa kunobva mumidzi
The beauty of the flowers comes from the roots

Proverbs also concern entomology and contain information about pests:

Ngezvemo zvimbutu zvamaonde
It is typical (like) insects inside a fig

Plant-animal associations in the ecosystem are conveyed in many proverbs:

Kure kwegava ndokusina mutsubvu
Far from a jackal is where there is no mutsubvu fruit tree

Gunguo rinodya mbamba/rino muchero waro
If there were no bambara nuts, what (fruits) would the crows eat?

Values contained in knowledge
In the affective domain, evaluative or connotive remarks are given in relation
to attributes of a natural resource referent in the proverb. The value of the
flower to the bee is in terms of pollen contribution; thus a wilted flower
does not attract the bee:

Ruva rasvava harikwedzi nyuchi
A wilting flower does not attract bees

Similarly, the chakata tree is evaluated first of all in terms of its productive
value to humans, and only thereafter aesthetically:

Totenda maruva/tadya chakata
We believe in the blossoms after eating the chakata (parinari) fruit

The following proverb conveys a sense of wonderment about how a
positively valued attribute can cause something which is regarded negative:

Matende mashava/aovazva doro
Beautiful red calabashes sour the beer

Phenomena observed in nature also helped to make sense of phenomena
that occurred in society:

Mukuwasha mukuyu hauperi kudyiwa
A son-in-law is a fig tree; he never stops being consumed

It can be seen from these proverbs that they are indeed an encapsulation
of ecological information born out of close observation and interaction
with the natural environment.

Use-values as contained in proverbs

Ecological basis

Knowledge of plant properties and of the wider environment in which these plants thrived informed such practices as designated methods of collection. The herbalist needing the bark of a particular tree would debark in a manner that ensured the tree or plant would not suffer from excessive sap loss (Makina, 1981). Similarly only dead wood on the ground was to be collected for firewood contrary to the present practices of tree cutting for firewood, caused by increased land pressure. Relevant proverbs referring to resource management practices that are informed by ecological knowledge are:

> *Gavi rakabva kumasvuuriro*
> A fibre came from its bark
>
> *Hombarume haiiti shura ne sango*
> A hunter has no mysterious notions about the forest.

Social basis

The following proverbs indicate the right of any member of the community to forest products like fruits and firewood:

> *Muchero wakurumbira wakuwa*
> The plentiful fruit is short-lived
>
> *Muchero wesango hauvimbwe nawo*
> The fruit of the forest cannot be relied on
>
> *Mukadzi isango rehuni/rino tsvakwa naani naani*
> A woman is like a forest in which anyone can look for firewood

In fact, the literal texts of these proverbs are reminiscent of Hardin's (1968) notion of the tragedy of the commons. According to Hardin, resources held in common will inevitably be over-exploited and degraded. While the literal meaning of these proverbs convey a similar impression, in actual practice the Shona situation was not one of open access. Common property as applied by the Shona had specified behavioural rules for resource utilisation. One of these rules is revealed in the following:

> *Matanda masairirwa/unosiya nerino muchenje*
> (Standing) logs are (to be) tested; you may leave the one eaten by ants

Hamutyinei and Plangger (1987, 24) explain the following proverb:

Maunga marema kudya muti waagere
Hairy caterpillars are stupid, they feed from the tree on which they
live

'This proverb may be used to dissuade boys from cutting a tree in order to
get its fruits which they cannot reach from the ground.' This and other
common property rules ensured equitable distribution of resources between
members of the community, and between them and the wild animals, at
levels that safeguarded natural reproduction. These rules and regulations
more often than not were backed by religious sanctions in the form of taboos.

Taboos

The presence of sacred forests is widespread in Zimbabwe. Sacred forest
taboos operated through limited access or stipulations of desirable behaviour
in those forests (Makina, 1981). Some trees were tabooed as firewood because
of their offensive smoke (to the eyes) or smell, or because of their specified
uses: for making tools (hoe or spear handles) or for ceremonial occasions
like burials or grave-side trees. There is a taboo against cutting fruit trees or
obtaining ripe fruit from a tree by throwing an object. This is the import of
the 'trick' aspect in the proverb:

Mashanje echakata/ kutsvukira mumuti kuti ndiposherwe
It's the trick of a chakata (parinari) fruit to redden (ripen) on the tree
and so become a target

The individual is enticed to break a regulation to obtain the ripened fruit
through prohibited means. Breaking a taboo was seen to have individual and
social implications. A violation of some taboo was supposed to affect the
whole community (for example, induce drought). The social implications
led to social control. Restitution rituals would involve the whole community.

Respect for nature

The following two proverbs sum up the general respectful attitude of the
Shona towards the natural environment. This attitude was based on the belief
of the pervasive God-Mwari of the cosmos (Mbiti, 1970).

Chakupa sango hachishorwe
What the forest gives you should not be despised

Tendai/ muchero ugowisa
Be grateful to the tree so that it may yield more fruit

Respect of what Musiki (the creator) had given was incorporated in what could or could not be said about a given aspect of the natural and physical world. To say this fruit is tasteless, or this mushroom is rotten, or that baboon is ugly was considered a serious spiritual offense. Punishment included loss of sense of direction while in the forest.

The natural resource management practices implied in the above proverbs ensured ecological stability and bio-diversity. The proverbs reveal information in cognitive, affective and skills domains related to natural resource utilisation and values. How can proverbs be used in environmental management education?

APPLICATION OF PROVERBS TO EXTENSION EDUCATION PROGRAMMES

Indigenous knowledge and the values contained therein can provide the framework for education programmes that are applicable to contemporary demands. Taboos and sacred phenomena, for example, have been affected by secularisation, Christianity and Western scientific influences. Fear of the bizarre or religious sanctions are no longer the basis of environmental related behaviour. The situation requires generation of new values based on what is familiar. Trees and forest protection and afforestation programmes could use concepts of taboos and stewardship to explain the negative long term effects of forest clearing. Government regulations and other similar measures could be translated as the taboos of the day. The proverb

> *Mbudzi kudya mufenje/kufana nyina*
> (If) a goat eats cabbage-tree leaves, it imitates its mother

can be used to explain that things can no longer be done according to tradition without modification. Knowledge is generated in response to prevailing conditions and demands. Chimhundu (in Hamutyinei and Plangger, 1987) gives proverbs based on modern imagery. For instance,

> *Saga reshuga/rakapera netisipunu*
> The bag of sugar/was finished by a teaspoon

> *Chabhenda chabenda/njanji haiswatudzwi* .
> What is crooked is crooked/a railway line cannot be straightened

These proverbs clearly depict prevailing conditions. Unlike Chimhundu, who does not encourage conscious coinage of proverbs, we suggest that such a conscious effort should be tried. The young people, for example, could be challenged to come up with proverbs or stories that convey environmental concerns from their perspective. Such a project could provide a vehicle for enjoyable learning in response to matters of serious national concern like conservation.

Proverb text is rich with material that can be used in developmental literature for formal and non-formal education. The proverb

> *Mwoyo muti unomera paunoda*
> The heart (of love) is like a tree, it germinates where it likes

compares the free spirit of love with the ubiquitous tree or forest. This proverb has been used by the author to write a page long teenage love story located in Mabvuku residential area of Harare, once surrounded by thick forest. Today the Mabvuku area is like a desert, trees no longer germinate where they like. The simple story was used as a basis for discussion and raising awareness on deforestation, causes of soil erosion, and the need to see economic demands of agriculture and fuelwood within a long term context of environmental impact. Culture based teaching-learning materials can contribute to reducing developmental conflicts of technical packages that are not based on emic perspectives.

Extension is almost synonymous with communication. It is common knowledge that young extension workers communicate in a mixture of Shona and English (colloquially known as Shonglish). The author listened to an extension worker give instruction on poultry and how small creatures called bacteria, which cannot be seen by the naked eye, can cause diseases. In Shona beliefs, disease can be caused by invisible people called witches. The equating of bacteria with witches can cause some 'noise' in communication (Rogers, 1983). However, if the concept of witches is used as a familiar frame of reference to explain bacteria or germ theory, there is likely to be 'similar comprehension of the object of communication'. Unfortunately communication in extension seems to be based on the premise that farmers must understand phenomena within the Western scientific framework. This hinders the learning process. The objective should be finding aspects within the culture that can be used as teaching and learning vehicles. Proverb text or other indigenous communication media can be used in this manner effectively.

Another important aspect of communication is the contextual use of proverbs. Proverbs in speech can be used to reinforce ideas or to correct behaviour. Both aspects can be applied in extension. The corrective use of proverbs is extolled for the indirect, diplomatic but terse way of pointing to negative behaviour without putting down the addressee (Penfield and Duru, 1988). An extension worker could use the proverb that follows, for example, in a humorous way to indicate that if soil conservation practices are not followed we might get to such low yields that it can be said of us,

> *Yaruma sei nzara/hurudza kurarira hute*
> How biting the hunger must be that (even) an expert farmer has wild plums for dinner

Proverbs may reinforce existing ideas. A spokesperson of a village farmers' group thanked the extension personnel at the end of a field day in Kachuta village, Guruve (May 1987). He referred to the following proverb:

> *Njere moto unogokwa*
> Knowledge is like live coals (in the traditional cooking fireplace), it can be borrowed (by a neighbour to start their own fire in the absence of lighters or matches)

This proverb conveys the central concern of this chapter, i.e. indigenous knowledge can be a basis of knowledge exchange and generation in which farmers participate as equals. Fortunately, there is some evidence of appreciation of people's knowledge as conveyed by Pito Shiri, a Zimbabwean extension worker, who said,

> The most important thing is to get to know people's traditions and humble yourself before them. Don't try to be different or act as a 'professional' from outside. If you can get farmers to understand that they know things you don't know, then you have got them on your side (Sofo *et al.*, 1980, 30).

CONCLUSION

Bown (1975, 1) aptly remarks that African policy makers

> may criticise foreign educational content, but do not criticise the structure or methodology of imported western systems of education, whether of formal schooling or of adult and continuing education.

This chapter is a response to Bown's challenge. Knowledge and values creation on the basis of indigenous knowledge is a form of mental and economic decolonisation given the colonial and racially dehumanising past. The material presented in this chapter shows that this is a fruitful exercise. It underscores Wilson's observation that

> local theories [on trees in the fields] are highly sophisticated and combine well with recent advances in scientific ecology, largely in opposition to the technocratic approach adopted by the state (Wilson, 1989, 380).

The analysis of some Shona proverbs on trees, forests and related taboos, indicate the possibilities of presenting indigenous knowledge in a form relevant and appropriate for environmental educational planners and teachers. Proverbs emerge as potential 'bridges' between conservation concerns by national planners and the perceptions and understanding of these problems by local people. Thus, generating and promoting relevant natural resource management practices through adaptation of cultural values and their synthesis with Western scientific ecology is a feasible project. Agricultural education and extension research and development conducted in an interdisciplinary framework with farmers engaged as equal partners could contribute to such a project.

Agriculture in indigenous knowledge systems is a way of life encompassing economic (food supply), social (land as common property implying common values and responsibility), religious (land as a God-given resource), and political dimensions (land administration as a responsibility of the chief and/ or government). This holistic approach could be used as the basis of an agricultural extension programme on land management and conservation.

Although the land tenure system and values have changed to a certain extent, there are social responsibilities for all landholders that are motivating factors for good land management. The present and future well-being of children are central values for parents and communities. This value-set encourages the conservation of resources for future generations, and is epitomised in the following lyric, which was composed by a group of women farmers during a Harare workshop in November 1987, and sung to the tune of 'What a friend we have in Jesus':

Makoronga oti fungisa   Gulleys (due to soil erosion) remind us

Kuti tigadzire minda   To conserve land/soil
Kana tazosiya nyika   So that when we depart from earth

Nyika isakukurwa.	The land will not be washed away.

Chorus:	Chorus:
Kana minda yatorwa nevana	When the children take over
Pane nyika tagara	The land we have been living on
Hapazowe nemibvunzo	There will not be such questions
'Kuti uyu munda ivu riripi.'	as 'Where is the soil of this land?'
Mombe ne hwai zvati vunza	The cattle and sheep will also ask
Tine vana toisepi	'Where do we raise our children?'
Tsuro ne mbeva zvinodaro	The rabbit and the mouse will say,
Huswa nemiti zviripi?	'Where is the grass and trees?'

Part IV
Encounters

Chapter 9

Farmer-extensionist encounters in Nyamaropa
From clashes to a negotiated order?

DUMISANI MAGADLELA

This chapter is about encounters between farmers and agricultural extension staff in Nyamaropa irrigation scheme in Eastern Zimbabwe. It shows how farmers perceive the role of Agritex,[1] and how Agritex staff are caught between administrative and advisory roles. The empirical evidence presented below is used to analyse farmer-extensionist relations as they change and shift in daily interaction between the two parties. The main argument put forward is that farmers enrol extension staff into their own projects to gain leverage in their group conflicts, and extension staff take advantage of farmers' divisions in attempts to enforce or impose specific ways of water management.

The first part presents encounters among Agritex staff, and between Agritex and farmers during 1994 when Agritex chose to take a not-so-active role in the way farmers dealt with their conflicts and water use problems. The second part looks at the 1995 season where the extension department decided to actively influence the state of affairs by forming Block Committees parallel to the Irrigation Management Committee[2] (sometimes referred to as the Management Committee). A third part is a presentation of an encounter between farmers and the Irrigation Supervisor in 1996 whereby the Supervisor ordered the closure of water to all farmers who had not paid their fees for that season, and crops were suffering. A fourth part introduces the crux of Agritex's problem in Nyamaropa from a farmer's perspective, being the confusion over what role the post-independence extension service has to play in an irrigation scheme. Finally, there is a discussion of implications and meanings of the presented encounters. The argument throughout the chapter is that people are manipulative of their relationships, and they strategise in their encounters with other actors in their social lives. Different farmers use their ethnic, kinship, political and other social resources to gain influence and have a say in irrigation management, or to upstage their opponents in the contests of control over the irrigation scheme.

In this chapter views of farmers and field staff are presented in long quotations, as they were expressed in specific encounters with each other, or with the researcher. The main purpose is to bring out dynamics of encounters as they occurred. I will argue that 'at the back' of these minute encounters is a process of change in Nyamaropa.

THE SETTING

Nyamaropa irrigation scheme is in Nyanga District of Manicaland Province. It was constructed between 1956 and 1960, and after several phases of development, now commands 450 hectares. During the period of study (1993–6) there were about 430 plotholders,[3] one Irrigation Officer, one Supervisor (both responsible for the irrigation and its surrounding dryland areas), four Extension Workers, and one Water Controller for each of the four Blocks making up the scheme. The four Water Controllers had their foreman, making a team of five in charge of water distribution. There were about 20 General Hands who helped maintain irrigation infrastructure. Most senior of all Agritex staff in the scheme was the Officer, followed by the Supervisor, Extension Workers, Water Controllers and General Hands.[4]

Extension Workers were supervised by the Supervisor who was in his sixties and was experienced in extension-related work. The supervisor said that there was need for both Extension Workers and Water Controllers to tighten the way they worked with farmers and be strict. Extension Workers said that he was sometimes too harsh with them (and farmers) and ordered them around. The Irrigation Officer, who was much younger than the Supervisor (in his thirties), was more reserved and preferred to carry out his duties quietly.

A cocktail of relations

The social situation among farmers in Nyamaropa presented a cocktail or relationships that characterised the heterogeneous nature of the population, derived, partly, from the nature of the project as a settlement scheme (see Reynolds, 1969; Magadlela and Hebinck, 1995). There were immigrants (mostly of the Manyika ethnic group) who came to join the scheme from around Manicaland Province and beyond, and locals (the majority of whom belonged to the Barwe ethnic group) who were resident in the area before the construction of the project. Most locals initially rejected the idea of joining the scheme, but decided to join when they realised 'the general good' of irrigating.

Unfortunately, most of the locals only managed to get smaller plots (an average size of two acres each as opposed to an average of four acres for those who joined earlier), and in parts of the scheme with *vlei* soils. The Irrigation Management Committee was formed by Derude[5] in 1983 as one of government's policy objectives to encourage or foster farmer participation as a step towards eventual farmer management of irrigation projects. The Irrigation Management Committee had a large Manyika immigrant following and Block Committees were backed mostly by local Barwe irrigators. Block Committees were formed in 1995 as a result of Agritex's dissatisfaction with the IMC together with other farmers' dislike of IMC leadership.

The basis of the disenchantment among the three parties was the displacement of locals from their land when the irrigation scheme started (they refused to join the project); the fact that locals who joined later got smaller plots; alleged lack of respect for traditional values by immigrant Manyika irrigators; and Agritex's resentment of IMC influence among some irrigators. Agritex staff purported to be neutral in dealing with all farmers, but sometimes got caught up in farmer-politics.

AGRITEX'S 'UNDERCOVER' STRATEGY

The 1994 winter season was one of the dry ones in Nyamaropa area, and this affected both irrigators and dryland farmers in such a way that they had to look to the little irrigation water to see them through to the next season's harvest. There had been talk of water being misused or stolen by some farmers in some blocks in the scheme. The Supervisor said that he would call a meeting with Water Controllers to thrash out once and for all the issue of how some farmers who had not paid up their fees after the June 30 deadline were being given water.

The meeting was attended by four Extension Workers and the four *nyamvura* (Water Controllers). The situation was tense (there were no farmers). The Supervisor was chairing the meeting. He went straight into the issue at stake,

Gentlemen I am disappointed with the work you are doing. Cows are all over the scheme, you do not do anything about it, they are walking all over the canals, and nobody has filed a report. Some of you want to be too nice, why? We never see you in the scheme, why? I don't want any comment from you. I am shifting you around. You have names of people who are not supposed to be given water, but you went ahead and gave them in other areas, how come? Why are you so supportive of farmers? You are selling us out, if I see this again, you will lose your jobs. You make us look evil and you look good. Why is it that people irrigate at will without anyone to monitor them? This is why so much water is lost. If you think you cannot manage any more, then leave. You are paid Government money to do this job, but you neglect it. You mess up your work and then ask for some days off. What is wrong with you?

When no response came from Water Controllers, Extension Worker Sithole came in with a question, 'How did some farmers get water when they had not paid up their fees as required?' Water Controllers pointed out that some farmers had proved to them that they had finished paying by showing them their receipts. Sawunyama, one of the farmers in question, and one who was said by Agritex to be a constant trouble maker, irrigated his crop and said that he was going to pay later, but the standing Agritex rule was that farmers in such situations should not be given water at all until they had paid up all they owed.

The Supervisor told Water Controllers to check with the irrigation clerk before giving water to anyone who was on the list of defaulters. He emphasised that 'If you open water for farmers you have to hang around until they are through with irrigating, they must be monitored to avoid losses and theft of water'. Water Controllers took all the criticisms quietly and left the talking to their superiors. Extension Worker Sithole then gave some advice to the Water Controllers on farmers. He said,

> Some farmers are very clever, you have to try and beat them in their attempts to beat you and your system, you have to join hands the four of you and work closely together, every now and then you have to check with each other which farmer appears in the other's papers so that you do not give the wrong farmer water when you are not supposed to.

The four Water Controllers acknowledged the piece of advice which they later admitted to me was nothing new since they argued, they worked closer with farmers and under more risky conditions than Extension Workers. One of them, Samunda, a local,[6] said that extension staff did not want to get into tough situations with farmers such as directly limiting water use or regulating and scheduling irrigation turns; 'they leave all the dirty work to water controllers who are always at the point of friction with farmers, while they sit in their offices'.[7]

Then there was mention of the problem of farmers having plots in two or three different blocks. When their names appeared on the list of those to be denied water, they acted as if it was only for one plot in one block, and continued to use the other one as if nothing happened. Agritex did not confront this situation directly, and quite rightly so I thought. After observing and listening to their problems, I came to realise that there was a critical link between staff and farmers which sometimes made it impossible for staff to directly confront farmers who flouted irrigation regulations and rules. The Irrigation Officer argued that they themselves were humans too and not blind and unfeeling enforcers of regulations. They were aware that they were an inherent part of the situation in Nyamaropa, and their lives were interlocked with the farmers.[8]

Water controllers apparently played an important (but not well-defined) role in promoting water piracy. After one of the Extension Workers had said that Water Controllers were sabotaging the work of extension staff, the Supervisor said,

> These Water Controllers are undermining us, they are weakening us in the eyes of farmers by working against our resolutions and the laws that we have put up. They are supposed to be promoting or supporting what we recommended and agreed upon, but they undermine it. They should be taught a good lesson too. What we need to do now is to work on them very carefully. What do we do now?

He went on to say,

> Let us shift them around, they have stayed too long at the same place and are feeling too warm. This may make them a bit tougher on farmers who demand favours. If they continue to flout the regulations like they have been doing, then they will have to join the general hands or leave.

The four Water Controllers were subsequently shifted around, but not before a little debate on which block and which Extension Worker was going to get which Water Controller. After being shifted, there were some complaints from some farmers about the 'new' Water Controllers who did not know which farmers were supposed to get water first. Later, it seemed farmers came to accept the Water Controllers in their blocks and worked with them.

Since the 1994 winter season was a dry one, farmers wanted to know how much water was available and how much acreage for winter crops was recommended by Agritex for the season in light of the available water. But no meetings were held early enough either by Agritex or the Irrigation Management Committee (or both as had been the case for the past seasons), to discuss the matter with farmers and possibly decide what acreage each farmer was to plant winter wheat and beans. When less water became available, and irrigation intervals increased from 14 days to over 25 days, farmers demanded a meeting with both Agritex and the Management Committee.

Agritex alleged that Management Committee members were undermining them. Previously, they would hold closed meetings with them before a general meeting with farmers, and agree on how to handle or present certain cases or issues to farmers. But during several meetings the Irrigation Management Committee changed what was agreed in the closed meeting to present a different picture altogether, thereby leaving Agritex in the cold. Agritex staff declared that they were not going to have any more meetings with the Irrigation Management Committee prior to a general meeting. They were not going to organise any meetings with or for farmers themselves since that was the task of the committee, and they were not going to sit in front of the audience with the Irrigation Management Committee during meetings any more. They were going to sit with the rest of the audience as mere observers.

When the meeting to discuss the water situation was held, the few (three out of six) Agritex staff present sat in the crowd and listened, but by the end of the meeting, only one of them was still present, the other two had gone back to their offices. The one left had to answer the question of how much water was left in the dam, and was asked by farmers to go and measure it with his colleagues and give feedback to farmers. This was done within the following week and the message was given to farmers who subsequently went on to use more land per farmer than recommended. There was nothing new in this show of defiance. It was an indication of farmers' own freedom to make their own choices and take risks in their farming outside Agritex realm of control or regulation.

AGRITEX, COMMITTEES AND MUTUAL ENROLMENT: THE DIVIDE AND RULE STRATEGY

Early in 1995 there was talk within Agritex circles that the Irrigation Management Committee was not working well with the department. Extension staff had initiated the formation of Block Committees with some local farmers who were not happy with the Irrigation Management Committee led by mostly immigrants. In a meeting with extension staff, the Supervisor and the Irrigation Officer were trying to sell the idea to the rest of their staff. The Supervisor said,

> Block Committees are strong. The Management Committee wants to hinder their progress. They have to form their top committee. Mpesa (the IMC chairman) is using other farmers as weapons in his fight with Block Committees, but it will not be long before the present Management Committee loses all its power. Its members are not attending Block Committee meetings, they say they want to see how the new structure works before becoming part of it. They (the chairman and his followers) have this misguided impression that they are legally entitled to their position, that no one here can dislodge them. They think it is their gazetted right to be there, but there is no irrigation policy that says they should be there.[9] They think that they can go up the ladder and have higher political offices deal with the issue on their behalf, we have to show them that they are wrong.

Extension Worker Sithole had his own interpretation of the whole situation. He said that the Management Committee could not be allowed to just stand as popular figures without working for the development of the scheme. Then he pointed out that Block Committees would work on getting more water from the catchment area with the guidance of Agritex, adding that, 'we are aware that the Management Committee wants to disrupt their progress. It's good that we are now working well with the Block Committees, and they appear to be working well with most farmers.' The Supervisor then asked when the next Management Committee elections would be held,[10] and no one seemed to know exactly when. Then he added that,

> We want to form a new Management Committee based on the present Block Committee members so that we talk the same language, right now there are two or more languages. As Agritex we have to support the new Block Committees because they seem to have a better grasp of the situation we are in today, and they are supported by the majority of farmers.

Extension Worker Sithole then pointed out that 'we have to tell the Irrigation Management Committee that things work out better if people participate in their own project', and the Supervisor added, 'we have to tell them that we are here for their good, for their interests, let us write a letter about our views to them'.

No letter was written though, but it was clear that there were serious dilemmas among Agritex staff about what steps to take next. They were obviously against the Irrigation Management Committee which they found to be uncooperative, and were in support of the new Block Committees which were seemingly easier to work with because they always consulted with Agritex in most of their activities. Agritex's enrolment of dissatisfied farmers in their strategy to unseat the Irrigation Management Committee could not be distinguished from local farmers' wish to remove the same committee from power. This seems to have been a well-hatched coup by both Agritex and some local and immigrant disgruntled irrigators to turn farmers against their 'popularly elected' body. For several months after the formation of Block Committees things seemed to be moving quite well and they were gaining support in the way they were *attempting to control* farmers' irrigation practices and fine those who broke the unwritten rules and the 1991 by-laws of the irrigation scheme.

For several weeks after that meeting there were discussions among farmers on the role of the two types of committees. I discussed the issue with some of them. One immigrant irrigator who joined the irrigation scheme in 1964, and one of the most outspoken irrigators, said that he hated the idea of seeing the irrigation scheme full of committees fighting over control of the scheme,

> There are too many committees here, there are just too many committees in this place. There should be one main committee in charge of everything. Too many people want power here, and this will not lead us anywhere. Where have you seen two bulls living peacefully in one kraal? If Agritex staff cannot do the job that they came here for, then they should leave!

But then there was an ethnic divide among irrigators which played a significant role in the way they aligned themselves, even temporarily, with one group or another. In this case the farmer cited above had the view that one committee would represent all farmers. As an immigrant himself, he supported the immigrant leadership of the IMC.

It seemed that there was no way that the Irrigation Management Committee would be removed from power without a struggle, with or without Agritex supporting the Block Committees which the IMC initially saw as just a 'rebellious faction'. The lines of division were not getting any clearer though.

On 21 June 1995, there was a meeting of Agritex, the Irrigation Management Committee and Block Committee members held at the Agritex offices. The meeting was convened by Agritex after discussions about the role of both the IMC and Block Committees in representing farmers' interests. Agritex staff said that there was a significant decline in farmers' production levels over the last couple of seasons. They said that they wanted to remind farmers that they were in a government irrigation project where they had to keep certain standards. The idea was to get the message across to farmers through their representatives that they have to raise their production. The Supervisor addressed them,

> We have problems with methods of farming and levels of production among farmers here. You are all irrigators here and must always know that this is a government project. Standards should be kept high. You must keep in mind that you can be evicted anytime. If you do not do the right thing, you will go back to where you came from. You are on state land all of you here. We will soon be grading all farmers in the irrigation scheme on how you produce, how you maintain your plots, and there will be a scale against which you will be measured, from good through average to poor. Some of you have too much land in the scheme, especially those who got two extra acres for good performance in the early days of the scheme. Some of your friends are now too old to work in the plots. This is the land that needs to be reallocated. You are in business here, and you have to do something to show that. As extension staff we are here to help you learn new and improved methods of farming, to preserve the soil, and live better lives. Lastly, as leaders in the community, you should lead everywhere. Why are some of you not paying your fees on time? The last date is June 30th of each year,[11] but few of you have paid. Do not prevent people from paying because when they suffer you will not be there. Let them come to me when they cannot afford the fees, then we can see what to do with their cases. You are the ones to lead the way, as staff we are neutral, we do not want to end up like prostitutes, going from place to place as if we do not know why we are here. Additionally, there is the issue of food aid[12] from the government. Should irrigators get a share of this or not? This is a

very big question and your answer will say a lot about yourselves and your performance as irrigation farmers. We have not signed the papers, and we are not going to sign them, especially where it says that we have to confirm that the person needs food aid, and is not able to provide for him/herself. We just have to send them back to the councillor. How can we sign them when the applicant is a 4-acre plotholder in an irrigation scheme? You should de-register and put your children in your plots.

This was quite a long speech, and everyone listened intently. Then it was Mpesa's turn to speak (the Management Committee chairman). Mpesa said, 'There is a lot of dirt blocking canals, whose responsibility is that?' and added, 'Block Committees should know who is doing what and where, they must monitor people in their areas closely, otherwise there is no point in having them.' On the face of it this was a surprising turn around which appeared, at least to me, as a way of accepting Block Committees by the Management Committee chairman who had been opposed to them all along. But it sounded like a ploy to get them to feel confident and then discredit them or give them tasks that they could not effectively perform. By the same token, Mpesa put himself in a commanding position, delegating what roles the Block Committee had to play. At the end of the meeting there was a feeling of co-operation between the IMC and the Block Committee. The IMC had not been as confrontational and defensive as some Agritex staff had predicted. Mpesa maintained his position that the Block Committees kept their new mandate to run their blocks and Agritex felt they had brokered the new spirit of co-operation between the two.

After two months, Agritex staff said that there were too many forces pushing Water Controllers to give or deny farmers water, and they ended up not knowing clearly who to listen to. They got instructions from Agritex, the Irrigation Management Committee, from Block Committees, and sometimes from powerful individual farmers. They ended up listening to whoever gave instructions.

One case that was spread round the scheme concerned a farmer, a young man, who was not in his plot to irrigate when water arrived in the sub-block. His plot was at the end of the canal, so water was running to waste. Someone reported the matter to the chairman of Block B committee, where the incident took place. He took a key and locked the gate that opened to the particular canal. The young farmer arrived just after the gate had been locked, and a nasty exchange ensued. He could not irrigate and had to wait for the next turn, approximately a week later. When news of this went round, it did not go down well with other farmers who felt that their freedom and entitlement to water was being threatened. Some of the older immigrant irrigators openly declared that the Irrigation Management Committee and its powerful chairman, Mpesa, should defend them and their irrigation scheme against the 'assault from Agritex and *their* Block Committees'. However, Agritex staff were not going to give up. They argued that,

> The Management Committee should not feel that we are trying to divide and rule them. This is democracy at the local level, and we are doing it ZANU(PF)[13] style all the way. Blocks and their committees are the constituents, and the new Management Committee will be elected from the constituents, just like Government Ministers from among elected Members of Parliament. If present Management Committee members are not popular in their blocks, too bad, they will be eliminated.

However, the wave of support for Mpesa grew, especially among more senior members of the irrigation scheme with four to six acres, most of whom cleared their own plots in the early 1960s, or those who inherited a plot from their relatives and believed that it was family property which should not be given up. Some of them argued that the Block Committees leader, Samunda, was being used by Agritex together with other 'new irrigators who joined yesterday'. An all-party meeting was set for 3 October 1995 at the Agritex offices, but Agritex staff decided to postpone it when they realised that their side was likely to be humiliated. The Management Committee and its supporters seemed to be on a victory match. They wanted to stop Block Committees and their supporters from dictating things in the scheme. Mpesa said that he would enjoy watching Agritex get embarrassed in front of farmers if they tried to impose the Block Committees' new IMC. He said that if farmers rose against Agritex in protest, he would enjoy watching the game, and they would beg him to rescue them. Agritex was not sure how to handle the situation.

'THERE SHALL BE GNASHING OF TEETH': FARMERS' RESPONSE TO AGRITEX

1996 had a stormy start in Nyamaropa irrigation scheme. Tension was brewing between farmers and Agritex — more specifically between farmers and the

Irrigation Supervisor who had ordered the closure of water to all farmers who had not paid their irrigation (or maintenance) fees for the 1996 (current) season. The story was that irrigation fee collectors from Mutare came and found only ±Z$14,000 out of a possible ±Z$60,000. The Supervisor said that he was told by the collectors that farmers in other irrigation schemes had almost finished paying their fees for the current season (citing Chibuwe irrigation scheme in Chipinge District in the same Province of Manicaland, as left with only about Z$300 to collect). He immediately called a meeting of Water Controllers and ordered them not to give water to anyone who had not paid fees for the current season.

He did not consult his extension colleagues, but took the decision alone. To compound the problem, he then left for a short training course in Harare. He left strict instructions that anyone who touched the gates and gave water to people who had not paid would pay the biggest price ever for the crime. Farmers came to the office to ask for water, but were told that no one would get it. They asked the Irrigation Officer who actually was superior to the Supervisor in authority in the scheme to intervene, but he allegedly refused to have anything to do with it because he had not been informed. He said, 'I want this whole thing to get to its rightful end.'

After almost a week, the situation was getting worse and crops were really stressed. Farmers were organising a protest demonstration at the Agritex offices. One elderly farmer friend of mine, who is quite fluent in spoken English openly expressed his excitement: 'there shall be gnashing of teeth . . .[and they shall be cast out to outer darkness. . .]', apparently quoting from the Bible in the book of Matthew about the hard times that were coming. Farmers then called for a meeting with the Supervisor. Before the meeting there was some talk among farmers, several of them calling for the dismissal of the Supervisor, and some saying that they wanted him to leave the irrigation scheme and be transferred to another place. They said that they were tired of his harshness and short temper, and blamed it on what they called his 'colonial training'.[14]

Irrigation Management Committee and Block Committee members were seemingly united in this cause to 'discipline' the Supervisor in the meeting. They selected some well-known outspoken farmers to attend the meeting where they would try and settle their score with the 'mean' Supervisor once and for all. This was a special task force meant to deal with the problem. The Supervisor, however, said that he was not going to hold an illegal meeting, and declared that he was going to meet committee members only. He went on to collect their names from Extension Workers to be able to screen non-committee members.

Some of the Extension Workers told him that he had made a mistake, but he turned on them, and said that he would 'sit on them' for being unpatriotic. He had asked for my opinion, and I urged him to go by the book and do what was right. From an economic point of view, he had the right idea and was only taking measures to get farmers to pay, but politically, and as far as producing food is concerned, many people, including his colleagues, thought he got it all wrong. Part of the seed and fertiliser used came from the Government Drought Recovery Programme, and he would be branded a sell-out who wanted the program to fail, and people to starve, in spite of his noble objectives meant to serve the same Government.

Before the meeting, the Supervisor called me into his office and showed me his job description and the by-laws that said farmers could pay their fees up to June 30 of every year, but he was changing that, saying things were too lax in the scheme. He then showed me a bunch of eviction forms that he said he could use if things got out of hand in the meeting. His resolution to get things done was admirable.

The meeting was attended by about 20 farmers, three out of six extension staff, two policemen invited by the Supervisor to come and monitor the proceedings just in case they got out of hand. It started with a prayer by Block D Committee chairman, Simoyi, of the Apostolic Faith Mission Church. He said, among other things 'Lord you are the organiser of things, may you help us know what we are saying . . .' This was followed by a traditional opening of clapping hands. Then the Management Committee's vice-chairman, Matiringe, opened the discussions going straight into the issue of the closing of water by Agritex for farmers who had not paid their fees for the current season. Below is an extract of exchanges from that meeting:

> Matombo (one of the specially invited vocal farmers): If water was closed by Agritex, we want to know where the others are. Are they united in this? Where is the Officer?
> Supervisor: Are you all committee members here? This may be an illegal meeting . . .
> Makanyanga (Management Committee Secretary): We have some committee members and selected irrigators representing everyone.
> Supervisor: Yes, but did you talk to the Management Committee chairman (Mpesa) about this? How will I answer him when he comes?

The vice-secretary of the IMC then said that he had talked to Mpesa who had given them the go-ahead. Farmers were getting anxious to discuss the issue at hand. They said that the IMC vice-chairman was good enough for them. Then the supervisor told them why he had decided to close water for all those who had not paid, citing the visit by fees collectors and referring to money being wanted by Treasury in Harare.

He told them that they were in a government scheme, and acknowledged their problems, but emphasised that they have to pay. He then told them a story of a woman who came to him crying but could not pay, and he had to pay her fees from his own pocket so that she would get water. He said,

> I have been listening to your problems. The laws of irrigating here have not changed. You still have to respect the right to irrigate, to reside and to depasture stock which you were given when you joined this project, and you can be removed if you do not perform your expected roles. If the law has changed, it has changed in your homes and not in our offices. You are not paying for water, but that may soon change too and you may be receiving monthly water bills.[15] We understand your problems. We are different from the White man who did not understand the role of the extended family. I have broken the law to help some of you [by giving farmers water when they had not paid up their fees just to save their crop]. But we have these eviction forms (waving a copy), and we can still use them. I receive all types of problems, and I have to do my job.

Kapadza, a well-known and outspoken businessman, popular for having once kicked a board used by the Irrigation Officer to illustrate something to farmers in a meeting where he repeatedly called for their ejection from Nyamaropa, then said that farmers were asking Agritex to take action for people who had arrears, but not during the current season.

The Supervisor said that that was not what Treasury and Central Government were thinking of. He added that he had met with Block Committees and they agreed to have people pay. Block Committees said that they indeed agreed that farmers had to pay, but did not agree with Agritex on the timing of the payment enforcement. The IMC vice-chairman said that there were too many committees, asked again why water was closed, and said,

> It seems it was out of spite, a way of fixing us, making us suffer. We were coming from a bad season of drought. We needed water to recover. If you were going somewhere you should have told us.

The Supervisor said: 'So you think I have *utsinye* (evil intentions)?' Matombo quickly replied, 'Yes you have'. Samunda, a local irrigator and a leader of the Block Committees, in a raised emotionally charged voice, said,

> We are not experts in farming, we all have to agree and co-ordinate. During the worst drought of 1992 we had to pay fees even when we had nothing from the fields. We asked for an exemption and were denied one. Why should we get such a penalty now for this, and at the time when a promising crop is suffering? We are still within our time to pay the fees, so why punish us when 'our deadline' [30 June] is not over yet?

The Supervisor responded by saying they all had to act like adults, not children. Then the Irrigation Management Committee secretary asked a 'simple' question,

> Secretary: When are we supposed to finish paying our fees for this season?.
> Supervisor: Harare said they want the money. . .
> Secretary: . . . our period from July to June is not over yet, are we not supposed to grow crops in order to be able to raise the fees?
> Supervisor: . . . but I talked to your chairman about all this and he agreed . . .

Several farmers in the crowd chorused, 'do you have the minutes of that meeting?' Then someone said that there was confusion between the Irrigation Management Committee and the new Block Committees. But Samunda immediately intervened: 'There is no confusion between the Management Committee and Block Committees, or among Block Committees themselves. We did not discuss the issue with the Supervisor. If we had been told we were certainly going to talk to farmers about it.'

Then Mautsa, who once advocated free-for-all irrigation during water shortage in 1994, asked, 'Who writes reports from this area to the top offices in Nyanga?' One committee member said that the Officer does it together with the Supervisor. The idea was to tease the Supervisor, by exploiting the fact that he had acted alone in ordering the closure of water. Farmers understood very well that there was no unity among Agritex staff.

Nyakatawa, a committee member, then asked, 'Was the closing of water legal? Why did it have to be done now, and which law was used in deciding on the issue?' The Supervisor, apparently upset, replied, 'You are free to go to anyone you like with this matter, you can go right up to the President if you so wish, but I have to go through my Department's channels, and I may never get to the President.' Then Manyanga, one of the Block Committee members said,

> You are here to help us, we are farmers, you and us have to help each other all the way, we are not fighting you. You should have called us to discuss the issue of water and fees, and what steps to take, or at least announce that things were going to be different from now on. Sudden changes like this can be very disruptive. Right now farmers are in the dark as to what exactly happened or is going to happen, they are busy trying to produce something for their families. You have to co-ordinate with the committees. It should not be like we are fighting. We do not want to fight, we have to teach each other. Government trucks came here to give us food, fertilizer and seed, and there are these grain loan schemes. How can people repay them when their crops are allowed to wilt when water is there? We have to teach each other.

The Supervisor consented, saying, 'Now you are talking sense, now we are understanding each other.'

After this exchange farmers pressed him to tell them when they should pay their fees. When no clear answer came, the vice-chairman then said to the Supervisor,

> When one makes a mistake, one should be man enough to admit it.
> It is not bad to admit that you were wrong, that you made a mistake,
> it is allowed to do that in a court of law or even in church.

The Supervisor said, 'Well, if you want me to . . .', but farmers cut in and said, almost together in a chorus, 'No!, it has to be from you, from your heart, for all of us . . . not for any one person'. He then admitted to having erred. The vice-chairman then thanked him for being brave enough to take the heat. Then one of the Block Committee members, Manyanga, pointed out that many people were now against Block Committees, and suggested that all committees, including the Irrigation Management Committee, be dissolved pending fresh elections.

The meeting then went into the issue of setting a date for the election of a new Management Committee, but the Supervisor and the vice-chairman were against the idea in the absence of the chairman. They promised to notify the chairman of the Management Committee when he came back. After the meeting some farmers could be seen talking and laughing with the

Supervisor, saying that they were hoping that something like this will never happen again, and that they have to work well with Agritex staff. It was agreed that farmers were going to be given water; but that after the deadline anyone still owing money to the department was not to be given water to irrigate.

REMINISCENCES OF A PRODUCTIVE PAST

The above unfolding of events during three dry seasons in Nyamaropa irrigation scheme seem to point at an existing confusion over how agricultural production could be organised most beneficially. The supervisor, realising his duties as an agriculturist, in the end tried to resort to rules and regulations, hoping to instill an atmosphere of production as he had known it in his days as Demonstrator under the colonial regime. This escalated the simmering conflict with different factions of farmers in the scheme, but also seemed to bring all parties together in a common definition of the aims of the irrigation scheme. It stressed the need for co-operation of all parties involved in the scheme.

The confusion over how to shape this mutually beneficial co-operation seems to have been triggered by a change in operations of extension agents in the scheme since independence. Illustrative of this change is what Mai Matata, wife of one of four brothers (all irrigators) whose father was one of the first irrigators in Nyamaropa, told me:

> We have our extension workers some of whom are our friends and maybe relatives [one of her daughters has a child with one of the extension workers]. We do not see much of them though, they just ride through the fields on their bikes, they say hello, and go on to wherever they may be going, probably to the shops, or to meetings. Extension workers of the olden days when the White Manager was here knew how to deal with farmers, they would stop, sit down or work with you while talking to you, they showed you how to do things. They had more commitment to their work. I could say they knew how to study farmers' thoughts and feelings, and would pre-empt farmers' actions and teach them well. These are different, but we are not complaining, and we do not want to kick them out of course. We need their services. Their assistance is vital for our survival especially now that things are more and more difficult, but not everyone listens to them all the time. It's like in a family, you get

some children who do not listen to their parents, but they cannot live well without the parents, can they?

One senior irrigation farmer, Manyuchi, who also ran a retail shop and a grinding mill in the area, made similar observations. Manyuchi was an immigrant, and a very enterprising farmer and businessman, employing casual labourers during peak labour periods. He said that he specialised in tobacco production because it gave him a lot of money. He actually rented other farmers' irrigated fields to grow tobacco. He said that he supported the Irrigation Management Committee which let Agritex staff do their job. He spoke softly and emphatically, and said,

> I know some farmers have been telling you a lot of wrong things about this issue, but we have to face the facts. Some are saying that during the Smith days[16] things were much better as far as extension services were concerned, especially in improving the lives of farmers. If I remember well, *madhumeni ekare* [agricultural demonstrators of the olden days, as Extension Workers were known then] knew how to do their work, they would scout your field even in your absence and when they met you they would tell you what your field looks like in terms of pests and diseases. We used to be the best cotton and tobacco farmers in the whole province. In agricultural shows we won a lot of prizes from the crops we produced.[17] Those men knew how to assist us to get the best out of the small plots. During our earlier days, extension workers used to be very frightening people, there was a lot of force used to get farmers to listen and follow, but this was not a constant thing, some of them were very nice and understood us. But most farmers were afraid of them. Some managers in the irrigation scheme were white, and they would punish farmers for not doing things right, such as not weeding or not planting on the recommended dates . . .

Then he went on to talk about the impact of national political changes on attitudes to extension services and extension staff.

> Today things have changed and the Government does not encourage that farmers be punished, people do what they want, and you can see what is happening with these committees who are taking over the irrigation scheme. Farmers are freer to do what they want, and you can say that there are times when they actually dictate to Agritex what they want, they are now powerful when it comes to dealing

with Agritex. Agritex seem to be ineffective when it comes to telling farmers what to do. Some of it is a result of fear of being reported to politicians, such as the case of 1992 when the Member of Parliament, the District Administrator and the Provincial Governor were called here by farmers who reported that Agritex was throwing away water when farmers' crops were wilting. Politicians here play an important role, and Agritex staff are just human beings too. They are afraid to be on the wrong side of political leaders.

He then shifted his focus to blaming the Government,

The Government made a mistake by putting too many people together to share small plots when they knew that the water source was not reliable, and that there was no lasting solution to the people in the catchment taking our water. It has actually become worse with time. They have killed the irrigation scheme, both Government and people in the catchment area. We are supposed to be modern farmers here, but we are not very different from farmers in the reserves.[18] Little education is not good, it is dangerous, *havanakufunda ba* (they are not educated, the farmers!). The manager [meaning the Irrigation Officer] is the father of water in the irrigation scheme, and is assisted by the Extension Workers and the Water Controllers. Block Committees should have nothing to do with water issues, they are just farmers, they should leave water management to Extension Workers and Water Controllers who are employed by government to do that job and are paid for it. If this goes on like this people will kill each other for water here. I cannot let someone else who is just a farmer like me tell me how to use how much water which we have the same entitlement to. Are they registered and known to Government as workers?

So why are they so busy organising us? Some of these Block Committee members really go out of their way, some of them go to the extent of buying keys and locking gates to stop other farmers from irrigating. How can that be justified, who gave them the right and the power to do that in the first place?

When asked about the belief among Agritex staff that farmers generally waste water, he said,

I will tell you something my friend, when you milk a cow, you cannot avoid spilling a drop or two of milk on the ground, and when you are weeding, chances are you will chop off one or two crops that you are weeding, we all make mistakes.

Our chat, or rather his talk, ended with him telling me that maybe if Agritex staff got a firmer grip of things in dealing with farmers, they would keep things under control. However, the events in 1996 showed that things were not that simple.

Still, Manyuchi's analysis, though not shared by all Nyamaropa farmers, touched on the crux of events in the three seasons described above. Agritex's position as an extension service has changed since independence. In trying to come to grips with the new situation, the agricultural extensionists followed three different subsequent strategies: first one of retreat (1994), then one of an active party on the scene supporting one group of farmers and by-passing the elected farmer representative body (1995) and finally one of true masters over farmers tapping from the arsenal of rules and regulations at hand (1996). All of these positions yielded a response by the farmers and resulted in a different dynamic of interactions. None of the strategies proved satisfactory, but it is important to note that with the Supervisor publicly offering his apologies for the way he handled the water fee issue, an opening seems to have been forged for a new style of operation in Nyamaropa: a type of management that is based on an openly negotiated order.

DISCUSSION AND CONCLUSIONS

This chapter dealt with encounters between Agritex extension staff managing a smallholder irrigation scheme and farmers living and working there. It is also about encounters between groups of farmers in the irrigation scheme, and how the different parties devise strategies to gain influence over others. The Nyamaropa 'saga' sheds light on a number of pertinent issues like the confusion over the role of the extension service in irrigation schemes; the need to appreciate the importance of negotiations in establishing good links between extension agents and their clients; the role of politicians in post-independence extension performance; and finally the need to view the work of extension agents in a wider network of social, kinship and political relationships.

The capriciousness of farmer-extensionist encounters should not confuse the wider pattern that emerges. My interpretation of the material presented here is that 'at the back' of these encounters unfolds a larger process of change concerning the relationship between local farmers and state agencies.

Extensionist-farmer relations in daily interaction in Nyamaropa had unpredictable twists and turns. The way Agritex staff were enroled into farmers' struggles for control of the irrigation scheme shows how cunning farmers can be in their quest to run their own affairs or at least circumvent institutionalised rules and regulations. Mutual enrolment by both Agritex and farmers in their respective 'projects' (such as the need for different groups to gain control in farmer representation) led almost directly to the creation of a tense political atmosphere where each party took advantage of the other to (temporarily) take charge of the scheme. This aspect of the politicised nature of extensionist-farmer relations apparently could not always be avoided. It seemed to be an inherent part of the field of extension. Drinkwater (1991, 226) says that '. . . Agritex's goal is highly political and not merely technical'. If indeed the main function of Agritex was 'to articulate farmer problems, to synthesise, distil, consolidate, adapt and disseminate the final research recommendations' (Nyathi, 1995, 14), then there was an inherent problem of thinking or assuming that farmers could not articulate their own problems. I see this type of thinking as a carry-over from the pre-independence era. The fundamental question is, how can Agritex and farmers disentangle themselves from this colonial legacy and forge a new way of dealing with each other. In this chapter we see the various actors grappling with this fundamental question, leading them into uncharted territory.

The role of Agritex in Nyamaropa was generally understood in Communal Areas as being to advise farmers and teach them improved and better methods of farming. This was for both irrigation and dryland farming areas. There seemed to be a mix-up in the way farmers perceived the role of Agritex in the government-run Nyamaropa irrigation scheme. This was based on the apparent confusion of whether to remain simply advisors to farmers, or to be administrators of a government project. This chapter did not attempt to define the role Agritex staff had to play in such contexts, but presented, raw as it were, situations where the role of extension staff came under the spotlight, and where farmers seemed to question the very presence of extension staff in their area.[19]

Farmers having found ways to steal water and get away with it, or having devised their own strategies to counter punitive measures taken against their failure to pay up irrigation fees, exposed Agritex's failure to handle forms of farmer resistance. What was of interest here was that there was a serious management problem of farmers doing most of what they wanted to do, and Agritex not taking decisive action. Some farmers wanted Agritex to take decisive action, otherwise they did not see the reason why extension staff should be stationed in the scheme in the first place, but there was no consensus on this. My feeling was that Agritex staff genuinely wanted farmers to be their own masters in irrigation affairs, but had to combine this with its official mandate to manage the project.

Thus, the official mandate and roles to be performed by Agritex do not match the expectation of farmers. Agritex hands are furthermore tied by the local political situation. The political element in farmer-agency relations could assume a more threatening posture than local level dynamics between rival parties can withstand. Such was the case in Nyamaropa when political heavy weights from District and Province came down on civil servants running the scheme and tried to solve their problems. Chambers says that '. . . no major redirection of extension activity is likely to achieve lasting success without sustained political support' (1974, 82). In a way, this was one of the strategies employed by farmers who brought in the politicians to win the battle and show extension staff that they had more power resources to fall back on or summon for assistance. Farmers enrol politicians in attempts to gain leverage in their conflicts with Agritex.

The environment in which Agritex has to perform its duties is thus very complex; compounded by the recurring and chronic water shortages in Nyamaropa. In such tense conflict situations when livelihoods are threatened, differential responses are likely to emerge. Conflict is part of the reality of interface encounters (Long, 1989) and of external interventions, and this is only one illustration.

The power game played by both Agritex and farmers in meetings as shown above exposes the difficult situation that extension staff are caught in daily. The strategic enrolment by farmers of Agritex into their leadership wrangles between the Management Committee and Block Committees, together with the way Agritex staff take advantage of the wrangles to initiate moves to dispose of a group of farmers they deem distractive of the project's development, help reveal the webs or networks of social, kinship and political relationships that exist among the different parties. The agency-farmer interface is not only a contest to determine whose view of the farming world is dominant, nor only a political power game to be understood in terms of ethnic, kinship and group divisions. Understanding such interface situations and their intricacies helps to unravel some of the problems that have beset the extension service for decades, in spite of marked successes (Rukuni, 1994a). It shows that local actors have come to realise that the future is theirs. They are now grappling with the various options open to them; the preferred development models and images of an idealised future. In Nyamaropa there is some evidence to say that a new, locally negotiated, order may gradually emerge; an order in which the roles of farmers and state representatives, and local and outside actors, will be defined in locally accepted terms.

NOTES

1. The Department of Agricultural and Technical Extension Services is responsible for agricultural extension and irrigation services to small-holder farmers in Zimbabwe.
2. The Irrigation Management Committee consists of a chairman, a vice-chairman, a secretary and a vice-secretary, a treasurer and about six committee members representing different villages around the irrigation scheme. Block Committees have the same structure except that they have only one village committee member each.
3. It is difficult to give an accurate figure on this because farmers continue to subdivide their plots and re-allocate them to their children, wives, and friends.
4. Water Controllers are officially graded as General Hands, but within the scheme the role of Water Controller is seen as a form of promotion from largely menial tasks performed by General Hands.
5. Derude (Department of Rural Development) was responsible for running small-holder irrigation schemes from 1981 to 1985. In 1985 Agritex took over.
6. Water Controller, Samunda, is a cousin to Samunda the leader of Block Committees in Nyamaropa (see below).
7. The Water Controller told me of one of his experiences where he refused to give water to a farmer who had not finished paying fees after the deadline; he was asked if the water was his, and was promised a good beating. Then he was banished from beer drinking places for months after being given threats of poisoning.
8. In some cases this interlocking took a very literal sense, like with extension worker for block B, Sithole, who had been wedlocked with one of the daughters of the Block Committee Chairman of that block.
9. The supervisor referred here to a fact that transpired during an irrigation management workshop for all Agritex staff in Manicaland, held in February 1995. Agritex's Deputy Director (Engineering) admitted in front of his staff that in actual fact the regulations they were being taught had never been gazetted.
10. The then committee had been in power for the last three seasons.
11. The main reason for highlighting this is to compare it with the case of the Supervisor closing water, in January 1996, to farmers who had not paid their fees when in fact the deadline had not passed (See below).

12. During dry seasons where farmers in Communal Areas do not get enough food from their harvests, the Government provides assistance by giving them food hand-outs. There is sometimes a form of selection whereby families with working members (or in areas where food is expected to be easy to produce, such as in irrigation schemes), are not given free food. Some irrigators in Nyamaropa applied for it and expected Agritex staff to sign the forms in confirmation of their plight. The latter refused.

13. This is the ruling political party in Zimbabwe.

14. The Supervisor was a respectable elderly man who had worked in extension for almost 40 years. He often referred to his experiences with White Managers and farmers. He said that lack of control in irrigation schemes had negatively affected production and irrigation performance.

15. Here the supervisor referred to the government's intentions to privatise the Department of Water Resources. As a consequence of this privatisation water bills would have to be paid by small-holder irrigators in future.

16. This is used to refer to the colonial period in general, but Smith was the last Prime Minister of the White Government that was removed from power by African nationalists in 1980.

17. This assertion is supported by letters written by White Nyamaropa Irrigation Managers in the sixties about the performance of African farmers in the irrigation scheme.

18. The word Reserve.here is used to refer to dryland farmers around the scheme. Irrigators regard themselves as living and working in a farm situation with a commercial or business orientation to it, while they regard dryland farmers as staying in the poor colonial reserves (now known as Communal Areas) with very little hope for development.

19. Extension staff, on their part, sometimes deliberately refrained from active interference by playing passive roles in the running of the irrigation project. This phenomenon probably has two sides to it: First, it could be that Agritex wanted to see whether farmers had the capacity to manage water on their own. Second, and most probably, they might have been aware of mistakes that farmers were making and wanted them to get a practical lesson on their own on how not to manage the scheme. They might not bear all the blame for the way farmers had punctured holes into the whole system of irrigating, but they surely were well-placed to know that joint appreciation of the project's problems would create better agency-farmer relations conducive to more efficient management.

Chapter 10

The uncooperative statistic
Farmers' response to experts' fact-finding mission

EMMANUEL MANZUNGU

INTRODUCTION

From time to time academic researchers and development practitioners take time to get information from 'grassroots' people on general and specific issues. In agriculture farmers constitute this grassroots level. Whether the purpose of the information is for academic or practical use, the common denominator is a supposed reliance on reputable and accurate data gathering methods. One of the methods that is commonly employed is the questionnaire survey. The idea quite often is to obtain data sets or statistics that can be used to generalise reality.

Chambers (1983) states four reasons why there has been a 'convergence on questionnaires'. In the first instance, surveys are ideal for planners who require planning data. Surveys constitute a ready tool to get information about the people for whom plans are being made. The second point is that natural scientists, with their mathematical orientation, find the method acceptable because of the possibility of doing statistical analyses which fulfils their scientific requirements. Thirdly, questionnaire surveys provide figures that are the hallmark in today's planning. Here economists and statisticians dominate the scene. Lastly, the convenience of the survey to process data away from source in a cosy environment has promoted its usage.

But how useful is this often-used tool? Chambers warns of the pitfalls of the method. He caricatures it this way:

> They (researchers) stare at print-outs and tables. Under pressure for 'findings', they take figures as facts. They have neither the time or inclination to reflect that these are aggregates of what has emerged from fallible coding of responses which are what investigators wrote down as their interpretation of their instructions as to how they were

> to write down what they believed respondents said to them, which
> was only what respondents were prepared to say to them in reply to
> the investigators' rendering of their understanding of a question and
> the respondent's understanding of the way they asked it; always
> assuming that an interview took place at all and that answers were
> not more congenially compiled under a tree or in a teashop or bar,
> without the tiresome complication of a respondent. The distortions
> are legion . . . These 'findings' are artefacts, a partial, cloudy and
> distorted view of the real rural world (Chambers, 1983, 53–54).

Chambers believes that there is a significant mutilation of reality that occurs because of the use, perhaps misuse, of the questionnaire survey. This chapter shows that there is indeed good reason to doubt the usefulness of this method. It will be shown that the basic problem seems to be a lack of understanding between the experts or planners and farmers which in turn is caused by different world views of the two groups of actors. This chapter concentrates on the stage of data generation by exploring how the process takes place. It does not cover analysis for reasons explained later. It will be shown that for various reasons farmers resent being treated as data sets and consequently play the role of the uncooperative statistic.

The chapter is based on a discourse between experts and farmers during socio-economic interviews conducted in Chibuwe irrigation scheme on Monday, 4 July 1994. On this day six officials of the Economic and Planning Unit in the Ministry of Agriculture came to the scheme to interview some 20 farmers. The author was present to witness the occasion. Before presenting the empirical material and the subsequent discussion, a methodological note is given.

A METHODOLOGICAL NOTE

An investigation into something 'unfamiliar' as farmers' response to experts' fact-finding mission demands a methodology that is equally 'unfamiliar'. By this I mean that a researcher interested in such an enterprise does not use formal or semi-structured interviews to ask farmers how they resist the experts' probings. One basic condition, it seems to me, has to be fulfilled if the investigation is to succeed. The researcher must be able to do his/her research in a situation where he/she is not a stranger and is accepted as an observer. In such a position, the researcher is likely to get the 'reality' as it is.

My on-going research in Chibuwe irrigation scheme made me to fit this description of an unobtrusive observer. By the time the said visit took place I had been working in the Chibuwe scheme since October 1993. This was my eighth research visit. By then I was familiar to Agritex officials in the scheme, and personal relations had been established. On the farmers' side I knew the Irrigation Management Committee members, most of whom I have had one or more interviews with. I had also been to different sections of the scheme and farmers, particularly from blocks A and D, where I did most of my research work, had seen me in their blocks. This interaction with Agritex and farmers put me in a position of advantage of the 'silent observer' during the deliberations.

Against this background I could easily sit among the farmers and be 'one of them' without causing any uneasiness among them. This was crucial for me as I could then indulge in some participant observation (Bernard, 1988, Chapter 7).

A word of caution needs to be said about the 'unit of analysis'. The discourse at the meeting provided the boundaries of the study. I was not afterwards able to look at how the farmers' responses were used, as the Ministry officials would not let me have access to the 'confidential' papers.

A DELAYED START, MISFIRES AND THE CONSEQUENCES

A fuel shortage-induced delay

Ministry officials had phoned the Irrigation Supervisor, Mr Chimhanda, and advised him of their impending visit. He in turn notified the Irrigation Management Committee. They also advised him that they wanted to interview 20 farmers. The selection of farmers was left to him. Since there were five blocks in the scheme it was decided to take four farmers from each block. After names of farmers to be interviewed were selected, water bailiffs were sent to notify them. As nobody knew what was required farmers were just summoned to the office with no explanation.

Farmers began to arrive on the day of the interview singly or in groups at the Agritex offices in time for the 11 am meeting. As they arrived farmers were told that Ministry officials, who were still to come, wanted to interview them. By 11 am all the farmers had congregated but there were no visitors in sight. Agritex officials and farmers alike began to wonder whether the meeting was going to take place. Two hours later the visitors were nowhere to be seen. The Irrigation Supervisor in consultation with the Irrigation Management Committee chairman agreed to let farmers, who by now were complaining of being famished, to go. They would then meet at 2 pm at the Kopa[1] buildings to again wait for the visitors.

Just as the farmers were dispersing, a Nissan diesel truck pulled up. The visitors had finally arrived. All the farmers were then called back. As farmers were returning the Irrigation Supervisor, Mr Mavhura, one of the three extension workers in the scheme, and myself, exchanged greetings with the visitors. One of the visitors explained that they were from the Economics Unit of the Ministry of Agriculture. They were going around monitoring government projects and Chibuwe had been drawn out in the sample. They were late because they could not find diesel at Birchenough Bridge. They had then travelled to Rimbi 20 km away only to find that there was also no fuel either. They had then proceeded to Checheche for another 20 km.

An attempt to gain lost time

The Irrigation Supervisor asked the visitors whether they were going to brief farmers first about their mission. The spokesman, apparently conscious of the journey ahead and anxious to regain lost time, replied that they just wanted to go straight into the interviews. They would just split up and each one of them would interview three to four farmers. The Irrigation Supervisor disagreed,

> It is not good to do that. You should not treat farmers as if they don't know anything. It is fair to give them an idea of what you want first. They may have also have some questions.

Left with no option the Ministry spokesman agreed.

Soon after everyone took seats in the shade outside the Agritex offices in preparation for the start of the meeting. Once everyone was seated the Irrigation Management Committee chairman began to chair the meeting. Oblivious of the anxiety of the visitors with regards to time, the chairman remained true to the usual etiquette. First, as is the routine, the meeting was opened by *hlombe*, a traditional clapping of hands organised such that men clap first and then women. *Hlombe* is a sign of homage to ancestral spirits for their guidance in the deliberations. Thereafter a prayer was offered also for guidance in the deliberations. The chairman then called upon the Irrigation Supervisor to proceed with the matter at hand.

The Tower of Babel[2] experience

The Irrigation Management Committee stood up to make introductions. I was also introduced. Before he invited the visitors to take the floor he remarked that in the interest of gender they (Agritex) had included some women in the survey sample. Thereafter one of the visitors was called upon to explain to farmers the purpose of the visit.

He apologised first for the delay. He then went on to explain the circumstances of the delay. After that he explained that they were from which government department. He went further and explained that they were on a countrywide tour of agricultural projects which numbered 25. The spokesman was interrupted by the Irrigation Management Committee chairman;

> Who is taking a record of the minutes. Our own secretary is not here. You (to one of the farmers with a note book) are you taking down the minutes otherwise all that we are saying here will be blown away by the wind . . . If you are not taking down the minutes and are taking down notes we need somebody to take the minutes.

It appeared that the lack of information about what the meeting was about had made the farmers come unprepared. The chairman's effort to organise minutes to be taken were 'helped' by one of the visitors. Apparently anxious to save time, one of the visitors offered to keep a record of the minutes. He promised to make them available after the meeting. The address then continued.

They were visiting a number of government projects throughout the country to find out how they performed. Such projects included resettlement schemes and irrigation schemes. They had already been to Nyamaropa, Chakohwa, Devure and Mutema. That is why they had come to Chibuwe. From there they would proceed to St Joseph and Rupangwana in Masvingo and other irrigation schemes in Midlands and Matabeleland provinces. Farmers could wonder how come Chibuwe was selected. This was done by a computer. They were going to interview the selected farmers individually so that nobody would hear what the other person was saying. They would process the data and in the report they would not mention individual farmers by name. The information that was given would strictly be confidential and thereafter they would submit the report to government. That is where their responsibility stopped. He continued to say it was a very worthwhile exercise although the farmers did not benefit materially. However, they would benefit

in the long run as the nation stood to benefit. It was also essential that
farmers gave truthful information. They should not expect any help be it
fertilizer or drought relief on account of having been interviewed.

Even before the spokesman finished his address I noticed that the chairman
of the Irrigation Management Committee looked very unsettled. I could
not think why. Barely had the spokesman sat down before a deluge of
questions began. Questions like this were asked:

> We heard that we were going to be asked some questions. Are we not
> allowed to ask questions in return?

> You said that you were not going to give any help. Does this mean
> that even if we are suffering we will be condemned to no help?

> You say you just write the report to government. What for? What is
> the point?

> You are behaving like the colonialists who had their favourite strategy
> of saying, We have been to this place and the other, they have already
> agreed to destock. This could be a ploy to raise maintenance fees.

> Why are they interested in my small plot? (This was directed to a
> fellow farmer)

The questions were coming thick and fast. Table 1 summarises the
questions/comments and farmers' interpretations.

On his part the spokesman tried to answer each question as cooly as he
could but each answer seemed to attract more questions. In some cases he
completely missed out on what was being said because of the heavy Ndau
language that was in use. The spokesman sounded a Zezuru speaker.

The situation deteriorated as farmers were clearly incensed by the whole
situation. One farmer, Madzonga, a member of the Irrigation Management
Committee, complained:

> It is now late and we are hungry. We have been here since morning.
> On top of that they say we just have to listen. Furthermore, they
> should have also told us what this was all about, then we could have
> brought our records. Personally I am going to answer all their
> questions, never mind, they do not want me to ask. It does not matter,
> this is just an interview. Whether you lie or not, it is just an interview.

Table 1: Experts' operational parameters and farmers' response

Experts' operational parameters	Farmers' responses
Individual interviews to safeguard confidentiality	Why the secrecy if it is just to find out about farming. Is this not a divide and rule technique?
Farmers should not expect any help because they have been interviewed	What is the point of all this? Even if we deserve help, are we going to be denied just because we have been interviewed
Records of each farm plot are needed	Why are they interested in my small plot?
We are going to ask some questions	That means we are not allowed to ask in turn
Other sites have already been visited	It is a ploy to dupe us into agreeing about things we do not want
We will submit the report to government, that is where our responsibility stops	So what?
Interviews are a means to arrive at the truth	It is only an interview. Even if you lie, it does not matter because it is just an interview
We are your children	But you are going to destroy your heritage. As for us we are near our graves, it is your fault
We will send you a report	We are being taken for a ride. It is what they do, it has happened to us before. We get promises and nothing happens

At this juncture the spokesman appealed to farmers to tell the truth.

As pressure mounted on the spokesman, one of the visitors came to the rescue of his embattled colleague. In order to diffuse the tension he said that they were their sons, they were not strangers. He then pointed at one of the woman farmers and said, 'I know you from Mutema Irrigation Scheme. I grew up there, so we are part of you.' The woman looked on impulsively either because she did not genuinely know him or she did not fancy falling for this 'gimmick'.

The Irrigation Supervisor realised that a deadlock had been reached. He suggested that the interviews should go on despite the fact that farmers were unsatisfied. I then suggested that because the farmers were not convinced of the motive of the interviews, would it not pay if after compilation of their report the visitors would send a copy to the farmers. The spokesman jumped at this. Yes, they would send, even last year's reports. He then said that they would leave a copy of the questionnaire with the Irrigation Supervisor. Madzonga was not impressed:

> This has been done to us before. People come and promise and nothing happens. We are not hopeful that anything will be sent. This game goes on all the time. Even you (pointing at me) did not tell us what you were doing around the scheme.

A RUSH TO BE INTERVIEWED AND STAGE-MANAGED RESPONSES

After the 'uproar' which should have left the experts dumbfounded, the group broke up and the interviews began. First the experts identified secure places where each interview could be conducted with maximum 'confidentiality'.

The first batch of interviews then began. Then farmers showed another interesting side of what they felt. There was bargaining around who should go first. It had nothing to do with feeling nervous. Practically all the farmers wanted to go and get over with it! As the first interviews were being conducted farmers complained that it was taking too long. The Irrigation Supervisor went to his office and got the questionnaire to check on its length. He brought it and confirmed the farmers' worst fears. It was just too long as it contained questions on farmers' biographical data and farming. Figures were asked for practically each question.

Then the conversation returned to these 'misinformed and unununderstanding' experts. Every farmer present took time to pour scorn on them. As that conversation continued one of the farmers who was being interviewed stood up and approached the group. The face of the farmers turned in his direction expectantly — had the interview been completed so that they could get over with it? It turned out that the farmer had come to consult about how many bags of maize were contained in a scotch cart, apparently because the question wanted yield in tonnes per hectare while farmers simply took the produce in a scotch cart home. Then the conversation went back to how much time it was taking for one person to be interviewed and how hungry they were. I decided to time the duration of some two interviews. Each interview took 25–30 minutes. As interviews were completed there was a visible relief on farmers' faces. Madzonga, being one not to miss a point of mischief, looked at the farmers who were yet to be interviewed and remarked, *Tava kundorivhumbura doro nemari yebhinzi?* (We are going to get stone drunk with cash from the sale of beans).

I left the interviews in progress and went to check the block D Master Farmer Training session which was being conducted by one of the extension workers. It turned out that I arrived before him. Just one woman farmer by then had turned up for the meeting. I expressed my doubts that the meeting would take place since the extension worker had at one time announced a postponement at the Agritex office because a number of the farmers involved had been summoned for the interviews. Another woman farmer arrived for the training. I then conversed with the farmers before their lesson. It was during this conversation that more insights about the whole exercise came to light.

The women began to talk about one farmer (it was the same farmer who wanted to know how many bags of maize were contained in a scotch cart) who, when he was told to report to Agritex, apparently with no explanation, had confided in an acquaintance,

> I have been called by Chimhanda. I don't know what's there. Maybe they want to take my field. I just thought I should let you know if anything should happen to me.

The women laughed at the fact as he was one of the farmers who did not attend any meetings. Today, they said, *angaawanikidzwa* (he had been fixed)!

CONCLUDING REMARKS

It is tempting to dismiss this group of farmers as unique, politicised or uncooperative. There is no guarantee, however, that a group of farmers who supply the required information without any protest are better informants. It is argued here that it is much more important to look at what happened as instructive.

It was clear that the experts were at a loss as how they had got it all wrong. The experts were surprised that this group of farmers were so critical of them while other farmers interviewed before were not. They were startled that the statistic was being uncooperative. But farmers did not consider themselves as mere statistics. Hence they were not concerned about conforming to the norm of other farmers. The other irritating point to farmers was that they were being exploited as sources of information, while not getting anything in return. The 'glib' talk about the nation benefitting was not impressive to farmers. It was thought that experts' reference that the reports would be sent to government would not raise any suspicions as the government was deemed to act for the good of the farmers. On the contrary farmers questioned the motives of government. The experts' lack of understanding of farmers was worsened by the fact that there was no dialogue being sought. Experts were interested in processing information of farmers to 'feed their hungry computer files', thereby keeping the industry going (Chambers, 1983).

One significant point was that farmers felt that they were not being taken as persons in their own right but as some property of Agritex. Consequently, experts promised to send previous reports to farmers through Agritex.[3] Experts did not seem to understand that they were one of the many people who sought information. The fact that the officers' normal duties did not include interacting with farmers worsened the situation. This would have meant, for instance, reducing the number of questions on the questionnaire. It would also appear that the time that was perceived by them as valuable was that of government officers and not that of farmers. But farmers had also other things to attend to.

Finally, the selection of farmers by local Agritex staff also made the data gathered prejudicial to 'objectivity'. Furthermore, a farmer who fears to be evicted by Agritex is not the person to tell you that they used less the amount of fertiliser recommended or that he got low yields. The timid farmer was the only one whose conversation was captured. This, however, does not mean that other farmers were not fearful too.

In summary, it is important not to consider farmers as merely data sets or sources of information. Even those farmers who may not be vocal can still silently resist and offer falsified information as a protest. This brief chapter has shown the need for experts to obtain a better understanding of farmers. The questionnaire survey, as employed by the officials, is not an effective data gathering tool to accomplish an understanding of the plight of the people which it is intended to achieve.

NOTES

1. Kopa is the Shona version of Cooperative. This building served as a Master Farmer training centre as well as an input and drought relief distribution point.
2. The tower of Babel in the Bible (Genesis 11) was the place where men fell into confusion because of a lack of a common language.
3. Eighteen months later the promised report had not been received.

Chapter 11

Getting back to the source
Farmer participatory research at the University of Zimbabwe

JEFF MUTIMBA

INTRODUCTION

In Chapter 2, we saw how technological interventions by the state have had little impact in communal areas, why farmers acted the way they did and how they used their own knowledge to deal with technological demands to meet their diverse needs. We also saw that their knowledge and sense-making had not formed part of the 'official' knowledge system. This chapter examines how the scientists' perspective of the technology development process in general, and of farmer participatory research (FPR) in particular, is shaped. As scientists are trained at universities, it is necessary to go back to the university and examine how they are trained. I will do this by examining two case studies of research projects that I participated in as a collaborator in the Faculty of Agriculture, University of Zimbabwe (UZ). The two case studies, that both focus on livestock, will demonstrate how research scientists are developed and moulded.

According to its mission statement, the Faculty of Agriculture sees its role as that of providing 'scientific leadership in Zimbabwe and the Southern African Development Community (SADC) region through teaching and research to generate new knowledge'. To this end, the Faculty sees need for 'regular consultation with the international community and with major agencies active in the agricultural community in order to better direct post-graduate research projects in areas where a wider section of the scientific community may cooperate and/or benefit'.[1] It is striking that the mission statement does not accord an important role to farmers in order to 'better direct research'.

Most of the graduates of the Faculty of Agriculture join the various sectors of the agricultural knowledge system as managers, researchers, extensionists or trainers (lecturers). Some remain at the UZ as trainers of other students. Until 1992, the UZ was the only local institution offering degree level training in agriculture. The majority of agricultural scientists in the country will, therefore, have gone through the UZ whilst only a few will also have received their training from universities outside the country, especially the United States of America and the United Kingdom. To fully explain the approach, attitudes and performance of research scientists in participatory development processes would require to analyse the philosophy, perception and practice of teaching at UZ. This ambitious task is beyond the scope of this chapter. Here, I will restrict myself to how the conceptualisation of farmer participatory research by UZ researchers is reflected in the two projects under scrutiny. Where possible, I will attempt to trace back where this conceptualisation comes from, and how it is sustained.

The first project I will discuss is called 'Improving crop productivity through animal draught power' and the second is called 'Improvement of research-extension-user linkages'. Each of the projects is led by a principal investigator. In the context of the Faculty of Agriculture, a principal investigator is usually a senior academic staff member, some with PhDs already, and looking for opportunities to further enhance their academic careers. Normally, they would have participated in developing the project proposal. As in the two case studies, the research projects usually involve students who do most of the actual research work in partial fulfilment of the requirements of their degree programmes.

In each case study, I will look at: the origin and objectives of the project; and follow briefly the process of implementation to capture the researchers' concept of farmer participation in terms of the steps that were taken to involve farmers and when in the research process; the rationale for bringing them in; the nature of farmer participation; attempts to understand farmers and their needs; attitudes towards indigenous technology; attempts (or lack thereof) of trying to understand FPR; and, the researchers' training background. My focus will be more on the researchers' concept of participatory technology development and less on the actual hard technologies emerging as a result of the initiatives. In any case, implementation of the projects had only just begun at the time of writing and no hard technological results had yet emerged. The discussion will cover the planning and the initial phases of implementation. This will be adequate to demonstrate the philosophical orientation of academic institutions and that of their resultant products — the graduates.

IMPROVING CROP PRODUCTIVITY THROUGH THE ANIMAL DRAUGHT POWER PROJECT

This project was looking at crop and livestock interaction with particular emphasis on factors that constrain the contribution of animal draught power to improved crop production. The main objective of the research project was to develop a model of draught animal systems in Zimbabwe which used oxen, cows or donkeys for crop production. The purpose of modelling was to assist extension workers and farmers to develop strategies for the most effective use of draught animals and of stockfeed resources in order to increase crop yields on smallholder farms. This was to be achieved by focusing on constraints that were known in Zimbabwe and making critical appraisal of their causes, the effects they initiated, and the interventions that would lead to reduction or avoidance.

Donor-imposed collaboration

As I was embarking on my PhD research programme in 1992, my supervisor (Dr Olivia Muchena) and I submitted a proposal to the Rockefeller Foundation for funding through their programme called 'Forum on Agricultural Resource Husbandry'. (My supervisor was going to be the principal investigator.) The objective of this programme was to encourage and support agricultural research by staff and graduate students in faculties of agriculture in selected countries in eastern and southern Africa. Agricultural resource husbandry was defined as the art and science of combining plants, land, water, off-farm inputs and labour to produce a crop or crops. Research in this area was expected to include agronomy, soil management, land management, cropping systems, and other economic, management and husbandry issues related to crop production.

After a year of deliberations, during which time there were encouraging signs that the Forum was really interested in our proposal, we were told that our research project could not 'stand alone' and could only be supported if it was done within the context of on-going crop or livestock research. We were advised to link up with a draught power project which was being evolved in the Department of Animal Science at the UZ. When I approached one of the planners of that project I was told that the proposal had already been submitted to the Forum so it was too late for me to be included in the project.

After a few months, one of the team members of the draught power project came back to me and said, 'Were you serious in wanting to team up with us on this project? Apparently, they (the Forum) are turning down our proposal because it lacks a farmer participatory perspective. Are you still interested?' I said, 'Of course, I am still interested', and he went on, 'OK then, give us the words and we will revise and re-submit it.' I then wrote a paragraph saying how participatory the research would be and the proposal was revised and sent back. After a few weeks, a reply came back saying the proposal had been approved.

The Forum programme was also intended to strengthen ties between university researchers, government policy, and national and international science while bringing the university closer to 'real world problem-solving opportunities'. Consequently, one of the conditions for funding was that a research project had to include collaboration with other major research agencies. Therefore, the proposal was that the draught power project was going to collaborate with the Agronomy Institute of DRSS (the Department of Research and Specialist Services in the Ministry of Agriculture), the Department of Agricultural Economics and Extension (through myself) at the UZ and the Centre for Tropical Veterinary Medicine, United Kingdom. These institutions were identified for their comparative strengths in aspects of the project: the Agronomy Institute was expected to provide inputs in terms of crop production since the project was looking at the contribution of draught power to crop production; the Department of Agricultural Economics and Extension was to provide inputs in terms of socio-economic issues (in particular, I was expected to provide the farmer participatory orientation); and, the Centre for Tropical Veterinary Medicine had developed computer models on draught power which the project would learn from.

Grappling with the concept of farmer participation

The project leader came to me and said 'We have now got the funding, and we are ready to get on with the job. What do we do now to show that farmers have participated?' Apparently, the methodology and research design had already been decided upon — the project had been designed to use mathematical modelling using secondary data. The draught power improvement was seen as merely a technical matter with research activities solely directed towards developing a model based on feed and work output of draught power animals. I suggested we needed to establish farmers' views on the draught animal power problems, their coping strategies and their needs for research in this area. I suggested that we needed this information for planning and implementation purposes and to ensure effectiveness of our research by addressing real issues of concern.

We agreed to do a participatory diagnosis for ten weeks using undergraduate agricultural students who had just completed their second year. The objectives of the diagnosis were re-defined as: (a) to establish the role of animal traction in smallholder farming systems and its inter-linkages with other subsystems; (b) to get a clearer understanding of the problem of draught power; and, (c) to identify possible areas of intervention. It was also hoped that the survey would provide reliable qualitative data on draught power that would assist in developing an animal traction model.

I drew up a checklist to be used by the students as a guideline only to allow them to investigate the draught power problems through field observations and discussions with farmers. When the checklist was given to a statistician who was going to assist in developing the model, he changed everything into a questionnaire with 'yes' or 'no' type answers. He said they had no computer programmes that would analyse the kind of data that my format would generate. We failed to reach a common understanding but agreed that the two instruments would be used together. I knew I had lost the argument and that the 'enumerators' would just use the questionnaires which were easy to complete and required little thinking on their part as they had nothing at stake in the project. My suggested instrument would have been better used by the actual (student) researchers who would, I assumed, see the need for more thorough understanding of the farmers, their views and their utilisation of draught power. This was not going to be possible by 'mechanically rushing' through a set of questions. We selected three communal areas in which to do the diagnosis. The selection was done on the basis of agro-ecological regions and farming systems to give us a representative picture of the draught power problems across the country.

By the time we wanted to do the diagnoses, the MSc students who were to work on the project had not yet been identified. So we decided on recruiting six undergraduate students (two for each area) to do the diagnosis — literally to be enumerators. By that time, most of the agricultural students had already found something to do during their vacation.² Therefore, out of the six we took, only two were agricultural students — two were from psychology and the other two were sociology students.

We agreed I should give a 'crash' orientation course on the FPR component of the project before sending them out in the field. I found that the sociology students were fairly familiar with farmer participatory concepts whilst the agricultural students were not. Agricultural students are taught about crops and livestock but not about the people (farmers) who produce those crops and livestock. A look at the courses taught in terms of research methods reveals fundamental differences between social science and agricultural students. For the crops and animal science students there is an optional course called 'experimental designs'. The few who choose the course are taught experimental designs, regression techniques, analysis over sites and statistical computing — and nothing about participatory methods. Agricultural economics students take a compulsory course called 'research in agricultural economics' in which the emphasis is on surveys and quantitative data analysis. The sociology students, on the other hand, take a compulsory course called 'social research methods' which covers field work (how to work with people), systematic observations and experimentation, participatory surveys and qualitative data analysis.

Whereas I had hoped that the students would make observations (on draught power utilisation and management) through repeated visits to, and discussions with, ten farmers each (60 farmers in all), the whole exercise was reduced to administering the questionnaire, and filling in the 'boxes', through single visits. Each pair of students had to interview at least 150 farmers in ten weeks (in all 452 farmers were interviewed). My intended sample of 60 was considered too small for statistical purposes.

Sociology students reported problems in using the questionnaires pointing out that it was 'a bit too technical and missed out on qualitative data such as that on coping strategies'. They also observed that the questionnaire assumed that farmers planted crops in pure stands. For example, there were questions on area for each crop and frequency of ploughing or cultivations for each crop, when in fact some of the crops were grown in associations. There was no provision for this. They suggested the questionnaire should have been pre-tested and modified before final use. Just before the field exercise commenced, I left for my own theoretical orientation training at Wageningen Agricultural University in the Netherlands. I was away for three months, and was therefore unavailable to suggest any remedial action in the field for the duration of the field exercise.

Getting back to the true aim of the project

The unwritten aim of the project was to provide an opportunity for researchers (myself included) to advance their academic careers. Two post-graduate students — one with a crop science background and another with animal science — were selected to work full time on this draught power project. They each developed a research proposal on an aspect of draught power and were both registered for an MSc degree. One student was to focus on the effect of draught power on some agronomic factors that affect productivity whilst the other was to look at animal factors affecting cattle draught power output.

The students proceeded to do a literature review and to analyse the data brought back from the diagnosis which was all quantitative. Having observed that the students were recent graduates of the University lacking the practical 'feel' of the situation they were planning for (they were even analysing data which they themselves had not collected), I volunteered to take them out for two days to talk with farmers. I thought we also needed explanations on the largely quantitative data that had been generated from the survey. At first, to the students, this was like an interesting break from their computer laboratory — otherwise it had not been part of the requirement for their degree programme. Once in the field, they began to get the feel of the nature and complexity of the problem. They were also able to follow up on issues that had not been clear from the survey data.

Back to the university, the students continued developing their research proposals. The one was looking at 'Some animal factors affecting draught power output in the smallholder sectors of Zimbabwe using models as aid to improved management'. The objectives of this part of the research were: to assemble literature on factors that affect draught animal power output with a view to integrating them into a model that would quantify the animal draught and crop production linkages; and, to develop cattle management packages that optimise draught animal power for improved crop productivity. The model was to be developed using concepts and quantitative values from literature. Whereas the farmers were looking for ways of improving the condition of their cattle (through the provision of water, disease control and feed — see survey results below), the student was going to produce sets of equations to represent the processes within the system and accompanying predictions. The other student was to look at 'The interrelationship between draught power supply, tillage practices and crop productivity'. The objectives were: to study how different tillage practices, their timing and other factors influence crop productivity; and, to develop packages that match tillage practice to available draught power in different soil types under different soil moisture to optimise crop yield. His assumption was that some recommendations on tillage had not been adopted by farmers because they were applied as individual components without looking at how other factors influenced their adoption. Again, the researcher's main interest was in modelling the processes that influenced, or were influenced by, different tillage practices in relation to draught power availability and output. The underlying assumption was that the project would produce models that could be used by extension workers in 'teaching' farmers.

Summary of survey results

Forty percent of farmers had adequate (at least two oxen) draught animals, 48 percent did not have any draught animals whilst twelve percent had inadequate draught animals. Farmers used oxen as a source of draught power to perform tasks like ploughing, planting, cultivation, ridging and transporting inputs (manure, fertilizer, seed) or outputs (crops and stover). To a smaller extent, cows were also used provided they were not milking. A few farmers used donkeys alone or in combination with cattle.

Farmers with adequate draught power were generally able to carry out key operations, like ploughing, planting and cultivation (to control weeds) in a timely manner. If they chose to, they could also perform other operations like winter ploughing, harrowing, marking planting furrows and ridging to conserve moisture. Availability of draught power also reduced demands for labour for operations like planting, weeding and transport (of inputs, of crops from the fields and to the markets). Draught animals, as with livestock in general, were an important source of fertilizer (manure). Those with inadequate draught power teamed up together and performed the various tasks in turns.

Those without draught power managed in different ways. Some had relatives with draught power who could make it available to them free of charge; some hired for cash; some did odd jobs (like weeding, harvesting, or digging in the garden) for others in exchange for draught power. In all cases, use of draught animals was restricted to the one most important operation (in terms of draught power requirements) — ploughing at planting time. Farmers could not afford to hire or beg for the other operations like winter ploughing (because they would sometimes need to plough again at planting time to control weeds), cultivation and transport. Even then, they could only plough after the owners were through with their own operations. Therefore, non-draught power owners usually ploughed and planted late resulting in low yields.

In a few cases where the farmers were either too poor to hire draught animals, too weak to give their labour in exchange for draught animals, or without relatives with draught animals, the farmers planted without ploughing. But this was like a desperate last resort. In general, farmers did not believe in zero tillage although it was being encouraged especially after the 1991/92 devastating drought that killed a lot of cattle. They had observed that crops planted on unploughed lands did not perform well generally and that there were much more weed problems than from ploughed lands. For those who did not have draught animals, this also meant increased labour requirements for hand-weeding to control the weeds. Only a few farmers had donkeys either as a sole source of draught power or which they used in combination with cattle. Generally, donkeys were considered to be a poor farmer's type of livestock. They were however considered hardy and more tolerant to drought conditions than cattle. They were also more tame than cattle and, for those who had a choice, they used them more for tasks that required precision like marking planting furrows and cultivation. They were slower than cattle but, because they were docile, even one person could do these operations where labour was a problem. Their other advantage was that they could carry loads on their back — reducing the need for scotch-carts (ox-carts). Cattle were preferred because they provided a wider range of benefits — higher quality manure, milk, and meat. At the end of their productive life — usually ten years — they could also be sold for cash.

In general, those who had draught animals where not really keen to share with those who did not have (even on a hiring basis). They felt that the animals were already working hard on their own lands given that the animals were generally weak most of the time. They only shared as a 'social obligation'. The nature of the draught power problem expressed itself in two forms — unavailability and inadequacy in terms of absolute numbers and in terms of the quality of the draught power available (i.e. weak draught animals). Forty eight percent of farmers did not have draught power at all. Some of them had never had (for a variety of reasons including poverty) whilst others lost their cattle during the 1991/92 drought and had not managed to re-stock. Some farmers had less than adequate numbers to make a span either because of poverty or they had lost some of their draught animals during the drought. The 1991/92 drought was followed by another two poor seasons in terms of rainfall so that building up the livestock numbers was slow.

In terms of cattle keeping, farmers mentioned three main problems — shortage of drinking water, diseases and inadequate grazing in that order of importance. The problem of drinking water was as a result of successive droughts. Dams and rivers had dried up and many boreholes were also drying up. Farmers suggested drilling more boreholes and the construction of dams.

The diseases were mainly tick-born (e.g. red water and heart water). Ticks were becoming a problem because the Department of Veterinary Services, which is responsible for providing dipping services to communal areas, was running out of funds to provide the services. As a result, cattle were going for months without dipping instead of the one or two weekly intervals. To alleviate the problem, the Department was introducing a cost recovery system. Farmers had already been informed that they were going to pay for the dipping services through an annual fee per animal. Despite their concern that the fee was too high ($3 per animal per year), they were supportive of the principle provided it led to an improvement in the services.

Shortage of grazing was largely due to pressure on the land resulting in more and more grazing land being turned into arable and residential land. The recent droughts had also exacerbated the problem in that the grazing lands were not producing enough herbage. The problems of grazing were more acute in winter when the veld would be dry. Farmers' solutions included supplementary feeding with (mainly maize) stover during winter; and, grazing their cattle on the arable lands during winter. This was one reason why farmers did not plough their lands in winter as recommended by the extension services (see also Chapter 2). At harvesting, cattle owners cut maize stover (and sometimes groundnut halms) and stored them on raised platforms for controlled feeding to their cattle during the dry period. They also cut grass from their contour ridges and garden plots and stored it in the same way. A few also gathered pods of *musekesa* (monkey bread or *Peliostigma thorningii*) and *mupangara* (*Acacia albida*) and fed their cattle. Draught animals got special preference during peak ploughing periods like October and November.

Usually, the cattle were given the feed after ploughing. Some farmers added salt to the stover or grass to make it more palatable.

The effects of diseases and poor grazing was that the draught animals were generally too weak for maximum production. Farmers mentioned that in the past when grazing was plentiful, they used to have big strong cattle which could do more work than the cattle they had now which were much smaller because of low levels of nutrition. As regards grazing, farmers suggested the establishment of controlled communal grazing schemes. The majority of those who made this suggestion also linked the 'schemes' to fencing. They felt that if the schemes were fenced, the need for herding (which is a labour-demanding activity) would also be reduced thereby releasing the scarce labour (with all children going to school) for more productive activities. The suggestion for grazing schemes was also laced with an underlying expectation that if farmers formed cohesive groups, they could attract donors who would provide fencing materials. Farmers also suggested planting and production of fodder crops. They wanted ideas on high yielding (in terms of herbage) fodder crops they could plant either in their arable or grazing lands. Farmers bordering large scale commercial farmers also suggested 'resettlement' into commercial farming land as a solution.

Other problems mentioned by farmers were: ophthalmia (for which they used local herbs to control); shortage of commercial drugs to treat sick cattle; shortage of money to buy the drugs; cattle rustling; high cost of hiring draught animals (at the time, it cost about Z$100 to plough a hectare); low calving rates; internal parasites; lumpy skin disease; and, slow response of veterinary staff when invited by farmers to attend to cattle health problems. Farmers thought the veterinary staff were elitist whilst they felt Agritex staff were more responsive to their needs.

'Marketing' the FPR concept

The Rockefeller Forum organised an in-country workshop for all researchers working on projects sponsored by the Forum. The principal investigator of the draught power project requested the organisers to invite me to the workshop as well. When I saw the invitation, I requested that I present a paper as well. The reply from the organisers was that my paper, which was on FPR, was 'not in line with the Forum workshop'. The organisers, who were also senior lecturers at the University, did not see the relevance of FPR and must have wondered what I was looking for in a purely technical workshop. I went back to the principal investigator of the draught power project to 'protest' and he suggested he would give me some of his time on the programme.

In preparation for the workshop, the students on the draught power project prepared presentations on their proposals and presented these to the draught power project team (of which I was a member) before the actual workshop — a kind of 'dry run'. We noticed that the students were not saying anything about farmers and whether they had any input. They were not even saying anything about the diagnostic survey and how the results were to be used. Considering that the project had been approved on the basis that farmers were going to be involved, we were concerned that the sponsors would not be happy with the presentations as they were. We therefore suggested that the students mention how the survey had clarified the nature and extent of the problem. We also advised the students to say that the models were going to be used to select recommendations that would be tested on-farm and with farmers and that any feedback would be used to modify the models. In other words, we were now strategising on how to relate the actual activities to the 'official' demands.

At the workshop, which was held in a three-star hotel — away from farmers — three people performed the official opening. Two of them were Rockefeller officials. In their speeches, the three mentioned, repeatedly, the need for technologies to be relevant to farmers' needs. However, there was no space in the programme to deal specifically with this issue of 'relevance' and how it could be achieved. There were 17 presentations covering: groundnut trials (2); soil and fertility management (5); tillage (3); draught power (3); dairy (1); and, weed management (3). I wondered where else FPR would be discussed if it was not discussed at such fora. I also wondered how the three 'officials' hoped to see relevant technologies emerging when the whole programme reflected a central concern only for biological issues relating to crops and livestock.

I managed to squeeze a paper on FPR using the draught power principal investigator's scheduled time. The paper appeared to have aroused immense interest with calls for more collaboration with those with this (FPR) philosophical disposition. However, there was observable unanimity among the researchers that farmers should not be exposed to technologies before the researchers worked on them first. 'What if it fails? What are you going to do?', interjected one prominent researcher to suggestions that farmers should be involved from the beginning and throughout the process. 'What if you work on the technology for two years, or five years, and you take it to farmers and they say they don't want it — what are you going to do?', I demanded, to some laughter. In all, it felt like a marketing exercise of the farmer participatory concept.

IMPROVEMENT OF RESEARCH-EXTENSION-USER LINKAGES PROJECT

Introduction

The International Livestock Research Institute (ILRI, formerly International Livestock Centre for Africa — ILCA), together with the Department of Animal Science, UZ, decided to launch a four-year research project on the 'improvement of research-extension-user linkages' with funding from the Danish International Development Agency (DANIDA). The overall objective of the project was to 'develop the capacity of national agricultural research system (NARS) scientists to a better understanding and conduct of research with smallholder farmers with a view to improving the research-extension-user linkages for the effective transfer of improved technology packages'.

This statement of objective already posed a conceptual problem. It sounded like the project was based on a 'transfer of technology' model. Its focus was on improvement of the flow process of 'improved' technology from research through extension to farmers. This reflected a top-down approach which contradicted the FPR philosophy espoused in the same statement of objective. This focus reflected a thinking in terms of a knowledge system that starts with the researcher who develops 'improved' technologies and hands them down to extension for onward transmission to farmers to replace their 'inferior' technologies.

The specific objectives of the project were:
i)   to develop training courses to give NARS scientists a better understanding of the farmers' problems, including the potential for human capacities and the research methods for solutions;
ii) to introduce and use the Farming Systems Research approach (FSR) as well as improved on-farm experimentation procedures;
iii) to develop and strengthen NARS capacity in the use of FPR;
iv) to develop cost effective farmer participatory methods.

As its major output, the project was expected to lead to sustainable increases in livestock production in Zimbabwe, particularly at the smallholder level, resulting from improvements in the technology development and transfer linkages (Olaloku, 1994).

A team of 28 experts (animal scientists and extensionists) was brought together at the UZ for a three-day pre-project implementation planning workshop to outline the project approach and underlying philosophy; and to identify the priority research and training issues and develop a programme of activities for the project. Invitations were also sent to ZFU (Zimbabwe Farmers Union, the union of smallholder farmers), but they did not send any representatives. Even if the Union had sent representatives it would have sent one or two who would most probably not have been effective — they would have simply been overwhelmed by the number of scientists present.

The workshop started with some lecture/discussion inputs on: the key issues and the underlying philosophy and approach of the project; farmer participatory approaches; research/extension/farmer linkages; and a method to describe the sustainability of livestock farming systems. This was followed by presentations by participants on their work experiences and their organisations' views/experiences in research, extension and farmer linkages. After participants shared their experiences, we then identified cattle keeping problems and potential solutions in communal areas through a brainstorming and ranking exercise. Led by the Zimbabwean participants (six of the participants came from countries in eastern and southern Africa and Denmark), participants identified shortage of grazing as the main problem. The problem was caused by lack of proper grazing management which, in turn, was caused by the communal land tenure system which was not conducive to proper grazing management. As a consequence of this grazing problem, farmers were experiencing serious shortages of draught power which in turn affected their crop yields.

To reduce the effects of insufficient grazing, the extension service was recommending collection, storage and controlled feeding of crop residues to cattle; production of improved fodder grasses; controlled (communal) grazing schemes; proper use of draught animals; and minimum tillage methods. However, these efforts were not having much impact as only few communities adopted, or had possibilities of adopting, grazing management schemes; yields of fodder grasses were generally low; lack of draught power led to poor crop performance which meant low stover yields.

After some lengthy debate on the problems, the workshop was informed that the planners of the project were, in fact, only interested in small-scale dairying and may be in short-term cattle fattening as well with particular emphasis on those cattle due for culling. We were taken aback as we thought we had been debating the real problems and we were now being asked to identify problems relating to small-scale dairying. What was not explained explicitly throughout was that ILRI's mandate was milk and beef production. There was not much small-scale dairy farming in the communal areas. There were only ten very small community projects throughout the country producing a total average of 6,000 litres of marketable milk per day nationwide.[3] We reluctantly accepted this re-orientation of focus. We had to, anyway, as we were interested in the funding that was coming with the project. If we wanted the funding, we had to find problems in small-scale dairy. So, again, we identified lack of appropriate dairy feed and lack of water as major limiting factors.

We then went on to debate in detail possible areas of intervention regarding feeding. We suggested the following as researchable areas: feed budgets; feed inventory; conservation methods; processing methods; utilisation methods; and inter-linkages of livestock and crops. Other problem areas that were identified were: unavailability of dairy stock (calling for breeding, calf rearing methods and herd health studies); and lack of enabling economic environment (calling for studies on viability models and rural milk processing). At some point, the group also tried to look at, and prioritise, researchable issues regarding researcher/advisor linkages; advisor/farmer linkages; and researcher/farmer linkages but failed to reach a common understanding on this approach. It was evident that the issue of linkages could not be dealt with in isolation. It was therefore agreed that this would be dealt with within the context of small-scale dairy research. The project would identify or develop linkages mechanisms that would facilitate the achievement of research objectives. However, four general problems were identified. These were (with possible solutions in brackets):

1) unclear policy direction (review and develop mechanisms for institutional linkages; develop mechanisms for incentives for research and advisory staff — like publications, competitions and honoraria);

2) inappropriate training (research and extension staff short term training on research methodologies, communication skills and technical subjects; farmer training programmes; individual higher degree training);

· 3) inadequate synthesis of information and publication (identify gaps in available information; produce working papers for the project; produce manuals for staff and farmers; publish papers in refereed journals); and

4) lack of financial resources for researchers (create income generating projects; sale of developed technologies; raise levies on produce).

The workshop ended with suggestions for a training course to sensitise scientists to the methodological issues and approaches for research on the linkages including the relevance and usefulness of FPR and on-farm livestock research methodologies. A coordinating team was set up to meet and formulate and prioritise the activities of the project. Subsequently, two researchers (PhD students), an animal scientist and a socio-economist, were selected to work on the project full time. A communal area, Gokwe South, was identified where small-scale dairy was being introduced by the Agricultural Rural Development Authority (ARDA), a parastatal, through its Dairy Development Project (DDP).

Background of the Dairy Development Project

Soon after independence, the Government of Zimbabwe, through the Ministry of Agriculture, announced that it wanted 'to see black and white cows throughout the communal areas' (a reference to the black and white Friesland dairy cows). This was part of the Government's broad objective of breaking the monopoly enjoyed by Whites in various sectors of the economy prior to independence. This posed a huge challenge given the circumstances in the communal areas. Communal farmers kept partially dual purpose cattle (for draught power and some milk) which were tolerant to the harsh conditions (limited feed, water and health care) found in these areas. In addition, the poor infrastructure (poor roads and long distance to markets) was not conducive to commercial milk production.

Money was sourced from the then European Economic Community (now the European Union — EU) to facilitate the introduction of dairy cattle in communal areas. ARDA was asked to spearhead the development of small-scale dairy in the communal and small-scale commercial farming areas. A Dairy Development Project unit (DDP) was then established within ARDA in 1983. Essentially, the activities of DDP included the provision, on soft loan basis, of proper breeds of milking cows and bulls, cooling and storage facilities. Interested groups of farmers approached DDP for assistance to go into dairying. The groups were expected to stand on their own after some time. Thirteen years later, ten projects had been started and none of them had yet been weaned although there was talk of weaning one or two of the first ones. Together with Agritex, DDP also provided normal extension services to dairy farmers.

The dairy project in Gokwe South was established in January 1994 on the initiative of the farmers. A group of farmers got together and agreed on a Z$400 membership fee and approached Agritex and DDP for guidance and assistance. It must be pointed out here that the group was composed of people who had worked in various sectors of the economy and had come to settle in Gokwe in search of economic opportunities in agriculture. Some even had formal agricultural training and had worked for many years in extension as advisors. Therefore, the group was composed of well informed people who knew about other initiatives and provisions within the public sector. At the time of writing, the group had mobilised 110 members, 69 of whom had already paid their membership fees with 35 of them already producing milk. The rest of the members were at various stages of procuring the milking cows. With DDP's assistance, a milk collection and distribution-cum-marketing centre had already been constructed and was functional.

Entry into the dairy project area

Following the recommendations of the pre-implementation workshop, a one week training workshop on 'Livestock Systems Research Methodologies' was organised and run in Gokwe. During the planning of the course, concern was raised whether scientists would not object to being 'stuck out' in a remote area like Gokwe for 'a whole week'. Many agricultural scientists in Zimbabwe are born and brought up in rural villages. However, as they grow up, they move away from the villages and the poverty associated with the villages. There is a general reluctance to go back to the villages for any length of time. Having studied 'their way out of poverty', they do not want to go back to it. We decided that if the concept of FPR was accepted, then we had to go where the farmers were. Even then, we selected the best hotel in the area for a venue. This attitude has the effect of maintaining the gap between farmers and researchers who cannot, therefore, participate as equal partners.

The training workshop, which covered FSR and FPR methods, was attended by two farmer representatives and 22 research and extension staff from the major potential collaborating agencies. At that time, the impression was given that the research agenda was yet to be evolved depending on what farmers saw as their problems. Whilst this was said, constant reference was made to 'dairy' throughout the discussions that it did not look likely that the project would look at anything else other than dairy. The whole tone of the workshop was on dairy — the selection of the participants was based on their interest in dairy; the selection of the research area was based on the dairy activities going on; and the selection of survey farmers was based on their interest in dairy. In addition, it had been decided from the onset that the research would be in two broad areas — animal production aspects and socio-economic aspects — hence researchers had been selected on that basis. One researcher remarked: 'We are not going to dictate to farmers what we want to do, but we are not going, all of a sudden, to pretend that we don't know anything either.' From this rhetoric, it was clear that all we needed here were 'diplomatic salesmanship' skills. The research agenda had already been set — but it still had to be sold to farmers in a way that they would feel they suggested it.

The concept of FPR was accepted at the philosophical level but so far, there had been no evidence of it in practice. At one of the earlier planning meetings, it was agreed that 'all the parties (research, extension and farmers) would participate at every stage until they came up with a research agenda'. But we had already come this far without having consulted a single farmer. The meetings we had held so far were considered too technical for farmers to be involved. Participation by those farmers present at the training workshop was minimal and limited to answering a few questions directed to them not because the training was too technical for them, but because they were too few compared to 'officials' present.

It was not going to be easy either for researchers to move away from their traditional methods of research. In discussing data handling and analysis, one resource person dealt with quantitative data only. When I asked about handling qualitative data he responded: 'What is that — is that when you go out with a tape recorder like a news reporter or journalist and you come back and write a composition — something like that?' The skills provided in this training workshop were expected to be used in the subsequent research programme of the project. Our failure to deal adequately with qualitative data handling methods would have implications on the extent to which the research process would be participatory with the tendency for researchers to look for quantitative data that would be able to withstand scientific rigour.

As an initial phase into the field, we also did a rapid rural appraisal (RRA) in the villages we were proposing to work. Our appraisal was focused mainly on dairy. This gave us some initial impressions about dairy farming in the area. I felt that the main strength of the RRA method was that it made the situation analysis more focused and systematic. We were able to get a lot of information in a short time. However, I observed several weaknesses in the way we used the approach. Firstly, we went into the field without prior adequate training on how to conduct dialogue with farmers. Therefore, in most cases, animation of discussions was poor with too much use of leading questions and less open-ended questions. What were supposed to be discussions, were reduced to question and answer sessions. I witnessed several cases involving especially junior scientists where farmers would be explaining something and the scientist would interrupt and say, 'No, I have not got to that yet, what I am asking is . . .' The farmers would then resort to answering strictly the questions asked and wait for the 'next question'.

Secondly, the questions we asked were too many and both farmers and researchers were exhausted at the end of each interview session. Because of the large number of questions, researchers were not able to internalise them well enough to be able to use them in the form of discussions. They therefore tended to refer to the many pages of written questions rushing through them without showing much interest in what the farmers were saying. The farmers were reduced to passive interviewees and respondents. Their participation was restricted to answering questions.

Thirdly, the time in the field tended to be too short for researchers to get to grips with the real issues. There was no time to develop trust and to understand how things actually happened on farms. Fourthly, whilst I found some of the techniques we used, like walking transects and household interviews, to be participatory though still extractive, I also found some of them, like mapping and matrix ranking, to be rather artificial, academic, and reflective of a Western way of thinking. Peasant farmers do not draw maps in their day to day communication. We drew the maps, disappeared with them and never really used them again.

The rapid appraisal was followed by a less 'rapid' and more thorough appraisal, by the two research students, a few weeks later designed to quantify findings of the rapid survey. In their order of importance, the following problems were identified:
1. low rainfall and shortage of water for domestic and agricultural use;
2. high costs of animal feeds;
3. problems of milk transport to the marketing centre;
4. cash shortage for agricultural purposes;
5. poor and limited availability of pastures;
6. high prevalence of cattle diseases and limited knowledge of them;
7. unavailability of acaricides and other veterinary drugs;
8. high prevalence of crop pests and diseases;

9. declining arable land;
10. shortage of fencing materials;
11. shortage of farm implements;
12. increase in gulley erosion;
13. lower producer prices;
14. shortage and high costs of transport of agricultural produce to markets.

The results of the survey were discussed with the dairy farmers at a report-back workshop organised for the purpose. The farmers confirmed the findings. Given these problems, farmers were asked what they thought should be done. They suggested that the problem of water was so serious that it needed to be attended to before all else. They emphasized that solving that one problem would also solve a few other problems on the list like pastures and feeds. They also pointed out that the problem of water was complex. It was exacerbated by growing pressure on the land with people coming to settle from other communal areas of the country resulting in unplanned settlements, shortage of grazing, accelerated soil erosion and rapid desertification. They therefore suggested that the district and local authorities (district administrator, district council, chiefs and village heads) should be involved as they were responsible for the welfare of the district and could control land utilisation. They could also deal with the influx of people coming in from other areas.

As researchers, we had no means of dealing with this problem. We suggested that this be referred to the local and district leadership at another meeting. Further discussion had to be steered towards researchable problems in dairying. In the end, farmers agreed feed shortage was an important problem that needed to be researched on. One farmer dramatised the problem with a reference to Farmers Co-op (a major source of commercial dairy meal): 'We have all the raw materials to make the (dairy) concentrate. We want to learn how Farmers Co-op makes it so we can mix and make it for ourselves'. The Department of Veterinary Services also came under attack for not responding to requests promptly, for not providing results of samples submitted for laboratory analysis and for poor dipping services.

Back to what we meant to do anyway

As was already evident from the pre-implementation workshop, the potential impact area where the project had relative advantage was in the area of feed given the expertise and technologies already available. The survey served to confirm what we had already identified and decided were the research issues. We were going to focus on the development of feeding systems based on home-grown feeds. We convinced ourselves that we needed not be ashamed of choosing the easier way as our key objective was to develop a linkage model that could be recommended for general application in research. The research on feeding systems was going to provide the context. We also convinced ourselves that if we were able to come out with fodder crops that could grow in this drought-prone area, we would have indirectly lowered the impact of low rainfall (and the consequent lack of water) — a key problem as seen by farmers. The key objective of developing a linkage model was not explained explicitly to farmers because there was nothing tangible in it that would benefit farmers directly.

The students then developed their PhD proposals based on this strategy. One student was going to focus on the technical aspects of selecting fodder crops suitable for the area; designing livestock feeding systems using selected fodder crops; and formulating and testing of rations based on farm-grown crops such as maize and sunflower. The other student was going to look at the socio-economic aspect and, in particular, he was going to analyse existing linkages and develop, test and adapt a linkage model during the process. The project hoped to develop the organisational means for interfaces that would enable knowledge to flow between the three sub-systems — research, extension and farmers. Interfacing and adaptive networking was to be done in a number of ways: through meetings with farmers; through the creation of teams composed of representatives from the various collaborating agencies and farmers; and through trial adaptation. The dialogue that had started to emerge, and was expected to intensify, was expected to allow knowledge to flow in both directions. An expected outcome of this was a linkage model that would be recommended for wider application.

On the ground the key collaborators were going to be selected farmers, Agritex and DDP. The last two were already working with farmers and they assisted the project in gaining 'entry' into the community. They identified the farmers, introduced the project to the farmers and to the local leadership and assisted in laying the ground work for field activities. They were expected to take up the technologies that would emerge from the project activities for wider dissemination. The Department of Veterinary Services was expected to provide advisory inputs on matters pertaining to animal health. At the information sharing level, other collaborators in the project were the Department of Agricultural Economics and Extension (UZ), DRSS, National

Institute of Animal Science in Denmark and ILRI in Ethiopia. Whilst the Department of Agricultural Economics and Extension was expected to provide inputs on socio-economic issues, the other three institutions were expected to share their experiences, expertise and findings as they were both involved in dairy research. The Department of Animal Science (UZ) managed and coordinated the project activities. Through their nominated representatives, the collaborating institutions also formed a coordinating committee which was to: monitor progress of the project; organise dissemination of project results and information; keep their institutions informed of progress; and become a model of functional research-extension-farmer linkage. Farmers were also to elect two representatives to this committee which would meet every three months.

The trials on fodder crops were going to involve legumes (fine stem stylo, siratro, lablab, silverleaf) and grasses (star grass, bana, napier, sabi panicum, giant rhodes) that had been researched on-station and used on large-scale commercial farms. These 'elite' pasture materials were going to be tried for their suitability (growth, productivity, persistence and quality) in this low rainfall smallholder area of Gokwe South. The local sources of forages like *musekesa* and *mupangara*, which were already being used by farmers, were not included in the trials because of, as one of the supervisors said, their low chances of success. One researcher remarked: 'If you are going to work with farmers you have to make sure you have technologies to sell them — otherwise you will be wasting their time — they have better things to do.'

The students then identified farmers to host the trials on their farms and explained to them: the terms of their participation; the experimental designs; treatments; inputs; roles; type of data to collect; steps and procedures. The process followed was very similar to the contract mode of participation as identified by Biggs (1989). Researchers were bringing proposals for testing. Selected farmers who could guarantee the conditions of the contract would participate — we wanted farmers who were willing to give up portions of their land to try fodder crops; we wanted farmers who could grow at least two hectares each of maize and sunflower to provide home-grown concentrate supplementation materials to be experimented with; we wanted farmers who had at least two dairy cows and were willing to make them available for the feeding trials. Reports would be produced and papers published (including PhD theses).

CONCLUSIONS FROM THE TWO CASE STUDIES

The two case studies demonstrate a growing realisation of the need for livestock research in smallholder agriculture. Traditionally, on-farm livestock trials were considered too difficult to conduct because of the level of risk involved (there were no mechanisms for compensating farmers in the event of the livestock dying because of research 'treatments') and because of the problem of maintaining farmers' enthusiasm due to the long-term nature of livestock research. On-farm research and extension was therefore more on crops where the risks were lower and the results were more immediate than in livestock. Researchers also shied away from on-farm livestock research because they did not have complete control of the animals resulting in what they termed 'noise' in the data.[4]

The following conclusions can be drawn from the two case studies presented in this chapter.

(a) Both projects were laced with an FPR label to make them attractive, may be to donors, when in fact the approach was top down. Researchers had already decided on what to research, the objectives, the treatments, the type of data to collect, the trial design, the type of trial, when to go to the farmers, what type of farmers to involve and the extent of their involvement before they even met the farmers. By the time the farmers were 'called in' and 'informed', it was too late for them to make any meaningful contributions. The farmers' contributions were limited to those activities that would ensure the successful implementation of the researchers' programmes or projects (like answering questions, providing their land and cows for the trials). The meetings with farmers were more like a marketing strategy either to show what the researchers had or to obtain farmers' commitment in what was to be done.

Farmer participation was relatively passive and restricted to the provision of information only. Farmers were mobilised to provide information which would (hopefully) be used by researchers to achieve their preconceived objectives. Farmers were not even told the full objectives of the research initiatives. The analysis of the information and decisions on how it would be used was at the discretion of the researchers despite the avowed intention of involving 'all actors throughout the process'.

Moreover, the concept of farmer participation itself was not well understood by the researchers. What we see are attempts to get farmers to participate in programmes designed by researchers and on terms laid down by researchers. This apparent ignorance stems from the fact that participatory research methods are neither known nor taught at the University. .

(b) The products (researchers) of the training process such as those described in the two examples are likely to come out with a 'transfer of technology' orientation that will be difficult to de-programme in their later working life. The graduates come out already conditioned to think they are the source of knowledge and that they have to take the initiative in the technological change process.

The notion that everything starts with the researcher needs to be changed. We should even change the way we talk about research-extension-farmer linkages and talk about farmer-extension-research linkages. This would make sense since everything, in fact, starts with farmers anyway. They have developed and accumulated knowledge and technology which they use and disseminate among themselves formally and informally. In the process, farmers encounter problems that need further research — including by formal research systems.

(c) The 'transfer of technology' (TOT) orientation is also reflected in the design of the learning environment at the University. 'Lecture' theatres are designed in such a way that the 'lecturer' *stands* and lectures from one side whilst students *sit* and listen from the other — giving little room for mutual sharing between the lecturer and the students and between the students themselves. The students, therefore, go out of the university with the impression that 'those who know' stand on one side and *lecture* to 'those who don't know' — an approach which is inappropriate for adults in general and FPR in particular.

(d) The products of this training process come out with the impression that 'good science' is about figures and statistics. They are not taught alternatives for different situations. This impression is further supported by the insistence on publications in scientific journals if one is to be promoted within the university system. In the two examples discussed above, the lecturers and supervisors involved were looking for materials acceptable for publication in their respective and specific disciplinary (refereed) journals — crop science or animal science journals. These refereed journals have their own rigorous 'scientific' demands which have very little to do with farmers' needs.

(e) As far as the staff and student researchers were concerned, their principal motivation and criteria for success in both examples was academic acceptability. If, along the line, farmers' needs were also addressed, it was more by coincidence than by design. Researchers' time was limited by funding and registration period. Hence their prime consideration was something that was academically acceptable at the particular degree level and not necessarily something practically useful to farmers. Even if farmers' needs were not addressed, that would not change anything — the researchers would still have accomplished their objectives — getting their degrees and/or having published. Whilst the above attitudes need to be changed right from source — at universities — there is need for bridging courses for those already in the field.

(f) The disciplinary specialisation of scientists and the accompanying organisational structures are inappropriate for a participatory approach as they are not flexible enough to meet farmers' diverse needs. Researchers are employed in particular disciplinary areas and they will always find research problems to work on, with a view to justify their existence even if the identified research problems were not the same as those felt by the farmers. In the second example, the planners of the project had no mandate to work on draught power, so they had to find something on small-scale dairy. That the researchers' interests coincided with the needs of even this small group of farmers was only a matter of coincidence tending towards 'arm-twisting' of farmers.

(g) In a donor-funded programme recipients are weaker partners who cannot determine the agenda. In addition, they are also in a hurry to get the funding and there is no time for proper consultations with farmers.

(h) The linkages and collaboration espoused in the two examples were donor-driven and were being imposed on institutions which had no culture nor provision for working together. They did not normally share the same vision. In addition, the linkages were exclusive and only limited to the 'projects' supported by donors when, in fact, the agricultural knowledge system was much wider. The linkages were created around the specific projects and there were no links to other structures with a life beyond the projects' life. Therefore the approach used by donors could in itself be misleading, and in any case difficult to sustain. Linkages cost money. Without donors, the collaborative activities discussed in the two examples would not have taken place at the same level using the local institutional resources — more so given that the institutions did not share the same vision. It was unlikely too that such activities would continue beyond the funding period.

A new paradigm for agricultural research

Agricultural researchers at the University of Zimbabwe are trained to be 'hard systems' thinkers (Checkland, 1989). In the two examples, researchers created systemic images to guide the construction of models representing transformation processes. That is, the way inputs are processed into outputs determines the function of the system. In the two examples, researchers specified relationships between feed (incoming) and production (outgoing) in their models. They were then going to compare the outcomes of the models so constructed to observations in the real world. The purpose of such hard systems learning is to achieve useful simulations of real world processes in order to be able to predict production either in the form of milk or work output.

Whilst we probably need thinkers that are conversant with 'hard systems', their proficiency should ideally broaden so as to incorporate the 'softer' aspects of these systems. Such researchers would be capable of handling different perspectives on technological transformation processes and thus be able to appreciate the various rationales and world views that can be found within an agricultural knowledge system. In theory, such a multiple perspective can more easily facilitate and accommodate learning processes and processes of technological adaptation that serve the needs of the different interest groups in the system.

To move from a production focus that essentially ignores people except as objective components, to one that recognises people and their relationships with their environment as the central concern of agricultural development, calls for a major shift in the world view of professionals who work with farmers. Indeed this is a call for a new type of professional agriculturist who in turn needs a new style of education — a new paradigm.

NOTES

1 Faculty of Agriculture Policy Paper, FAG/33/93 (mimeo), University of Zimbabwe.

2 BSc agricultural students are required to spend at least ten weeks on an internship programme during their long vacation after their second year.

3 *The Herald,* May 11, 1995.

4 The story was told of a researcher who arranged to conduct feeding trials using two farmers' cows. They put ear tags on the cows for identification. On one of his regular visits, and half way through the experiment, the researcher was presented with the two ear tags — the cows had been disposed of in a marriage settlement.

Chapter 12

Conclusion

ALEX BOLDING, JEFF MUTIMBA AND PIETER VAN DER ZAAG

From the case studies discussed in this book emerge the contours of two separate worlds that operate simultaneously in the field of smallholder agriculture. The duality is exemplified by, on the one hand the world of smallholder farmers, and the perspectives of the national extension and research agencies on the other. This duality is continuously reproduced in the agricultural interventions described in this book, and can be captured by opposing terms such as indigenous versus modern livelihoods; farming practice versus scientific expertise; contextual knowledge versus disciplinary dogma; risk-spreading versus economic gain; performance versus potential; evolution versus revolution; and the view of an insider as against that of an outsider. At the heart of the dichotomy lies the puzzle of agricultural innovation or stagnation that affects the well-being of people in Zimbabwe's communal areas. This important insight, we believe, must inform any venture that seeks to bridge the gap between these two worlds. In this brief concluding section, we focus on the founding pillars of the two bridge heads, before suggesting ways to link the two together.

THE COLONIAL LEGACY: SEGREGATED DEVELOPMENT

The present-day agricultural extension service in Zimbabwe is rooted in a long and tenacious tradition. For the commercial sector this tradition goes back to the end of the last century, while that of the communal sector is over 70 years old. Little wonder that nearly two decades after Independence, agricultural extension for the smallholder sector is still grappling with the new disposition, while falling back on extension practices developed during the pre-independence era. Three legacies from the past stand out.

One deeply ingrained conviction holds that the communal areas as they are today can, in actual fact, be transformed into commercially sound agricultural ventures. But most communal areas are at a disadvantage, for

308

instance with respect to soil quality, water resources (low or erratic rainfall, downstream of most river systems, far removed from storage facilities) and the availability of credit (lack of collateral). Even in the most marginal districts, projects based on alternative arable techniques, such as low external input agriculture, are still being pursued. This is despite the fact that those areas would only be viable for extensive grazing schemes, requiring the out-migration of large numbers of households to other, better endowed, areas where arable agriculture is a viable enterprise. The question arising is: what are the real and realistic options of transforming communal area farming households into viable agricultural enterprises (viz. Agritex mission statement). With most of the agricultural research stations located in well-watered commercial farming areas there is little to ensure that this question will be answered in a satisfactory manner.

A second important legacy still informing present-day agricultural extension is that most technologies and techniques that are proposed hail from either the temperate regions of the northern hemisphere or from the well-watered central 'watershed' in Zimbabwe, i.e. the main commercial farming areas. These proposed novel technologies may not be compatible with the local conditions (e.g. contour ridge), presuppose the availability of other resources (e.g. cash), and may force farmers into taking larger risks than they are able or prepared to take (securing food for their own family is a first). A pertinent question is: to what extent have research efforts seriously addressed the real (as against the perceived) technology needs of communal farmers?

Communal farmers are thus to be found on the subsistence-commercial agriculture continuum. They are neither of the two, and both at the same time. They are farmers, but possibly also temporary city migrants, artisans, mothers, etc. The formal agricultural extension service is not flexible enough to address the specific, and diverse, needs emerging from the heterogeneous communal area population. Agritex has the mandate to promote the commercialisation of communal agriculture, and this is clearly a third legacy of the past. That this mandate may imply different strategies for communal farmers at different points of the commercial-subsistence continuum is a confusing reality, for which the extension service appears ill-equipped. Farmers weigh up the various constraints and opportunities related to various crop scenarios, labour demands, capital requirements and the vagaries of the market. They may therefore sometimes opt not to follow Agritex advice. An extension of this point is that parts of the Master Farmer curriculum are irrelevant to some or most communal farmers. Given the multifaceted demands of communal farmers, how can agricultural extension best service this sector?

TECHNOLOGY: ADOPTION THROUGH ADAPTION

The past weighs heavily on the present-day extension practice. The case studies presented in the preceding chapters demonstrated the variegated response of communal farmers to the opportunities offered by the extension service and the wider economy. Farmers critically pick those items they deem beneficial, re-work them if necessary, simply ignore (or pay lip-service to) the rest, while holding on to the proven aspects of their original farming system. In Southern Africa it did not require an extension service to introduce a completely new staple crop, maize. The introduction of the plough, it appears, would have happened with or without Mr. Alvord's gospel. The case studies are replete with examples of how farmers adapt and re-define recommended techniques and practices to suit their own needs. Thus farmers critically evaluate proposed recommendations, and learn. But do researchers and extensionists similarly learn from this process of selective adoption and adaptation?

The answer to this question appears to be that at the field level, at the front line, agricultural extension workers and officers indeed have a wealth of experience from which the higher echelons of the extension service, and researchers, might benefit. In practice such positive feedback hardly ever occurs. The reward system within the extension service does not stimulate frontline staff to report the trials and tribulations of real-world extension. So, on paper at least, the extension service does not recognise farmer practices and sticks to 'proven' recommendations. The re-working of packages by farmers is considered non-adoption, and thus a 'failure' of the extension service. Official reports are polished up, and as a result, it is the extension service and the research institutions that fail to learn and update their agendas. They instead appear to stick to a paradigm which gradually gets outdated. They have become institutionalised to an extent that they cannot possibly meet the varied, multi-faceted and changing demands brought forward by farmers. Whereas in the past the linear transportation of new technologies from research institutes to the field has resulted in some notable successes (e.g. introduction of hybrid maize varieties), it is questionable whether such successes can be brought about in a more democratic environment. Since communal area people have proven themselves capable of weighing and choosing alternatives, Agritex should explore ways of proving itself as a facilitator, much like a supermarket where communal people come and take their pick.

KNOWLEDGE AND RESEARCH: DICHOTOMISED MINDS

The evidence provided in this book points out that the formal research and extension services availed to farmers are frequently inadequate. It is important to fully appreciate why this is the case, before being able to suggest solutions. There may be different reasons why these services fall short of farmers' expectations, but the general thread emerging from the case studies is that researchers and extensionists sometimes fail to fully appreciate the reality and constraints faced by smallholders. Hence their advice may not address the farmers' problem. At the root of this lack of appreciation may lie constraints in the field of (a) gender, (b) education, (c) evaluation/research, and (d) voice. Let us attempt to substantiate these constraints.

(a) The first constraint is simple and straightforward: the majority of smallholder farmers is female, while a large majority of extensionists is male. This point is probably not the most important constraint, but it may hamper good communication between farmers and extensionists, and thus jeopardise the important and positive role women farmers play in agricultural development. The on-going process of feminisation of the agricultural field in Southern Africa has not been recognised by extensionists, who, when talking of farmers, still have a male head of household and a male agriculturist in mind.

(b) The second constraint relates to the incompatibility of the 'modern' education and training of researchers and extensionists, with that of farmers. They may thus interpret similar physical processes differently, which will make 'shared problem appreciation' difficult to accomplish, let alone finding mutually agreed solutions. This incompatibility is, in theory, surprising since most of the researchers and extensionists themselves have a rural background. It would appear that the 'modern' educational system fails to equip graduates to look critically at their 'modern' knowledge, transcend it, and find original and new ways to bridge the perception gap between farmers and themselves. The modern education system seems to do just the opposite, and widen the gap. Or, in the words of Professor Rukuni, 'education only serves to escape the reality of the communal areas.'[1] Encounters between farmers and extension workers therefore sometimes appear to be mere role plays, after which both parties return to their usual business. There is urgent need to address this apparent paradox, and adjust the education system. The suggestion made by Dr. Muchena, namely to stimulate what she calls the 'affective' aspects of a formal training through using, for instance, local proverbs, requires serious attention.

(c)   The third constraint is about evaluation and research. The research and extension services do not precisely know, and monitor, the economic and social value of so-called small-scale crops. They tend to belittle their value, which is often much higher than assumed. Likewise, in the statistics these same crops feature dismally. It is believed that both acreage and yields of the 'small-scale crops' are severely under-estimated, reinforcing the impression that these crops are not important. Another information gathering tool, viz. questionnaire surveys, often fails to capture farmers' real concerns. In order to avoid further bias in our databases, the research and evaluation tools should be sharpened, for instance through combining different research methods, collecting both quantitative and qualitative data. In addition, it will be necessary for researchers and extensionists to spend more time with the farmers when collecting data, so that a social relationship can be built based on mutual respect. Until there is a reward system that rewards time in the field as against time in the office, researchers, and even extension workers, may not receive the incentives to spend more time with farmers.

(d)   The fourth constraint relates to the lack of 'voice' of smallholder farmers and their inability to make claims on the research and extension agendas, and influence them. If smallholder farmers would have this voice and the power to actively set the agendas of research and extension, all other constraints would most likely also be solved. This is shown by the commercial farming sector where the linkages with research and extension appear to be effective and efficient. The poverty of the smallholder farmers is a major underlying factor. Until this is solved, other mechanisms should assist farmers in making their voice known in the research and extension arenas. It is unfortunate, however, that the efforts by ZFU and government so far (for example through COFRE and the Farming System Research Unit within DR&SS) have yielded disappointing results.

STATE-FARMER INTERFACE: FROM CONTROL TO FACILITATION

The relationship between 'the state' and the smallholder farmer is a tense one. In Zimbabwe, this can be partly explained as a colonial legacy. Agricultural extension features centre-stage in this relationship. The extension approach developed from persuasion and demonstration in the early Alvord days, to force (Native Land Husbandry Act) and back to negotiated development and expert advice after 1980. A general trend is that the extension service first of all manages to reach the resource-rich farmers, and since the days of Alvord the Master Farmer badge is considered a social, and even a political, asset more than anything else. Apart from a potential alliance between local extension workers and resource rich farmers, extension staff have to fulfil roles that central government assigns them. A major role of Agritex in the present era is to assist government programmes to reach the grassroots, for instance distributing drought relief packages (seed and fertiliser). Then there is the institutional reality: the extension worker has to comply with the institutional rules and regulations, reporting procedures, policy guidelines which often are of little help, a modest salary, and cope with a situation where resources (for instance transport) are scarce.

The manifold roles puts the extension worker in an awkward position: she may aim to reach the resource poor farmers, end up assisting resource rich farmers, lack clear guidelines, rewards and transport, while being entrusted with an additional role with important political connotations. A typical trajectory would thus be that of an extension worker who starts her career energetically but after some years of frantic efforts resigns to the easy target groups. However, in some cases, the present lack of resources to make the wheel go round leads to unexpected forms of ingenuity and cross-cutting action, as has been the case in forestry extension, where Agritex and Forestry Commission staff have linked up in a coordinated effort to address the de-forestation challenge.

In irrigation schemes, Agritex officers suffer from their dual role as managers and advisors, creating conflicting allegiances. In rainfed agriculture such a conflictive situation does not occur; here farmers tend to ignore Agritex. In both situations Agritex could benefit from a reorientation in focus. As a facilitator and information supplier (e.g. of market prices and opportunities) it could play a meaningful role in future. However, as it stands today, Agritex staff are frustrated in their efforts by a myriad of departmental rules and regulations that date from a time when political backing for strong state interventions was there. A lack of policy instruments that take cognisance of the post-independence system of governance has left the Department in a political vacuum, especially in irrigation contexts.

BRIDGING THE GAP

Researchers and donors alike have a habit of looking at the state or government affiliated institutions as the prime movers in agricultural intervention. In doing so, the existing imbalances and approaches are often reproduced, albeit in a different complexion. In the process political decisions that affect the fundamentals of agricultural production and its resource base are normally avoided or ignored (Ferguson, 1990). Zimbabwe in particular has a long standing tradition of state driven interventions in rural areas that are informed by a desire to control the aspirations of its inhabitants. However, some recent developments seem to indicate that this situation is changing.

The land designation exercise that started in 1997 can potentially even out some of the resource inequalities that have hampered the emergence of a vibrant smallholder agricultural sector. The move to acquire land for smallholder farmers provides a big opportunity to close the gap between agricultural performance and potential. However, the modalities of the resettlement programme are as yet unknown. The data presented in the preceding chapters suggest to evaluate the implications of the above described dichotomies and come up with a different programme than the resettlement models of the early eighties. To us this implies a radical break with the role of Agritex as 'the master of smallholder farmers'. The case studies in this book suggest that a role as mediator and facilitator is more likely to produce results.

Recent public sector reform programmes also seem to point to a new configuration of the three sub-systems of the Research-Extension-Farmer continuum. A bird-eye look at the Agricultural Sector Investment Programme (ASIP) and the Water Sector Investment Programme (WSIP) suggests that these reform initiatives aim at reviving the 'prime movers' of the Transfer of Technology (TOT) complex according to a new institutional agenda. By improving the structure and efficiency of public research institutions and increasing the accountability of public research and extension to its clients, it is hoped to get the machinery of the TOT paradigm in moving order for a third 'agricultural revolution' (Eicher and Rukuni, 1994, 406). The

Department of Research and Specialist Services is currently being restructured through privatisation of some services and a new emphasis on cost recovery. However, the complex needs of smallholder farmers and the circumstances in communal areas of today do not lend themselves for quick fixes that can be addressed by a restructured and financially sound public system of research and extension. More hopeful aspects of the current reforms concern efforts to improve the accountability of the research system to its clients. The reconstitution and revival of the Agricultural Research Council (ARC) seems promising. The new ARC is taking half of its membership from farmer associations and has established provincial committees aiming at improved dialogue between research, extension and farmers. The ARC will manage an Agricultural Research and Extension Fund (AREF) to channel all international donor and local private resources to research priorities that have been identified by their committees (Beynon *et al.*, 1998). However, it is not clear whether Agritex and DR&SS will continue to get separate funding or will fall under the new ARC structure. Reforms in the water sector also seem to suggest that policy formulation and implementation is becoming subject to user influence, instead of being prescribed to the users by the state. The mobilisation of stakeholders at catchment level in so-called catchment councils embodies the first attempt in Zimbabwe to create local platforms for decision making on rural development initiatives (Manzungu *et al.*, 1999). These reforms provide opportunities for public agencies like Agritex and DR&SS to learn about and 'grow' in their new role as facilitators and mediators. Of crucial importance for the success of these policy reforms will be whether the strong private research institutes can be enticed to address the research needs of the smallholder sector (e.g. by developing new hybrid millet varieties). This might require a re-orientation of the Commercial Farmers Union that has to date mainly invested in its own network of private and semi-public research institutions.

After independence and increasingly so after Zimbabwe embarked on a programme of economic liberalisation, new players have entered the field of agricultural extension. This is reflected in the surge in contract farming arrangements between horticultural companies and smallholder growers concerning predominantly export crops (such as sweet peas, baby corn, tomatoes and paprika). Simultaneously new buyers have appeared on the formerly state controlled cash crop market. This has promoted competition and led to increased prices (for example, in cotton) and better service provision to growers. In most cases this development is accompanied by crop specific extension services provided by private companies. Paprika contractors, for instance, normally supply their own advisory agents in areas where their clients are concentrated. These agents operate parallel to Agritex extension workers. Their operations are client driven and are not confused by Agritex's obligation to implement government policies and programmes (like drought recovery). This development has received scant attention in this book. Nevertheless we consider it important, since it opens up ways for smallholder farmers, and particularly future resettlement farmers, to choose and even hire the expertise and services they require. This can help in closing the gap between what Agritex has on offer and what smallholder farmers require. Of course we are not blind to the possible negative effects of unscrupulous companies recommending their own services or products at the expense of other considerations like household food security. We believe Agritex and research institutions have a role to play in levelling the playing field, by means of provision of market information, quality control, and identifying opportunities and constraints for sustainable smallholder livelihood strategies.

In theory there is also a role to play for farmers' unions, like the ZFU, and other non-governmental organisations. So far the ZFU has been heavy at the top and thin on the ground, as is reflected by the conspicuous absence of the ZFU and its activities at the grassroots level in the preceding case studies in this book. International donors, however, have increasingly shifted their attention to organisations like the ZFU in order to develop the countervailing power of the small-holder clientele of Agritex. Consequently the ZFU has been involved in quite a number of policy (re-)formulation exercises and resource intensive extension projects. Unfortunately most of these initiatives are donor rather than client driven and prove unsustainable in the long run. Nevertheless the ZFU and various NGOs at national and local levels have become a force to reckon with for Agritex and other government institutions. The main reason that Agritex is drawn into their activities, though, is the fact that these NGOs have increasingly succeeded in diverting donor funds from government channels into their own hands (Vivian, 1994). Besides the adoption of a new development discourse (participation, sustainability), this development has neither resulted in a diminishing influence of Agritex nor in a wide-spread change of approach on the ground. The latter is possibly related to the fact that NGOs mainly draw their personnel from former Agritex employees. Still, the fact that small holder farmers are taken serious in their own right by NGOs and donors is a positive development that in the long run might produce a stronger articulation of smallholders' needs. Zimbabwe presently harbours a strong and growing NGO sector that could play a meaningful role in any future development initiatives.

Despite the noble intentions of the above initiatives, the material presented in this book shows that the future of agricultural extension is not just a matter of getting your finances right and improving feed-back linkages in the old Technology of Transfer model. As the dialogue between the medical doctor and rain doctor during the days of Livingstone (see the Introduction) demonstrates, the main challenge is to bring two different world views together. The issue at stake is to get professionals with the right mindset to appreciate the different aspects that make up a sustainable livelihood in communal and resettlement areas. This requires a focus on practices rather than ideal models of how things should be. The strength of any future approach will be tested by the degree to which it is locally embedded and succeeds in capturing the ideas and conditions embedded in local practices. We believe an emphasis on interdisciplinary training and intervention practice is essential in such a new approach. By means of this book we hope to have contributed to the development of a new curriculum for future extension practitioners.

ENDNOTE

1. Public statement by Professor Rukuni during the 'Water development for diversification within the smallholder farming sector' workshop organised by the ZFU and the Friedrich Ebert Stiftung in Harare, 30 May 1995.

Bibliography

AGRITEX, 1990, Mission statement and philosophy. Agritex, Harare.

——, 1993, Mission statement and philosophy. Agritex, Harare.

——, 1994, Irrigation manual. Second Edition. Agritex, Harare.

——, n.d., Master farmer training record; advanced. Government Printer, Harare.

ALLAN, W., 1945, *African land usage. Human problems in British Central Africa,* Rhodes-Livingstone Institute 3: 13-20.

——, 1949, *Studies in African Land Usage in Northern Rhodesia.* Rhodes-Livingstone Papers No.15.

ALEXANDER, J., 1995, Things fall apart, the centre can hold: processes of post-war political change in Zimbabwe's rural areas. In: N. Bhebe and T. Ranger (eds.), *Society in Zimbabwe's Liberation War, Volume 2.* University of Zimbabwe Publications, Harare.

ALVORD, E.D., n.d., *The Gospel of the Plow or A Guided Destiny.* Autobiography of E. D. Alvord, O.B.E., M.Sc.

——, 1928, The great hunger. *NADA* 1(6): 35-43.

——, 1929, Agricultural life of Rhodesian Natives. *NADA* 2: 9-16.

——, 1930, Agricultural demonstration work on Native Reserves. Department of Native Development, Occasional paper No.3, Southern Rhodesia.

——, 1948, The progress of Native agriculture in Southern Rhodesia. *The New Rhodesia,* 6 August: 18-19.

——, 1958, Development of native agriculture and land tenure in Southern Rhodesia. Unpublished manuscript. University of Zimbabwe Library, Harare.

ANDERSON, D.M., 1984, Depression, dust bowl, demography and drought: the colonial state and soil conservation in East Africa in the 1930s. *African Affairs* 83 (332): 321-43.

ANDERSSON, J.A., 1999, The politics of land scarcity: land disputes in Save Communal Area, Zimbabwe. *Journal of Southern African Studies* 25(4): 553-578.

——, 2001a, Administrators' knowledge and state control in colonial Zimbabwe; the invention of the rural-urban divide in Buhera district, 1912-80. *Journal of African History.*

——, 2001b, Re-interpreting the rural-urban connection; migration practices and socio-cultural dispositions of Buhera workers in Harare. *Africa* 71 (1).

AQUINA, SR. MARY, 1963, *The social background of agriculture in Chilimanzi Reserve. Human problems in British Central Africa,* Rhodes-Livingstone institute) 33(4): 68-79.

——, 1965, The tribes of Victoria reserve. *Native Affairs Department Annual (NADA)* 9(2): 6-15.

ARNAIZ, MARIA E.O., MERRILL-SANDS, DEBORAH M., AND BEREAN MUKWENDE, 1995, The Zimbabwe Farmers' Union: Its current and potential role in technology development and transfer. Overseas Development Institute, London.

ATKINSON, N.D., 1974. *A History of Educational Policy in Southern Rhodesia.* Unpublished PhD, University of London.

AURET, DIANA, 1990, *A decade of development: Zimbabwe 1980-1990.* Mambo Press, Gweru.

AYLEN, D., 1939a, Soil and water conservation. *Rhodesia Agricultural Journal* 36(1): 12-30.

——, 1939b, Soil and water conservation, part II; chapter 4 - Construction of soil conservation works. *Rhodesia Agricultural Journal* 36 (6): 452-484.

——, 1939c, Soil and water conservation, part II; chapter 5 - Contour ridges (continued); construction and maintenance of contour ridges. *Rhodesia Agricultural Journal* 36(7): 534-552.

——, 1940a, The ridger grader. *Rhodesia Agricultural Journal* 37(3): 151-156.

——, 1940b, Contour planting and terracing of orchards. *Rhodesia Agricultural Journal* 37(4): 196-208.

——, 1940c, Soil and water conservation, part IV; Prevention and control of gullies. *Rhodesia Agricultural Journal* 37(6): 330-348.

——, 1941, Who built the first contour ridges? *Rhodesia Agricultural Journal* 38(3): 144-148.

——, (with K.J. MACKENZIE AND E.D. ALVORD), 1942, Conserving soil in the Native Reserves. *Rhodesia Agricultural Journal* 39(3): 152-160.

AYLEN, D., AND R. HAMILTON ROBERTS, 1937, Soil conservation. Bulletin No. 1019. Reprint from *Rhodesia Agricultural Journal* March 1937. Ministry of Agriculture and Lands. Govt. Stationery Office, Salisbury.

BAGCHEE, ARUNA, 1994, *Agricultural extension in Africa.* World Bank Discussion Paper no. 231, Africa Technical Department Series, Washington.

BAN, A.W. VAN DEN, AND H.S. HAWKINS, 1988, *Agricultural extension.* Longman, Burnt Hill, Harlow.

BANKS, P.F., 1980, The indigenous woodland - a diminishing source of fuelwood. Energy Symposium 80. Forestry Commission, Harare.

BATES, D.B., 1980, Important innovations introduced by agricultural extension to the tribal farmer over the last twenty years. *The Zimbabwe Science News* 14 (7): 187-190.

BAWDEN, R., 1994, *A learning approach to sustainable agriculture and rural development: Reflections from Hawkesbury.* Hawkesbury College, Australia.

BEACH, D.N., 1994, *A Zimbabwean past.* Mambo press, Gweru.

——, 1995, Archaeology and history in Nyanga, Zimbabwe. Paper for presentation to the Tenth Pan African Archaeological Congress and History Seminar Paper No. 97. Dept of History, University of Zimbabwe, Harare.

BEINART, WILLIAM, 1984, Soil erosion, conservationism and ideas about development: a Southern African exploration, 1900-1960. *Journal of Southern African Studies* 11(1): 52-83.

BEINART, W., AND COATES, P., 1995, *Environment and history. The taming of nature in the USA and South Africa.* Routledge, London.

BENOR, D. AND J.Q. HARRISON, 1977, *Agricultural extension: training and visit system.* World bank, Washington.

BERNARD, B. RUSSEL, 1988, *Research methods in cultural anthropology.* Sage Publications, London.

BERNSTEIN, HENRY, 1977, Notes on capital and peasantry. *Review of African Political Economy* 10: 60-73.

BEVAN, E.W., 1924, The education of Natives in pastoral pursuits. *NADA* 1(2): 13-16.

BEYNON, J., S. AKROYD, A. DUNCAN, S. JONES, 1998, *Financing the future: options for agricultural research and extension in Sub-Saharan Africa.* Oxford Policy Management, Oxford.

BIGGS, S.D., 1989, Resource-poor farmer participation in research: a synthesis of experiences from nine national agricultural research systems. OFCOR Comparative Study Paper No. 3. ISNAR, The Hague.

BOSERUP, ESTHER, 1970, *Woman's role in economic development.* Earthscan Publications, London.

BOURDIEU, PIERRE, 1990, *The logic of practice.* Polity Press, Cambridge.

BOURDILLON, M.F.C., AND E. MADZUDZO, 1994, Report on small-scale irrigation schemes in Zimbabwe. Unpublished report prepared for Danida, Harare.

BOWN, L., 1975, *A rusty person is worse than rusty iron.* Syracuse University, Syracuse.

BROKENSHA, D., D.M. WARREN, O. WERNER, 1980, *Indigenous knowledge systems and development.* University Press of America, Washington.

BROMLEY, D., D. FEENY, M.A. MACKEAN, P. PETERS,1992, *Making the commons work: theory, practice and policy.* ICS San Fransico.

BROMLEY, K.A., R.C. HANNINGTON, G.B. JONES, AND C.J. LIGHTFOOT, 1968, Melsetter Regional Plan. Department of Conservation and Extension, Salisbury.

BULMAN, E.M., 1970, The Native Land Husbandry Act - A Failure in Land Reform. Unpublished MSc thesis.

BUNDY, C., 1988, *The rise and fall of the South African Peasantry* (second edition). David Philip, Cape Town.

CALLEAR, D., 1985, Who wants to be a peasant? Food production in a labour-exporting area of Zimbabwe. In: J. Pottier (ed), *Food systems in Central and Southern Africa*. School of Oriental and African Studies, London, pp.217-230.

CARNEY, DIANA, 1998, *Changing public and private roles in agricultural service provision*. Overseas Development Institute, London.

CASEY, J., AND K. MUIR, 1986, *Forestry for rural development in Zimbabwe*. ODI-Social Forestry Network Paper 3c. Chameleon Press, London.

CHAMBERS, ROBERT, 1974, *Managing rural development: ideas and experiences from East Africa*. The Scandinavian Institute of African Studies, Uppsala.

———, 1983, *Rural development: putting the last first*. Longman, New York.

———, 1993, *Challenging the professions. Frontiers for rural development*. Intermediate Technology Publications, London.

CHAMBERS, R., AND B.P. GHILDYAL, 1985, Agricultural research for resource poor farmers: the farmer first model. *Agricultural Research and Administration* XX: 1-30.

CHAMBERS, R., PACEY, A., AND L. A. THRUPP, 1989, *Farmer first. Farmer innovation and agricultural research*. Intermediate Technology Publications, London.

CHECKLAND, P.B., 1981, *Systems thinking, systems practice*. John Wiley, Chichester.

———, 1989, Soft systems methodology. *Human Systems Management* 8: 273-289.

CHECKLAND, P., AND J. SCHOLES, 1990, *Soft systems methodology in action*. John Wiley and Sons, Chichester.

CHEAL, D., 1996, 'Showing them you love them'; gift-giving and the dialectic of intimacy. In: A.E. Komter (ed), *The gift, an interdisciplinary perspective*. Amsterdam University Press, Amsterdam; pp. 96-106 (originally published in 1987).

CHEATER, ANGELA, 1981, Women and their participation in commercial agricultural production. *Development and Change* 12(3): 349-377.

CHIKUKWA, T., RWAMBIWA, CHAPARAPATA, GODOBO, L. NYAGWANDE AND J.A. BOLDING, 1996, Summary of outcomes from kuturaya meetings in Tiya and Shinja Resettlement. Paper presented at Agritex Engineering Branch conference, Hwange, December 1996.

CHIMEDZA, R., 1989, The impact of irrigation development on women farmers in Zimbabwe: The case of Mushandike and Tagarika irrigation schemes. Department of Agricultural Economics and Extension working paper. University of Zimbabwe, Harare.

CHRISTOPLOS, I., 1999, Not Quite facing the abyss?, Book review of B.E. Swanson, R.P. Bentz and A.J. Sofranko (eds.), 1997, *Improving agricultural extension: a reference manual*, FAO, Rome, *Journal of Agricultural Education and Extension*, 6(1), pp. 63-65.

CHRISTOPLOS, I. AND U. NITSCH, 1996, *Pluralism and the extension agent*. (Publications on Agriculture no.1, Department for Natural Resources and the Environment). Sida, Sweden.

CHUMA, E., AND J. HAGMANN, 1995, Summary of results and experiences from on-station and on-farm testing and development of conservation tillage systems in semi-arid Masvingo, In: S. Twomlow, J. Ellis-Jones, J. Hagmann and H. Loos (eds.), *Soil and water conservation for smallholder farmers in semi-arid Zimbabwe, Transfers between research and extension*, Proceedings of a technical workshop, 3-7 April 1995, Masvingo, pp. 41-60.

CLIFFE, LIONEL, 1988, The conservation issue in Zimbabwe. *Review of African Political Economy* 42: 48-58.

CLOSE, A.M., 1943, Soil erosion. *Rhodesia Agricultural Journal* 40(3): 193-198.

COFRE, 1990, Research and extension linkages for small holder agriculture in Zimbabwe. In: E.M. Shumba, S.R. Waddington and L.A. Navarro (eds.) *Proceedings of a workshop on assessing the performance of the committee for on-farm research and extension (COFRE)*. Kadoma, Zimbabwe, 7-9 May 1990.

COMAROFF, J. and J. COMAROFF, 1992, *Ethnogra· · and the historical imagination*. Westview Press, Boulder.

CORMACK, R.M.M., 1951, The mechanical protection of arable land. *Rhodesia Agricultural Journal* 48(2): 135-162.

CORMACK, R.M.M., AND ISABEL WHITELAW, 1957, *Conservation: a guide book for teachers*. Government Printer, Salisbury.

CSO, 1964, Final report of the April/May 1962 census of Africans in Southern Rhodesia. Central Statistical Office, Salisbury.

——, 1989, Manicaland Province, comparative tables; district population indicators and information for development planning. Central Statistical Office, Harare.

——, 1994a, Census 1992; Zimbabwe national report. Central Statistical Office, Harare.

——, 1994b, Census 1992; Manicaland Profile. Central Statistical Office, Harare.

DAVIS, B., AND W. DÖPCKE, 1987, Survival and accumulation in Gutu: Class formation and the rise of the state in colonial Zimbabwe, 1900-1939., *Journal of Southern African Studies* 14 (1): 64-98.

DERUDE, 1983, Policy paper on small scale irrigation schemes. Ministry of Lands, Resettlement and Rural Development, Harare.

DIKITO, M.S., 1993, Women and irrigation. Unpublished report. Double Day Consult, Harare.

DRINKWATER, MICHAEL, 1989, Technical development and peasant impoverishment: Land use policy in Zimbabwe's Midlands province. *Journal of Southern African Studies* 15 (2): 287-305.

——, 1991, *The state and agrarian change in Zimbabwe's communal areas*. MacMillan, London.

DUNLOP, H., 1971, The development of European agriculture in Rhodesia. Occasional paper No.5, Department of Economics, University of Rhodesia, Salisbury.

EDQUIST, CHARLES, AND OLLE EDQVIST, 1978, Social carriers of techniques for development. Discussion paper no. 123. Research Policy Programme, University of Lund.

EICHER, CARL E., AND MANDIVAMBA RUKUNI, 1994, Zimbabwe's agricultural revolution: lessons for southern Africa. In: M. Rukuni and C.K. Eicher (eds.), *Zimbabwe's agricultural revolution*. University of Zimbabwe Publications, Harare, pp. 393-411.

ELWELL, H., 1989, Soil erosion. In: Post production activities and marketing of food crops in small-scale farming areas of Zimbabwe. Department of Agricultural, Technical and Extension Services (Agritex), Harare.

FARRINGTON, J.A., 1994, *Public sector agricultural extension: Is there life after structural adjustment?* (Natural Resource perspectives No.2) Overseas Development Institute, Sussex.

FAULKNER, E., 1945, *Ploughman's folly*. Michael Joseph Ltd., London.

——, 1948. *Ploughing in prejudices*. Michael Joseph Ltd., London.

FAO, 1990a, Irrigation subsector review and development strategy. Technical report and Annexes. Food and Agriculture Organisation, Harare.

——, 1990b, Forest resource assessment. FAO Forestry Department. Food and Agriculture Organisation, Rome.

FERGUSON, JAMES, 1990, *The anti-politics machine. Development, depoliticization and bureaucratic power in Lesotho*. Cambridge University Press, Cambridge.

FLOYD, B.N., 1959, Changing patterns of African land use in Southern Rhodesia. Human Problems in British Central Africa. *Rhodes-Livingstone Journal* 25: 21-39.

——, 1960, *Changing patterns of African land use in Southern Rhodesia*. Volume 2, Rhodes-Livingstone Institute, Lusaka.

FORESTRY COMMISSION, 1982, The rural afforestation project. Volume 1: Main Report. Forestry Commission, Harare.

FREIRE, P., 1970, *Cultural action for freedom*. Center for the Study of Development and Social Change, Cambridge, MA.

——, 1973, *Pedagogy of the oppressed*. Herder and Herder, New York.

FRIEDMANN, H., 1980, Household production and the national economy: Concepts for the analysis of agrarian formations. *Journal of Peasant Studies* 7(2): 158-184.

GASPER, D., 1988, Rural growth points and rural industries in Zimbabwe: Ideologies and policies. *Development and Change* 19 (3): 425-66.

GATTER, PHILIP, 1993, Anthropology in farming systems research, a participant observer in Zambia. In: J. Pottier (ed), *Practising development: Local science perspectives*. Routledge, London; pp.153-186.

GITTINGER, J.P., 1982, *Economic analysis of agricultural projects*. World Bank, Baltimore, Maryland.

GOODENOUGH, W.H., 1981, *Culture, language, and society*. The Benjamin/Cummings, London.

GORE, C., Y. KATERERE, S. MOYO, G. MHONE, D. MAZAMBANI, P. NGOBESE, D. GUMBO, AND L. LEIKVOLD, 1992, The case for sustainable development in Zimbabwe: Conceptual problems, conflicts and contradictions. A report prepared for the United Nations Conference on Environment and Development. ENDA and ZERO, Harare.

GRANT, P.M., 1976, Peasant farming on infertile sands. *Rhodesia Science News* 10 (10): 252-254.

——, 1981, The fertilization of sandy soils in peasant agriculture. *Zimbabwe Agricultural Journal* 78 (5): 169-175.

GRUNDY, I.M., B.M. CAMPBELL, S. BALEBEREHO, R. CUNLIFFE, C. TAFANGENYASHA, R. FERGUSSON, AND D. PARRY, 1993, Availability and use of trees in Mutanda Resettlement Area, Zimbabwe. *Forest Ecology and Management* 56: 243-266.

GUBBELS, PETER, 1994, Populist pipedream or practical paradigm? Farmer-driven research and the project agro-forestier in Burkina Faso. In: I. Scoones and J. Thompson (eds.), *Beyond farmer first. Rural people's knowledge, agricultural research and extension practice*. Intermediate Technology Publications, London, pp. 238-243.

HAGMANN, J., AND K. MURWIRA, 1995, Indigenous soil and water conservation in southern Zimbabwe: A study on techniques, historical changes and recent developments under participatory research and extension. Paper presented at the UZ/ZIMWESI Workshop 'Extension intervention and local strategies in resource management: New perspectives on agricultural innovation in Zimbabwe'. Harare, 10-12 January 1995.

HAILEY, 1938, *An African Survey*. London (cited in Floyd, 1960).

HAMILTON ROBERTS, R., 1930, Soil erosion: notes on contour ridging. *Rhodesia Agricultural Journal* 27(8): 841-845.

HAMILTON, P., 1964, Population pressure and land use in Chiweshe reserve. *Human problems in British Central Africa* 36: 40-58 .

HAMUTYINEI, M.A., AND A.B. PLANGGER, 1987, *Tsumo-Shumo*. Second edition. Mambo Press, Gweru.

HANYANI-MLAMBO, B.T., 1995, Farmers, intervention and the environment: A case analysis of government initiatives, local level perceptions and counter strategies on conservation forestry in Chinyika Resettlement Scheme, Zimbabwe. Unpublished MSc thesis. Wageningen Agricultural University.

HARDIN, G., 1968, The tragedy of the commons. *Science* 13: 1243-48.

HAUG, R., 1999, Some leading issues in international agricultural extension, a literature review. *Journal of Agricultural Education and Extension* 5 (4): 263-274.

HAVELOCK, R.D., 1969, Planning for innovation: A comparative study of the literature on the dissemination and utilization of scientific knowledge. Ann Arbor.

HAVILAND, P.H., 1934, Soil erosion. *Rhodesia Agricultural Journal* 31(6): 420-450.

——, 1947, Annual Report of the Irrigation Department, Southern Rhodesia, for the year ended 31st December 1946. *Rhodesia Agricultural Journal* 44(4): 327-347.

HOLLEMAN, J.F., 1958, *African interlude*. Nasionale boekhandel, Cape Town.

——, 1974, Marriage, bridewealth and women in African society. In: J.F Holleman, *Issues in African law*. Mouton, The Hague.

——, 1969, *Chief, council and commissioner; some problems of government in Rhodesia*. Van Gorcum, Assen.

HOWES, M., 1985, The uses of indigenous technical knowledge in development. In: D. Brokensha, D.M. Warren and O. Werner (eds.), *Indigenous knowledge systems and development*. University Press of America, Lanham, MD, pp. 335-52.

HUDSON, N.W., 1957, Soil erosion and tobacco growing. *Rhodesia Agricultural Journal* 54(6): 547-555.

HUGHES, A.J.B., 1974, *Development in Rhodesian tribal areas, An overview*. Tribal Areas of Rhodesian Research Foundation, Salisbury.

HUIZER, GERRIT, 1995, Indigenous knowledge and popular spirituality: A challenge to developmentalists. In: *Agriculture and spirituality: Essays from the Crossroads conference at Wageningen Agricultural University*. Crossroads, The Hague.

HULME, D., 1991, Agricultural extension services as machines: The impact of the T&V approach. In: W.M. Rivera and D.J. Gustafson (eds.), *Agricultural extension: Worldwide institutional evolution and forces for change*. Elsevier, Amsterdam.

HUNT, A.F., 1958, Manicaland irrigation schemes; economic investigations. Department of Native Economics and Marketing. Salisbury.

IFAD, 1996, IFAD's evolving experience with extension support in Eastern and Southern Africa. Paper presented at the second informal consultation on international support to agricultural extension systems in Africa. International Fund for Agricultural Development, Rome, October 8-9.

IVY, P., 1978, The dryland cash-crop production potential of natural region 2. *Zambezia* 6 (2): 147-160.

JABAVU, D.D.T., 1921, *The Black Problem. Papers and Addresses on various Native Problems. (Second Edition)* The book department, Lovedale.

JANSEN, DORIS J. (WITH D. MFOTE AND K. KASERARAH), 1993, Economics of irrigation; A modular approach for comparing the benefits with the costs. Final report. Paper prepared for the Planning and Research Unit, Ministry of Lands and Agriculture., Harare.

JOHNSON, D., 1992, Settler farmers and coerced African labour in Southern Rhodesia, 1936-46. *Journal of African History* 33 (1): 111-128.

JORDAN, J.D., 1963, Zimutu reserve; A land-use appreciation. *Human problems in British Central Africa* 36:59-61.

KAIMOWITZ, D., M. SNYDER, P. ENGEL, 1989a, *A conceptual framework for studying the links between agricultural research and technology transfer in developing countries.* ISNAR, The Hague.

KAIMOWITZ, D., D. MERRILL-SANDS, K. SAYCE, AND S CHATTER, 1989b, *The technology traingle: Linking farmers, technology transfer agents, and agricultural researchers: summary report of an international workshop held at ISNAR,* The Hague, 20-25 November 1989. ISNAR, The Hague.

KAJEMBE, J.C., 1994, Indigenous management systems as a basis for community forestry in Tanzania: A case study of Dodoma Urban and Lushoto Districts. Unpublished Phd dissertation. Wageningen Agricultural University.

KATERERE, Y., 1985, The fuelwood crisis in Zimbabwe - some possible solutions. *The Zimbabwe Science News* 19: 71-3.

KATERERE, Y., S. MOYO, AND L. MUJAKACHI, 1993, The national context: Land, agriculture and structural adjustment, and the Forestry Commission. In: P.N. Bradley and K. McNamara (eds.), *Living with trees: Policies for forest management in Zimbabwe.* The World Bank, Washington, D.C., pp. 11-27.

KENNAN, P.B., 1980, Agricultural extension in Zimbabwe, 1950-1980. *The Zimbabwe Science News* 14 (7): 183-86.

KOMTER, A.E., 1996a, Introduction. In: A.E. Komter (ed.), *The gift, an interdisciplinary perspective.* Amsterdam University Press, Amsterdam, pp. 3-12.

——, 1996b, Women, gifts and power. In: A.E. Komter (ed), *The gift, an interdisciplinary perspective.* Amsterdam University Press, Amsterdam, pp. 119-31.

KOPYTOFF, I., 1986, The cultural biography of things: Commoditization as process. In: A. Appadurai (ed.), *The social life of things; commodities in cultural perspective.* Cambridge University Press, Cambridge, pp.64-91.

KUNAKA, A.M., 1990, Gums and re-afforestation in communal lands: A radical approach. A paper presented at the Third Agritex Manicaland Province Annual Technical Workshop. Kariba, August 6-10, 1990.

LAL, R., 1979. Soil tillage and crop production., International Institute of Tropical Agriculture, Ibadan, Nigeria (Proceedings Series No.2).

——, 1986, No tillage and surface tillage systems to alleviate soil-related constraints in the tropics, In: M.A. Sprague and G.B. Triplett (eds.), *No tillage and surface tillage agriculture: The tillage revolution.* Wiley & Sons, New York, 261-317.

LEEUWIS, CEES, 1995, The stimulation of development and innovation: Reflections on projects, planning, participation and platforms. *European Journal of Agricultural Education and Extension* 2 (3): 15-27.

LIVINGSTONE, D., 1857, *Missionary travels and researches in South Africa.* Murray, London.

LLOYD, B.W., 1962. Early history of Domboshawa School, Period 1920-1939. *NADA* 1962: 4-13.

LONG, ANDREW, 1992, Goods, knowledge and beer; The methodological significance of situational analysis and discourse. In: N. Long and A. Long (eds.), *Battlefields of knowledge.* Routledge, London, pp. 147-170.

LONG, NORMAN, 1989, Introduction: *the raison d'tre* for studying rural development interface. In: Norman Long (ed.), *Encounters at the interface: A perspective on social discontinuities in rural development.* Wageningen Studies in Sociology 27. Wageningen Agricultural University, pp. 1-10.

——, 1992, From paradigm lost to paradigm regained? The case for an actor-oriented sociology of development. In: Norman Long and Ann Long (eds.), *Battlefields of knowledge: The interlocking of theory and practice in social research and development.* Routledge, London, pp. 16-43.

——, 1994, Agrarian change, neoliberalism and commoditization: An actor oriented perspective on social value. Keynote lecture, XVI Colloquium 'Las Disputas por el Mexico Rural: Transformaciones de practicas, identidades y proyectos'. El Colegia de Michoacan, 16-18th November 1994.

LONG, NORMAN AND JAN DOUWE VAN DER PLOEG, 1989, Demythologizing planned intervention: An actor perspective. *Sociologia Ruralis* 29(3/4): 226-249.

LONG, NORMAN, AND MAGDALENA VILLARREAL, 1993, Exploring development interfaces: From transfer of knowledge to the transformation of meaning. In: Frans J. Schuurman (ed.), *Beyond the impasse, new directions in development theory.* Zed Books, London.

LOVELACE, G.W., 1984, Cultural beliefs and management of agro-ecosystems. In: T.A. Rambo and P.E. Sajise (eds.), *An introduction to human ecology research on agricultural systems in southeast Asia.* University of the Philippines, Los Baòos, pp. 194-205.

LOW, A.R.C. 1994, Environmental and economic dilemmas for farm-households in Africa: When 'Low-input sustainable agriculture' translates to 'high-cost Unsustainable livelihoods'. *Environmental Conservation* 21(3): 220-224.

LUE-MBIZVO, C., AND J.C. MOHAMED, 1993, The institutional and legal framework for natural resource management. Local Level Natural Resource Management

Project (Makoni District) Working Paper No. 3. Stockholm Environment Institute and ZERO, Harare.

MACKAY, HUGHIE, AND GARETH GILLESPIE, 1992, Extending the social shaping of technology approach: Ideology and appropriation. *Social Studies of Science* 22: 685-716.

MACKENZIE, F., 1992, Development from within? The struggle to survive. In: D.R.F. Taylor and F. Mackenzie (eds.), *Development from within: Survival in rural Africa*. Routledge, London, pp. 1-32.

MADONDO, B.B.S., 1992, *The role of women in male-headed households in the management of smallholder irrigation schemes in Manicaland Province, Zimbabwe*. Agritex, Mutare, Project No. 1/92.

——, 1995, Agricultural transfer systems of the past and present. Paper presented at a technical workshop 'Soil and water conservation for smallholders in semi-arid Zimbabwe: Transfers between research and extension', 3-7 April 1995. Masvingo.

MAGADLELA, DUMISANI, AND PAUL HEBINCK, 1995, Dry fields and spirits in trees: A social analysis of development intervention in Nyamaropa irrigation scheme, Zimbabwe. *Zambezia* 22(1): 43-62.

MAGWA, 1995, The future of master farmer training programme. Paper presented at an Agritex workshop for Chimanimani District Staff, 22-24 March 1995, Mutare.

MAINWARING, C., 1921, The Advantage of Autumn and Early Winter Ploughing. *Rhodesia Agricultural Journal* 18 (4): 387-388.

MAKADHO, J.M., 1994, An analysis of water management performance in smallholder irrigation schemes in Zimbabwe. PhD thesis, University of Zimbabwe, Harare.

MAKINA, J.C., 1981, Some traditional wildlife conservation practices and their role in the planning of conservation education programmes using mass media with special emphasis on radio and television. Department of Adult Education, University of Zimbabwe, Harare.

MAKONI, I.J., 1990, *National survey of biomass: Woodfuel activities in Zimbabwe*. SADCC, Luanda.

MANZUNGU, EMMANUEL, 1994, The interplay between technical and social factors in smallholder irrigation schemes in Zimbabwe. Unpublished PhD proposal, May 1994, Wageningen Agricultural University.

——, 1995a, Design and practice: From the drawing board to the farmers' field; The case of Fuve Panganai irrigation scheme. Paper presented at the UZ/ZIMWESI workshop 'New perspectives on agricultural innovation in Zimbabwe', 10-12 January 1995, Mandel Training Centre, Harare.

——, 1995b, Engineering or domineering? The politics of water control in Mutambara irrigation scheme. *Zambezia* 25(2): 115-136.

MANZUNGU, EMMANUEL, AND PIETER VAN DER ZAAG (eds), 1996, *The practice of smallholder irrigation; Case studies from Zimbabwe*. University of Zimbabwe Publications, Harare.

MANZUNGU, EMMANUEL, AIDAN SENZANJE AND PIETER VAN DER ZAAG (eds), 1999, *Water for agriculture in Zimbabwe. Policy and management options for the smallholder sector.* University of Zimbabwe Publications, Harare.

MARSDEN, D., 1989, Indigenous management and the management of indigenous knowledge. Unpublished draft paper. Technology and Social Program, Iowa State University, Ames Iowa.

MASHAVIRA, T.T., P. HYNES, S. TWOMLOW AND T. WILLCOCKS, 1995, Lessons learned from 12 years of conservation tillage research by Cotton Research Institute under semi-arid smallholder conditions, In: S. Twomlow, J. Ellis-Jones, J. Hagmann, H. and Loos (eds.), *Soil and Water Conservation for smallholder farmers in semi-arid Zimbabwe, Transfers between research and extension. Proceedings of a technical workshop, 3-7 April 1995*, Masvingo, 22-31.

MBITI, J.S., 1970, *African religions and philosophies*. Doubleday, Garden City, NY.

MCGREGOR, JOANN, 1991, Ecology, policy and ideology: An historical study of woodland use and change in Zimbabwe's communal areas., PhD thesis, Loughborough University of Technology, UK.

——, 1995, Conservation, control and ecological change: The politics and ecology of colonial conservation in Shurugwi, Zimbabwe. *Environment and History* 1(3): 257-79.

MCNEIL, J.D., 1985, *Curriculum: A comprehensive introduction.* Third edition. Little, Brown and Co., Boston.

MEHRETU, ASSEFA, 1994, Social poverty profile of communal areas. In: Mandivamba Rukuni and Carl E. Eicher (eds.), *Zimbabwe's agricultural revolution*. University of Zimbabwe Publications, Harare, pp. 56-69.

MEINZEN-DICK, RUTH, GODSWILL MAKOMBE AND MARTHA SULLINS, 1993, Agroeconomic performance of smallholder irrigation in Zimbabwe. Paper presented at the University of Zimbabwe/AGRITEX/IFPRI workshop 'Irrigation performance in Zimbabwe'. Juliasdale.

MILLER, DANIEL, 1987, *Material culture and mass consumption.* Basil Blackwell, Oxford.

MOYO, P.H., 1925, Native life in the Reserves. *NADA* 1 (3): 47.

MOYO, S., 1995, *The land question in Zimbabwe.* SAPES Books, Harare.

MPOFU, T.P.Z., 1987, History of soil conservation in Zimbabwe. In: Coordination Unit, SADCC Soil and Water Conservation and Land Utilization Programme, History of Soil Conservation in the SADCC Region. Report No. 8. SADCC, Maseru.

MTETWA, R, 1976, The political and economic history of the Duma people of South-Eastern Rhodesia, unpublished PhD, University of Rhodesia.

MUCHENA, O.N., 1990, An analysis of indigenous knowledge systems: Implications for agricultural extension education with particular reference to natural resource management in Zimbabwe. Unpublished PhD dissertation. Iowa State University, Ames, Iowa.

MUIR, KAY AND MALCOLM J. BLACKIE, 1994, The commercialization of agriculture. In: M. Rukuni and C.K. Eicher (eds), *Zimbabwe's agricultural revolution*. University of Zimbabwe Publications, Harare, pp. 195-207.

MUNDY, H.G., 1921, The interdependence of crop rotation and mixed farming. *Rhodesia Agricultural Journal* 18 (4): 343-352.

———, 1928, *Sub-tropical agriculture in South Africa with special reference to Rhodesia*. Rhodesian Printing and Publishing, Bulawayo, Southern Rhodesia.

MUNDY, H.G., AND WALTERS, J.A.T., 1919, Rotation experiments, 1913-1919. *Rhodesia Agricultural Journal* 16 (6): 513-520.

MUTIZWA-MANGIZA, N.D., 1985, Community development in pre-independence Zimbabwe: A study of policy with special reference to rural land. Supplement to *Zambezia*, University of Zimbabwe Publications, Harare.

MVUDUDU, S.C., 1993, Gender issues, constraints and potentials in Agritex, an overview. Unpublished draft report. Agritex, UNDP, FAO, Harare.

NADA, 1963, Obituary: Herbert Stanley Keigwin. *NADA*: 122.

NHIRA, C., AND L. FORTMANN, 1991, Local control and management of forest and environmental resources in Zimbabwe: Institutional capacity. Centre for Applied Social Sciences Occasional Paper. CASS, Harare.

———, 1993, Local woodland management: Realities at the grass roots. In: P.N. Bradley and K. McNamara K. (eds.), *Living with trees: Policies for forest management in Zimbabwe*. The World Bank, Washington, D.C., pp. 139-155.

NIEMEIJER, D., 1996, The dynamics of African agricultural history: Is it time for a new development paradigm? *Development and Change* 27(1): 87-110.

NJOROGE, R.J., AND G.F. BENNARS, 1986, *Philosophy and education in Africa*. Transafrica Press, Nairobi.

NORTON, A.J., 1984, Tillage practices and their effects on soil conservation. Paper presented at the World Bank Project Review Conference, 15-18 June 1984, Kariba.

NYAMAPFENE, K., 1982, Some perspectives on soil erosion and conservation. *Zimbabwe Science News* 16(12): 286-288.

———, 1987, A short history and annotated bibliography on soil and water conservation in Zimbabwe. SADCC Coordination Unit, Soil and Water Conservation and Land Utilization, Report No. 12. SADCC, Maseru.

———, 1989, Adaptation to marginal land amongst the peasant farmers of Zimbabwe. *Journal of Southern African Studies* 15(2): 384-389.

NYAMUDEZA, P., 1996, Agronomic practices for the low rainfall natural region 5 of Zimbabwe. Paper presented at the UZ/ZIMWESI workshop on Water for agriculture. Current practices and future prospects. 11-13 March 1996, Mandel training centre, Harare.

NYAMUDEZA, P., 1999, Agronomic practices for the low rainfall regions of Zimbabwe. In: E. Manzungu, A. Senzanje, and P. Van der Zaag (eds.), *Water for agriculture in Zimbabwe. Policy and management options for the smallholder sector.* University of Zimbabwe Publications, Harare, pp. 49-63.

NYATHI, M., 1995, Research linkages: Analytical description of Zimbabwe's agricultural knowledge and information systems. *News and Views* 63(October-December 1995): 12-15.

NYONI, J., 1990, The determination of plot size in small scale irrigation schemes in Zimbabwe. In: George M. Ruigu and Mandivamba Rukuni (eds.), *Irrigation policy in Kenya and Zimbabwe. Proceedings of the second intermediate seminar on irrigation farming in Kenya and Zimbabwe* held at Juliasdale, Harare, 26-30 May 1987. Institute of Development Studies. University of Nairobi, pp.173-185.

NZIMA, M.D.S., 1990, A cropping systems research approach to cropping programmes in small scale irrigation schemes. In: George M. Ruigu and Mandivamba Rukuni (eds.), *Irrigation policy in Kenya and Zimbabwe. Proceedings of the second intermediate seminar on irrigation farming in Kenya and Zimbabwe* held at Juliasdale, Harare, 26-30 May 1987. Institute of Development Studies. University of Nairobi, pp. 375-384.

OLALOKU, E.A., 1994, *Pre-project implementation planning workshop on improvement of research-extension-user linkages: An overview. Paper presented at the Pre-Project Implementation Planning Workshop on Improvement of Research-Extension-User Linkages.* University of Zimbabwe, Harare, August 2-4, 1994.

OLDRIEVE, B., 1993, *Conservation farming for communal, small-scale, resettlement and co-operative farmers of Zimbabwe. A farm management handbook.* Prestige Business Services Ltd., Harare.

————, 1995, Sustainable land utilization. Paper presented at a technical workshop on soil and water conservation for smallholder farmers in semi-arid Zimbabwe: Transfers between research and extension, 3-7 April, 1995, Masvingo.

ONSELEN, C. VAN., 1976, *Chibaro: African mine labour in Southern Rhodesia, 1900-1933.* Pluto press, London.

OWENS, L.B., 1979, Landscape reduction by weathering in small Rhodesian watersheds. *Geology* 7: 281-284 (cited in Nyamapfene, 1982: 287).

PAGE, S.L.J., C.M. MGUNI AND S.Z. SITOLE, 1985, *Pests and diseases of crops in communal areas of Zimbabwe.* British Overseas Development Administration Technical Report, London.

PAGE, SAM L.J., AND HÉLAN E. PAGE, 1991, Western hegemony over African agriculture in Southern Rhodesia and its continuing threat to food security in independent Zimbabwe. *Agriculture and Human Values,* Fall 1991, pp.3-18.

PALMER, ROBIN, AND NEIL PARSONS (eds.), 1977, *The roots of rural poverty in Central and Southern Africa.* Heinemann, London.

PALMER, R., 1977a, The agricultural history of Rhodesia. in: Robin Palmer and Neil Parsons (eds.), *The roots of rural poverty in Central and Southern Africa.* London, Heinemann, 221-254.

———, 1977b, *Land and racial domination in Rhodesia.* Heinemann, London.

PARKER, C., 1984, Control of Striga and other weeds in Mali. Weed Research Division, Internal report, Long Ashton research Station, U.K.

PAZVAKAVAMBWA, S., 1994, Agricultural extension. In: M. Rukuni and C.K. Eicher (eds.), *Zimbabwe's agricultural revolution.* University of Zimbabwe Publications, Harare, pp. 104-13.

PEACOCK, T., 1995, Financial and economic aspects of smallholder irrigation in Zimbabwe and the prospects for further development. Paper presented at the Zimbabwe Farmers' Union/Friedrich Ebert Stiftung conference 'Water development for diversification within the smallholder farming sector', 30 May 1995, Harare.

PENFIELD, J., AND M. DURU, 1988, Proverbs: Metaphors that teach. *Anthropological Quarterly* 61(3): 119-28.

PHIMISTER, IAN, 1984, Accommodating imperialism: The compromise of the settler state in Southern Rhodesia. *Journal of African History* 25(3): 279-94.

———, 1986, Discourse and the discipline of historical context: Conservationism and ideas about development in Southern Rhodesia 1930-1950. *Journal of Southern African Studies* 12(2): 263-275.

———, 1988, *An economic and social history of Zimbabwe: 1890-1948. Capital accumulation and class struggle.* Longman, London.

———, 1993, Rethinking the reserves: Southern Rhodesia's land husbandry act reviewed. *Journal of Southern African Studies* 19(2): 225-239.

PLOWES, D.C.H., 1980, The impact of agricultural extension in the Eastern Tribal Areas of Zimbabwe. *The Zimbabwe Science News* 14 (7): 197-200.

PLUMWOOD, V., AND R. ROUTLEY, 1982, World rainforest destruction - the social factors. *The Ecologist* 12(1): 4-22.

PRETTY, J.N., 1995, *Regenerating agriculture: Policies and practice for sustainability and self-reliance.* Earthscan, London.

PRETTY, J.N., AND R. CHAMBERS, 1993, Towards a new learning paradigm: New professionalism and institutions for agriculture. Discussion paper No. 334, IDS/IIED, Brighton.

RAFTOPOULOS, B., 1999, Nationalism in Salisbury, 1953-65. In: B. Raftopoulos and T. Yoshikuni (eds), *Sites of struggle; Essays in Zimbabwe's urban history*. Weaver Press, Harare.

RANGER, TERENCE, 1970, *The African Voice in Southern Rhodesia*. Heinemann, London.

——, 1978, Growing from the roots; Reflections on peasant research in Central and Southern Africa. *Journal of Southern African Studies* 5(1): 99-133.

——, 1985, *Peasant consciousness and guerilla war in Zimbabwe; A comparative study*. James Currey, London.

RENNIE, J.K., 1973, Christianity, colonialism and the origins of nationalism among the Ndau of Southern Rhodesia, 1890-1935. PhD Thesis, North Western University, USA.

REIJNTJES, C., B. HAVERKORT, AND A. BAYER-WATERS, 1992, *Farming for the future: An introduction to low-external-input and sustainable agriculture*. MacMillan, London.

REYNOLDS, NORMAN, 1969, A socio-economic study of an African development scheme. Unpublished PhD thesis. University of Cape Town, Cape Town.

RHOADES, R., AND A. BEBBINGTON, 1995, Farmers who experiment: An untapped resource for agricultural research and development. In: D.M. Warren, L.J. Slikkerveer and D. Brokensha (eds.), *The cultural dimension of development: Indigenous knowledge systems*. Intermediate Technology Publications, London, pp. 296-307.

Rhodesia, 1945, Report of Secretary, Department of Agriculture and Lands, for year ending 31st December 1944. *Rhodesia Agricultural Journal* 42(4): 323-344.

——, 1952, *Official Yearbook of Southern Rhodesia*, No. 4, 1952. Central African Statistical Office. Rhodesian Printing and Publishing Company, Salisbury.

RICHARDS, PAUL, 1985, *Indigenous agricultural revolution: Ecology and food production in West Africa*. Hutchinson, London.

——, 1986, *Coping with hunger: Hazard and experiment in an African rice farming system*. Allen & Unwin, London.

——, 1991, Experimenting farmers and agricultural research. In: M. Haswell and D. Hunt (eds.), *Rural households in emerging societies: Technology and change in sub-Saharan Africa*. Berg, Oxford and New York.

RIVERA, W.M., 1991, Agricultural extension worldwide: A critical turning point. In: W.M. Rivera and D.J. Gustafson (eds.), *Agricultural extension: Worldwide institutional evolution and forces for change*. Elsevier, Amsterdam.

ROBERTSON, C. L., AND A.D. HUSBAND, 1936, Results from Glenara Soil Conservation Experiment Station. *Rhodesia Agricultural Journal* 33(3): 162-172.

RODER, WOLF, 1965, The Sabi valley irrigation projects. Dept. of Geography Research Paper No. 99. University of Chicago Press, Chicago, Illinois.

ROEP, DIRK, AND RENÉ DE BRUIN, 1994, Regional marginalization, styles of farming and technology development. In: Jan Douwe van der Ploeg and Ann Long (eds.),

Born from within: Practice and perspectives of endogenous rural development. Van Gorcum, Assen, pp. 215-227.

ROGERS, E.M., 1962, 1971, 1983, *Diffusion of innovations.* The Free Press, New York.

RÖLING, NIELS, 1982, Alternative approaches in extension. In: G.W. Jones and M.J. Rolls (eds.), *Progress in rural development and community development (Vol. 1).* John Wiley and Sons, New York.

———, 1988, *Extension science: Information systems in agricultural development.* Cambridge, Cambridge University Press.

———, 1994, The challenging role of agricultural extension. In: *Agricultural Extension in Africa: Proceedings of an international workshop,* January 1994. Technical Centre for Agricultural and Rural Co-operation, Yaounde, pp. 7-20.

RÖLING, NIELS, E.VAN DE VLIERT, AND J. PONTIUS, 1995, Searching for strategies to replicate a successful extension approach: Training of IPM trainers in Indonesia. *European Journal of Agricultural Education and Extension* 1(4): 41-63.

RÖLING, NIELS, AND JULES N. PRETTY, 1997, Extension's role in sustainable agricultural development. In: B.E. Swanson, R.P. Bentz and A.J. Sofranko (eds.), *Improving agricultural extension. A reference manual.* Food and Agriculture Organization, Rome.

ROTH, R.E., 1987, Environment management education: A model for sustainable natural resource development. In: D.D. Southgate and J.F. Disinger (eds.), *Sustainable resource development in third world.* Westview Press, London, pp. 129-38.

ROWLAND, J.W., 1974, *The conservation ideal.* SARCUSS, Pretoria (cited in Stocking, 1985, footnote 4, p.47).

RUKUNI, MANDIVAMBA, 1984, An analysis of economic and institutional factors affecting irrigation development in communal lands in Zimbabwe. PhD thesis. University of Zimbabwe, Harare.

———, 1988, The evolution of smallholder policy in Zimbabwe: 1928-1986. *Irrigation and Drainage Systems* 2(1988): 199-210.

———, 1994, The evolution of agricultural policy: 1890-1990. In: M. Rukuni and C.K. Eicher (eds.), *Zimbabwe's agricultural revolution.* University of Zimbabwe Publications, Harare, pp. 15-39.

RUKUNI, MANDIVAMBA, AND JOHANNES MAKADHO, 1994, Irrigation development. In: Mandivamba Rukuni and Carl E. Eicher (eds.), *Zimbabwe's agricultural revolution.* University of Zimbabwe Publications, Harare, pp. 127-138.

RUKUNI M., M.J. BLACKIE, AND C.K. EICHER, 1998, Crafting smallholder-driven agricultural research systems in Southern Africa. *World Development* 26(6): 1073-1087.

RUTTAN, V.W., 1988, *Cultural endowments and economic development: What can we learn from anthropology?* The University of Chicago Press, Chicago.

SANDERS, H.C., 1966, *The cooperative extension service.* Prentice-Hall, New York, USA.

SARDAR, Z. (ed.), 1988, *The revenge of athena: Science, exploration and the third world.* Mansell, New York.

SCHRIJVERS, JOKE, 1985, *Mothers for life; Motherhood and marginalization in the North-Central Province of Sri Lanka.* Eburon, Delft, The Netherlands.

SCOONES, I., C. CHIBUDU, S. CHIKURA, P. JERANYAMA, D. MACHAKA, W. MACHANJA, B. MAVEDZENGE, B. MOMBESHORA, M. MUDHARA, C. MUDZIWO, F. MURIMBARIMBA AND B. ZIREREZA, 1996, *Hazards and opportunities. Farming livelihoods in dryland Africa. Lessons from Zimbabwe.* Zed Books Ltd., London.

SCOONES, I. AND J. THOMPSON, 1994, *Beyond farmer first: Rural people's knowledge, agricultural research and extension practice.* Intermediate Technology Publications, London.

SCOTT, R.V., 1970. *The reluctant farmer. The rise of agricultural extension to 1914.* University of Illinois Press, Urbana.

SINCLAIR, SHIRLEY, 1971, *The story of Melsetter.* M.O. Collins, Salisbury.

SITHOLE, N.P., 1995, The impact of smallholder irrigation on household food security in marginal areas. Unpublished MSc thesis, Department of Economics, University of Zimbabwe, Harare.

SOFO, M., D. TOPOUZIS, S. HORST, S. AMAKAJE, N. KARRITHI, S. OWALTARA, C.L. MORRA AND L. KILIMVIKO, 1980, Making extension work: The key to Africa's future. *African Farmer* 4: 28-33.

STACK, JAYNE l.,1994, The distributional consequences of the smallholder maize revolution. In: Rukuni, M. and C.K. Eicher (eds.), *Zimbabwe's agricultural revolution.* University of Zimbabwe Publications, Harare, pp. 258-269.

STOCKING, MICHAEL A., 1978, Relationship of agricultural history and settlement to severe soil erosion in Rhodesia. *Zambezia* 6: 129-45 (cited in Stocking, 1985, footnote 5, p.47).

——, 1985, Soil conservation policy in colonial Africa. In: Douglas Helms and Susan L. Flader (eds.), *The history of soil and water conservation.* The Agricultural History Society. University of California Press, Washington D.C.

SUTTON, J.E.G., 1984, Irrigation and soil-conservation in African agricultural history, with a consideration of the Inyanga terracing (Zimbabwe) and Engaruka Irrigation works (Tanzania). *Journal of African History* 25: 25-41.

SWANSON, B.E., R.P. BENTZ, A.J. SOFRANKO, 1997, *Improving agricultural extension: A reference manual.* FAO, Rome.

TAPERERWA, S., AND J.B. CHIVIZHE, 1994, A review of master farmer training and group extension approaches as well as individual visits as vehicles for technology transfer and knowledge exchange in agricultural extension in Manicaland since 1980. Paper presented at the Agritex National Technical Conference, 21-25 February 1994, Harare.

TAWONEZVI, P.H., 1994, Agricultural research policy. In: Rukuni, M. and C.K. Eicher (eds.), *Zimbabwe's agricultural revolution*. University of Zimbabwe Publications, Harare, pp. 92-103.

TAYLOR, G., 1925, Review (Agriculture in East Africa). *NADA* 1(3): 86-90.

TEMPANY, SIR HAROLD A., 1949, The practice of soil conservation in the British colonial empire. *Technical Communication* No. 45. Commonwealth Bureau of Soil Science, Harpenden.

THEISEN, R.J., 1979, Socio-economic factors involved in the profitable cropping of worked out granite soils in tribal communities, *Rhodesia Agricultural Journal*, 76 (1): 27-34.

THRUPP, L.A., 1988, Legitimizing local knowledge: "scientized packages" or empowerment for third world people. In: D.M. Warren, L.J. Slikkeveer and S.O. Titilola (eds.), *Indigenous knowledge systems: Implications for agriculture and international development. Studies in Technology and Social Change* No. 11. Iowa State University, Ames, Iowa, pp. 138-53.

TIFFEN, MARY, MICHAEL MORTIMORE AND FRANCIS GICHUKI, 1994, *More people, less erosion; Environmental recovery in Kenya*. John Wiley & Sons, Chichester.

VAN DER ZAAG, P., 1992, The material dimension of social practice in water management, a case study in Mexico. In: Geert Diemer and Jacques Slabbers (eds.), *Irrigators and engineers*. Thesis Publishers, Amsterdam, 73-83.

VAN DER ZAAG, P., AND RÖLING, N., 1996, The water acts in the Nyachowa catchment area. In: Manzungu, E. and P. van der Zaag (eds.), *The practice of smallholder irrigation. Case studies from Zimbabwe*. University of Zimbabwe Publications, Harare.

VELSEN, J. VAN, 1964, Trends in African nationalism in Southern Rhodesia. *Kroniek van Afrika* 2(2).

VERSCHOOR, GERARD, 1992, Identity, networks and space; New dimensions in the study of small-scale enterprise and commoditization. In: Norman Long and Ann Long (eds.), *Battlefields of knowledge*. Routledge, London, pp. 171-189.

VICKERY, K.P., 1996, The Rhodesia Railways African Strike of 1945, part II: Cause, consequence, significance. *Journal of Southern African Studies* 25(1): 49-72 .

VIJFHUIZEN, CARIN, 1995, Leaders in a *musha*; A case study of the engendered organisation of food in irrigated agriculture. Paper presented at a UZ/ZIMWESI workshop 'Extension, intervention and local strategies in resource management; New perspectives on agricultural innovation in Zimbabwe'. Harare, 10 and 11 January 1995.

——, 1996, Who feeds the children; Gender ideology and the practice of plot allocation in an irrigation scheme. In: E. Manzungu and P. van der Zaag (eds.), *The practice of smallholder irrigation; Case studies from Zimbabwe*. University of Zimbabwe Publications, Harare, pp.126-147.

VINCENT, V., 1962, *Extension methods in Southern Africa.* Government Printers, Pretoria.

VINCENT, V. and R.G. Thomas, 1957, *Agricultural survey of Southern Rhodesia: Part I: Agro-ecological survey.* Federal Printer, Salisbury.

VIVIAN, J., 1994, NGO's and sustainable development in Zimbabwe; No magic bullets. *Development and Change* 25: 167-93.

WARREN, D.M., 1989, The impact of nineteenth century social science in establishing negative values and attitudes towards indigenous knowledge systems. In: D.M. Warren, L.J. Slikkeveer and S.O. Titilola (eds.), *Indigenous knowledge systems: Implications for agriculture and international development. Studies in Technology and Social Change* No. 11. Iowa State University, Ames, Iowa, pp. 171-84.

WASHINGTON, BOOKER T., 1901, *Up from Slavery.* Doubleday and Company, New York (reprinted in 1965 by Avon Book division, New York).

WATSON-VERRAN, HELEN, AND DAVID TURNBALL, 1995, Science and other indigenous knowledge systems. In: Sheila Jasanoff, Gerald E. Markle, James C. Petersen and Trevor Pinch (eds.), *Handbook of science and technology studies.* Sage Publications, Thousand Oaks, London and New Delhi.

WATT, W. MARTIN, 1913, The dangers and prevention of soil erosion. *Rhodesia Agricultural Journal* 10(4): 667-675.

WEINMANN, H., 1972, Agricultural research and development in Southern Rhodesia under the rule of the British South Africa Company, 1890-1923. Occasional paper No. 4, Department of Agriculture, University of Rhodesia, Salisbury.

———, 1975, Agricultural research and development in Southern Rhodesia, 1924-1950. *Series in Science* No. 2, University of Rhodesia, Salisbury .

WEINRICH, A.K.H., 1973, *Black and white elites in rural Rhodesia.* Manchester University Press, Manchester.

———, 1975, *African farmers in Rhodesia.* Manchester University Press, Manchester.

WHITLOW, J.R., 1979a, *The household use of woodland resources in rural areas.* Natural Resources Board, Harare.

———, 1979b, Deforestation — Some global and national perspectives. *Geographical Proceedings of Zimbabwe* 12: 13-30.

———, 1980, Deforestation in Zimbabwe — Some problems and prospects. Natural Resources Board, Harare.

WHITLOW, RICHARD, 1985, Conflicts in land use in Zimbabwe: Political, economic and environmental perspectives. *Land Use Policy* October 1985: 309-322.

———, 1988, Soil erosion and conservation policy in Zimbabwe: Past, present and future. *Land Use Policy* October 1988: 419-433.

WHITLOW, R., AND L. ZINYAMA, 1988, Up hill and down vale: Farming and settlement patterns in Zimunya communal land. *Geographical Journal of Zimbabwe* 19: 29-45.

WILKEN, GENE C., 1987, Good farmers: Traditional agricultural resource management in Mexico and Central America. University of California Press, Berkeley.

WILSON, K.B., 1989, Trees in fields in southern Zimbabwe. *Journal of Southern Africa Studies* 15(2): 369-88.

——, 1990, Ecological dynamics and human welfare: A case study of population, health and nutrition in Zimbabwe, PhD Thesis, University of London.

——, 1995, 'Water used to be scattered in the landscape': Local understandings of soil erosion and land use planning in southern Zimbabwe. *Environment and History* 1(3): 281-96.

WILSON, N.H., 1923, The development of Native Reserves. One phase of native policy for Southern Rhodesia. *NADA* 1(1): 86-94.

WOLF, ERIC, 1966, *Peasants. Foundations of modern anthropology series*. Prentice Hall, Englewood Cliffs, New Yersey.

WOLMER, WILLIAM, AND IAN SCOONES, 2000, The science of 'civilized' agriculture: The mixed farming discourse in Zimbabwe. *African Affairs* 99: 575-600.

WORBY, E., 2000, Discipline without oppression: Sequence, timing and marginality in Southern Rhodesia's post-war development regime. *Journal of African History* 41(1): 101-125.

WORTHINGTON, E.B., 1938, *Science in Africa,* London (cited in Floyd, 1959).

ZIMBABWE, 1987, National conservation strategy, the Zimbabwe road to survival. Ministry of Natural Resources and Tourism, Harare.

——, 1994, Report of the Commission of inquiry into appropriate agricultural land tenure systems. Volume one: Main Report; Volume two: technical reports. Government Printer, Harare.

ZINYAMA, L.M., 1993, The evolution of the spacial structure of greater Harare: 1890-1990. In: L.M. Zinyama, D.S. Tevera and S.D. Cumming (eds), *Harare, the growth and problems of the city*. University of Zimbabwe Publications, Harare, 7-32.

ZVOBGO, C.J.M., 1996, *A history of Christian missions in Zimbabwe, 1890-1939*. Mambo Press, Gweru.

ZWEERS, WIM, 1995, Ecological spirituality as a point of departure for an intercultural dialogue. In: *Agriculture and spirituality: Essays from the Crossroads conference at Wageningen Agricultural University*. Crossroads, The Hague.

Index

AUTHOR INDEX

Agritex, 111, 112, 114, 128, 135, 149
Alexander, J. 23
Allan, W. 18, 50, 58
Alvord, E.D. 14, 29, 36, 41-62 *passim*,
65-70 *passim*, 112, 164, 310
Anderson, D.M. 36
Andersson, J.A. 10, 20, 30, 67,
Aquina, M. 18
Arnaiz, M.E.O.29
Atkinson, N.D. 37, 39
Auret, D. 114
Aylen, D. 188, 189, 190, 192

Bagchee, A. 3, 9
Bates, 23
Bawden, R. 7
Bebbington, A. 109
Beinart, W. 12, 36, 67, 133, 185, 186,
191, 192, 200, 234
Bernard, B.R. 273
Beynon, J. 7, 9, 26, 28, 315
Biggs, S.D. 303
Blackie, M.J. 21, 23, 25, 28
Bolding, 29, 133
Boserup, M.F.C. 223, 224
Bourdillon, 113
Bown, L. 241
Brokensha, D. 5
Bromley, K.A. 6, 198
Bulman, E.M. 200
Bundy, C. 55

Callear, D. 134
Carney, D. 8
Casey, J. 158

Chambers, R. 4, 5, 9, 232, 268, 271,
272, 280
Chavunduka Commission 25
Cheal, D. 220
Cheater, A. 134
Checkland, P.B.7, 306
Chikukwa, T. 52
Chimedza, R. 224
Chivizhe, J.B. 115
Christopolos, I. 8, 9
Chuma, E. 58
Cliffe, L. 28, 191, 234
Coates 12, 67
Cold Storage Commission 11
Comaroff, J. 31
Cormack, R.M.M. 189
Cotton Research and Industry
Board 11
CSO, 137

Dairy Marketing Board 11
Davies, B. 57
de Bruin, R. 126
Derude, 114
Dikito, M.S. 224
Dissanayake, 201
Dopcke, W. 57
Drinkwater, M. 12, 25, 36, 67, 180,
267
Dunlop, H. 18, 20
Duru, 234, 241

Eicher, C.E. 314
Elwell, H. 101

FAO, 112
Farrington, J.A. 7
Faulkner, E. 58
Ferguson, 314
Floyd, B.N. 18, 50, 51, 52
Forestry Commission 161, 162
Fortmann, L. 158
Freire, P. 229, 232

Gasper, D. 22
Gatter, P. 143, 149
Gittinger, J.P. 116
Goodenough, W.H. 229
Gore, 157
Grant, P.M. 51-54 passim
Gubbels, P. 6

Hagmann, J. 26, 58, 189
Hailey, 50
Hamilton, P. 188
Hamilton, R.R. 18
Hamutyinei, M.A. 234, 237, 239
Hanyani-Mlambo, B.T. 30
Hardin, G. 237
Harrison, 3
Haug, R. 8
Havelock, R.D. 4
Haviland, P.H. 188, 189
Hawkins, 2
Hebinck, P. 248
Holleman, J.F. 24, 135
Howes, M. 229
Hudson, N.W. 191
Hulme, D. 3

Jackson, C.N.C. 43, 44, 47
Jansen, D.J. 112, 113
Jordan, J.D. 18

Kaimowitz, D. 4
Kajembe, J.C. 158

Katerere, Y. 161
Keigwin, H.S. 35, 36, 38-43 passim, 70
Kennan, P.B. 18, 20, 24, 63
Komter, A.E. 220
Kopytoff, L. 219, 220
Kunaka, A.M. 165

Lal, R. 58
Leeuwis, C. 7
Levi-Strauss, 229
Livingstone, D. 31, 32
Lloyd, B.W. 39
Long, N. 160, 201, 209, 220, 268
Lovelace, G.W. 230
Low, A.R.C. 7
Lue-Mbizvo, C. 158

Madondo, B.B.S. 26, 27, 115, 224
Madzudzo, E. 113, 126
Magadlela, D. 32, 248
Magwa, 115, 118
Mainwaring, C. 57
Makadho, J. 112
Makina, J.C. 234, 237, 238
Makotholiso, 46
Manzungu, E. 1, 32, 62, 112, 113, 115, 315
Marsden, D. 231, 232
Mashavira, T.T. 58
Maslow, A. 126
May, 241
Mbiti, J.S. 238
McGregor, J. 36, 50, 52, 55, 59, 61, 67
McNeil, J.D. 231, 232
Meinzen-Dick, R. 112
Mehretu, A. 112
Mohammed, J.C. 158
Moyo, S. 25
Mtetwa, R. 15, 50
Muchena, O. 30, 126, 284, 312
Muir, K. 21, 23, 25, 158

Mundy, H.G. 12, 13, 41, 43, 44
Murwira, K, 26, 32, 189
Mutimba, J. 32
Mutizwa-Mangiza, N.D. 20
Muzorewa, A. 22
Mvududu, S.C. 224
Mvundhlama, 44

Natural Resources Board 11
Nhema, Chief 48
Nhira, C. 158
Nitsch, U. 8
Nobbs 12
Norton, A.J. 58, 101
Nyamapfene, K.52
Nyamapfene, 188
Nyamudeza, P. 26, 52
Nyathi, M. 267
Nyoni, 112
Nzima, 113

Olaloku, E.A. 295
Oldrieve, B. 58, 101
Onselen, C.V. 10

Page, H.E. 58, 59, 115
Page, S.L.J. 44, 51, 58, 59, 101, 115
Palmer, R. 10, 11, 14, 15, 37, 49
Pazvakavamba, S. 27, 36
Peacock, T. 113, 114
Phimister, I. 11, 12, 13, 15, 18, 21,
    185, 192, 199, 200
Pig Industry Board 11
Plangger, A.B. 233, 234, 239
Plowes, D.C.H. 23, 63
Pole-Evans, 53
Pretty, J.N. 6, 79

Raftopoulos, B. 20
Ranger, T. 10, 11, 13, 13, 16, 22, 23,
    24, 37, 49, 57, 134

Reijntjes, C. 6
Rennie, J.K. 13, 55, 60
Reynolds, N. 248
Rhoades, R. 109
Rhodesia Agricultural Journal 13, 187
Richards, P. 5, 109, 229, 232
Rivera, W.M. 2, 8, 9
Robertson, C.L.191
Roder, W. 113
Roep, D. 126
Rogers, E.M. 3, 62, 240
Röling, N. 4, 6, 7, 109
Roth, R.E. 233
Routley, 157
Rukuni, M. 8, 11, 22, 26, 28, 113, 114,
    268, 312
Ruttan, V.W. 230

Sardar, 229
Scholes, J. 7
Schrijvers, J. 225
Scoones, I. 5, 7, 12, 17, 25, 27, 52, 54,
    55, 59
Senzanje, 1
Shantz, 50
Sithole, S.Z. 219
Smith, I. 21
Sofo, M. 241
Stack, J. 28
Sugar Industry Board 11
Swanson, B.E. 8, 9

Tapererwa, S.115
Tawonezvi, P.H. 17, 22, 23, 25
Taylor, G. 44, 50
Tempany, H. 190, 191, 197
Theisen, R.J. 52
Thompson, J.5
Thrupp, L.A. 231
Turnball, D. 126

Vambe, 66
Van der
    Ploeg 201, Zaag, P. 1, 30,
    den ban 2
Velsen, J.V. 20
Vickery, K.P. 20
Vijfhuizen, C. 30, 223, 224
Villarreal, 201
Vincent, V. 18
Vivian, J. 29

Walters, J.A.T. 13
Warren, D.M. 228

Washington, B.T. 38
Watson 185-Verran 126
Weinmann, 13, 17, 18, 23, 55, 190
Weinrich 11, 24
Whitelow, I. 200
Wilken, G.C.197
Wilson, K.B. 39, 52, 242
Wolmer, W. 12, 27
Worby, E. 23

Zheve, 121
Zimbabwe, 9, 234
Zweers, W. 126

SUBJECT INDEX

Afforestation 179
African
    education 39
    reserves
        Agricultural extension, in 18,
        23, 36-48; crop rotation
        practices, in 52-54, 96-97;
        manure use, in 51-52; mono
        cropping, 95-96; ploughing
        methods, in 16-18, 22, 25-26,
        240; research 16-18; state
        interventions, in 11-21 *passim;*
        weeding in 58-59; zero tillage,
        in 58 *see also* communal
        agriculture
Agricultural
    demonstrator
        policy 36, 43
        programme 40, 60-62
            African farmer request, for
            46; launch, of 41-49
    Improvements, Christianity and
    59; knowledge systems 7; labour,
    crop choices and 146-148; past

and present 35-129; Rural
Development Authority 85, 88,
91, 297
Research 16-18, 22, 25-26, 290
    and Extension Fund 315
Sector Investment Programme
314; training in Africa 39-41
Agriculture
    African
        agriculturalist 49; changes, to
        49-50; decline, of 10, 15, 16,
        20-24 *passim*, 132-134;
        development, of 38; merits,
        of 50-51
    communal
        organisation, of 147-149;
        success, of 28
    Finance Corporation 91, 167
    extension
        constraints, of 311-312; crisis,
        of 8-9; curriculum 229, 310;
        definition, of 2; paradigms, in
        2; post-independence 26-27,
        35

research 22
    small-scale farmers 23-25
    NGO's role of 28, 317;
    success, of 28

Bench terrace 30, 184-188, 192-199
    labour requirements, of 196-198;
    use, of 198-199
Biriwiri
    bench terraces, in 193-198; soil
    conservation techniques, in 184,
    186-188

Cattle Levy Act 67, 132
Chibuwe irrigation scheme 115
    Agriculture extension, in 173-
    1179 passim; forestation issues, in
    178; grain production, in 168-170;
    tobacco farming, in 167-168
Chiduku Reserve 63, 171
Chinamora Reserve 62
Chinyika Resettlement
    Area 163, 177-178; Scheme 157-
    181
Cold Storage Commission 106
Collective action for sustainable
    agriculture 6
Cotton
    Company of Zimbabwe 122;
    Marketing Board 28
Commercial agriculture
    expansion, of 14; extention
    services, to 13; labour
    mobilisation, for 10; state
    intervention, in 11-14
Committee of Enquiry on Native
    Education 39
Communal
    agriculture
        commercialisation, of 24-25,
        110-127; labour migrations, in

110-127; post-colonial
    developments, in 134;
    revolutions, in 16-17; risk
    management, in 123; state
    intervention, in 16, 132-135
farmers 29, 172-173
Conservation
    politics of, 185-186; technology
    184-199
Contour ridges 189-200

Danish International Development
    Agency 294
Dairy Development Project 297, 298
Deforestation 163-164, 177 see also
    afforestation
Domboshawa Native Industrial
    School 39

Economic Structural Adjustment
    Programme 27
The Education Ordinance 37

Farmer
    first paradigm 4-7
    participatory research 5, 282-288
    passim, 298
        marketing, of 292-294
    women 209-226 passim
farming
    African 14, 49-51; commercial 20,
    27; systems research 4
Food security 87
    constraints 89-90
Forestry
    conservation
        resettlement, and 158 -160 see
        also rural reforestation
        Programme
        extension 160-167

grassroots 166, 173-174;
limitations, of 180-181
research 161
Fuve Pangani Irrigation Scheme 118,
119, 126

Green revolution 3
Grain Marketing Board 87
Groundnut
farming 210-222
tomatoes, and 222-223; value,
of 211-222
harvesting, of 211-212; use 211-
222
Gums
problems, with 165-166

Household modernisation 64-65

Indigenous
farming systems 15
knowledge 228
curriculum 230-233, elements
of 232-233; extension work,
and 239-241; sources, of 233-
241
International Livestock Research
Institute 294, 296, 303
Irrigation
schemes 225; technology 223

Keigwin Plan 37

Labour migration 136-149 passim
expenditures, of 143
Lancaster House Agreement 22
Land
Apportionment Act 11, 15, 60,
132; Husbandry Act 61
Local knowledge and networks 209-
226

Maize Control Act 67, 132
Master farmer 15, 24, 63-66 passim;
66, 70, 309
criteria, for 64; training
programme 115-118, 127
Missionary Conference of Southern
Rhodesia 39
Morris Carter Land Commission,
The 11
Murambinda area 140
Mutambara Reserve 62

Native
Farmers Association 62
Land Husbandry Act 16, 20-21,
63, 313
opposition, to 20
Production and Trade
Commission 16
National Farm Irrigation Fund 28
Natural Resources Board 16, 165,
167, 176, 177
New institutionalism paradigm 7-9
Nyamaropa Irrigation Scheme 248,
267
Nyanyadzi Irrigation Scheme 63, 69

Participatory
rural appraisal 5; technology
development 5

Rapid rural appraisal 299, 300
Reforestation 164
Reserve
modernisation 61; types, of 62-63
Rhodesia National Farmers Union 22
Rural
Reforestation Programme 158,
159, 161-164 passim, 174; urban
migration 138, 139, 141-146

Selukwe Reserve 62
Shiota Reserve 63
Smallholder
    agriculture
    constraints, in 112;
    commercialisation, of 112;
    farmers, technology needs of 84;
    farming methods 51-63;
    improvements, to 35-44, 46, 90-
    136 *passim*
    irrigation 112-114
        gender, and 209-226
    revolutions, in 16-17, 28
Soil conservation 184, 187
Southern
    African Development Community
    282
    Rhodesia Native
        Affairs Committee 37;
        Association 62

Squatters 171-172

Tawona smallholder irrigation 209-
    217
    Groundnut production, in 210-
    212, 216
Technology transfer 157-181
    NGO's, and 5
Tobacco
    Marketing Board 15, production
    14
Tomatoes 218
Transfer of technology 3-5 *passim*, 18,
    30, 305, 314
Trees 157 *see also* forestry
Tribal Trust Land
    Authorities Act 21, 24;
    Development Corporation 22

Unilateral Declaration of
Independence 21, 22

Water Sector Investment Programme
    314

Zimbabwe
    agricultural
        parastatals 11; research
        stations 13
    agro-ecological classification 17
    Department of
        Agriculture 12-13 and
        Extension Services 9-12, 85,
        110, 127, 128, 131, 249, 253,
        255, 257, 258, 265-268 *passim*,
        312, 313; Conservation and
        Extension Services 18, 20, 22,
        26, 36; Native Affairs 37;
        Research and Specialist
        Services 16, 22, 25, 26, 84,
        312, 315
    Farmers Union 27, 29, 295, 316;
    segregated agricultural
    development 10-12, 20-22, 36,
    308
Zimuto Reserve 62
Zimunya Reserve 62

www.ingramcontent.com/pod-product-compliance
Lightning Source LLC
Chambersburg PA
CBHW060805220326
41598CB00022B/2540